FROM METEORITE IMPACT
TO CONSTELLATION CITY

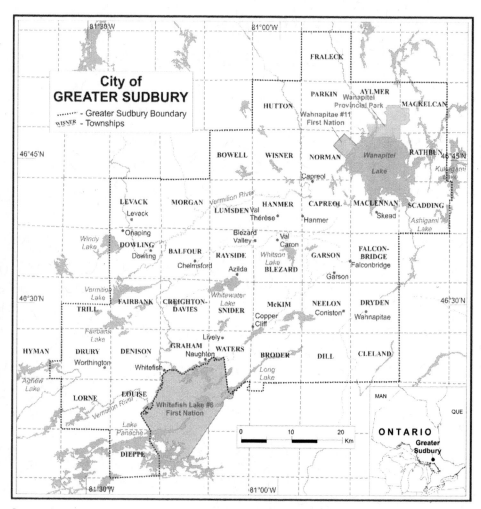

City of GREATER SUDBURY

.... - Greater Sudbury Boundary
WISNER - Townships

Source: Growth and Development Department, City of Greater Sudbury.

FROM METEORITE IMPACT
TO CONSTELLATION CITY

A Historical Geography of Greater Sudbury

Oiva W. Saarinen

WILFRID LAURIER
UNIVERSITY PRESS

This book has been published with the help of a grant from the Canadian Federation for the Humanities and Social Sciences, through the Awards to Scholarly Publications Program, using funds provided by the Social Sciences and Humanities Research Council of Canada. Wilfrid Laurier University Press acknowledges the financial support of the Government of Canada through the Canada Book Fund for our publishing activities.

Library and Archives Canada Cataloguing in Publication

Saarinen, Oiva W., 1937–
From meteorite impact to constellation city : a historical geography of Greater Sudbury /
Oiva W. Saarinen.

Includes bibliographical references and index.
Issued also in electronic format.
ISBN 978-1-55458-837-4

FC3099.S83S33 2013   971.3'133   C2012-907130-7

Electronic monograph in multiple formats.
Issued also in print format.
ISBN 978-1-55458-874-9 (PDF).—ISBN 978-1-55458-875-6 (EPUB)

1. Greater Sudbury (Ont.)—Historical geography. 2. Greater Sudbury (Ont.)—History. I. Title.

FC3099.S83S33 2013   971.3'133   C2012-907131-5

Cover design by Martyn Schmoll. Front-cover image adapted from Canada-maps.org. Text design by Janette Thompson (Jansom).

This book is printed on FSC recycled paper and is certified Ecologo. It is made from 100% post-consumer fibre, processed chlorine free, and manufactured using biogas energy.

Printed in Canada

Every reasonable effort has been made to acquire permission for copyright material used in this text, and to acknowledge all such indebtedness accurately. Any errors and omissions called to the publisher's attention will be corrected in future printings.

Contents

List of Illustrations

TABLES

AERIAL PHOTOGRAPHS

List of Biographies

Preface and Acknowledgements

Time and space. These are the two defining features that give meaning to places such as the City of Greater Sudbury. In this book, *history and geography* provide the context for a journey that began billions of years ago and is still ongoing. It is a fascinating odyssey, encompassing dramatic physical and human events. Among these can be included volcanic eruptions, meteorite impacts, the ebb and flow of continental glaciation, Aboriginal occupancy, exploration by the Europeans, the presence of fur traders, lumbermen, and Americans, the rise of global mining giants such as Inco/Vale Canada Limited and Falconbridge/Xstrata Nickel, unionism, environmental pollution and recovery, and the creation of a constellation city of some 160 000 people. The story includes the history of more than 88 000 departed souls who, at one time or another, chose Sudbury as their home and whose remains now reside in one of its twenty-eight cemeteries.[1]

I have often been asked how long it took to write this book. The simple answer is—*a lifetime.* The book flows from all that I have experienced over a lengthy lifespan that dates back to 1937 when I first saw the light of day at St. Joseph's Hospital in downtown Sudbury. From there, my sense of community belonging and understanding was enhanced by living in several areas of the city, including the West End, Minnow Lake, New Sudbury, Long Lake, and the Lockerby area. More than forty years of tenure as a professor in the Department of Geography at Laurentian University, and ongoing interactions with other faculty and their varying perspectives, provided me with an academic and theoretical vantage point to assess the place where I grew up. My stint at the University of London, and travels throughout Finland and other parts of Europe from the 1960s and beyond were also significant, as they provided me with a comparative stance that honed my appreciation of Sudbury's "sense of place" within a broader geographical context. This perspective is evident in the book's reach, which frequently extends beyond Sudbury to encompass Northeastern Ontario, Canada, North America, and even the entire globe. In short, the book reflects my personalized view of the Sudbury area as I have experienced it over more than seven decades.

The title deserves comment. It sets the framework for the book through its emphasis on two major themes, one physical in nature, the other human. The physical theme involves the great meteorite impact that took place in Sudbury some 1.85 billion years ago. This cataclysmic incident, called the "Sudbury Event" by geologists, transformed the area into a mineral-rich zone that gave rise to two of the world's largest mining companies, and profoundly shaped the regional settlement pattern. Had this impact not occurred, the area would have evolved simply as another lonely wilderness tract in the vast Precambrian Shield. The human theme centres on the creation of the City of Greater Sudbury, a municipal entity that came into being on January 1, 2001, encompassing 3 354 square kilometres of territory, a geographical area more than five times the size of the City of Toronto (but with only slightly more than 1 per cent of the latter's density) and two-thirds the size of the province of Prince Edward Island.[2] In addition to being the largest municipality in Ontario, and one of the largest in Canada by size, the city has the unique distinction of not developing in a normal metropolitan settlement pattern. In contrast to other large Canadian municipalities that exhibit continuous habitation from their downtown cores to outlying suburbs, Sudbury is made up of a complex of settlements consisting of the former Town/City of Sudbury and a surrounding assemblage of some fifty built-up areas, each with a population of more than two hundred people. Thus, Sudbury can be viewed as a "constellation city" made up of individual, smaller communities, each with its own attributes, forming a whole that is greater than the sum of its parts.[3] The importance of this distinctive human setting cannot be over-emphasized; it has shaped much of the history of the area. Unless otherwise noted, the use of the terms "Sudbury," "Greater Sudbury," and "region" in this book will normally refer to the area encompassed by the present-day City of Greater Sudbury.

The book is framed within the context of historical geography, divided into fifteen chapters. While some chapters are chronological in nature, others have been arranged within a thematic context. The material in the book has been largely acquired from existing sources. In the introductory chapters, the historical and geographical perspective has been widened to place Sudbury into broader provincial, national, and global contexts that have been rarely examined elsewhere. Chapter 1 outlines the unfolding of the physical landscape, setting the framework for a better understanding of what transpires in the rest of the book. This is followed, in Chapter 2, by a treatment of the original occupancy of the land by Aboriginals following the retreat of the last glaciers from the area some 10 000 years ago. In Chapter 3 I deal with the manner in which the land was transformed into reference lines known as meridians, base lines, and township grids, a necessary prerequisite for the subsequent settlement of the territory by white people. Chapter 4 explores the discovery of minerals associated with the Sudbury Structure and the formative years of mining exploration and development. Chapter 5 continues with a look at the evolution of Sudbury from a CPR townsite first into a town, and then a city by the Second World War. I examine Copper Cliff's changing setting from that of a mining camp to a corporate town in Chapter 6. The focus of the book returns to mining in Chapter 7, which outlines how two competitive mining giants, The International Nickel Company of Canada and Mond Nickel, found it necessary to merge into a global giant known simply as Inco. Chapter 8 delves into the expansion of the pattern of settlement beyond Copper Cliff and Sudbury, and reviews outlying communities associated with railway stations, mining camps, smelter sites, and other company towns. Chapter 9 begins with a look at other phases of outlying settlement associated with forestry and agriculture. While never located within Sudbury's municipal boundaries, the Whitefish Lake and Wahnapitae Indian Reserves, as well as the Burwash Industrial Farm are included here because of their effects on the area. The story of the evolution of Falconbridge Nickel and Inco and their later transformations into Xstrata Nickel and Vale Canada is the subject of Chapter 10. Chapter 11 illustrates how the Sudbury area evolved from a company-town setting into a new constellation entity known as the Regional Municipality of Sudbury that dominates much of Northeastern Ontario. In Chapter 12 I look at Sudbury's transition from a regional constellation into the current City of Greater Sudbury. Chapter 13 examines the history behind the area that served as one of the world's main centres of industrial unionism. Chapter 14 deals with the area's best-known international achievement: recent efforts to heal the landscape after years of resource extraction. Finally, Chapter 15 completes the book with a reflection on Sudbury's past, present, and future. I conclude that the community is poised to move in a new direction, one that lies "beyond a rock and a hard place."

I would like to acknowledge all the people and organizations who have contributed to the voluminous list of historical and geographical publications pertaining to the Sudbury area. I have found it a real pleasure in researching this story to see the spirit of cooperation shown by the people who worked with me throughout this undertaking. Their willingness to share information and constructive criticism proved to be invaluable, and contributed significantly to the completion of the book. To all of them I express my deep and sincere gratitude.

I would, however, like to extend special thanks to a few individuals. First, I must acknowledge the cartographic and computer wizardry provided by Léo Larivière, technologist with the Department of Geography and the Library Department at Laurentian University. As well, I extend my appreciation to former academic colleagues, including P.J. Barnett, Matt Bray, P.J. Julig, D.H. Rousell, Robert Segsworth, Gerald Tapper, and Carl Wallace. Other contributors include Dick DeStefano, Narasim Katary, Bill Lautenbach, Gerry Lougheed, Jr., and Bruno Pollesel. I wish to give my thanks also to those external advisors and editors at Wilfrid Laurier Press who read drafts of the manuscript and made cogent suggestions. Financial assistance for this undertaking was provided by Laurentian University and the Finnish National Society. Finally, I would like to express my deep and sincere appreciation to my wife, Edith, for her ongoing support, which included the onerous task of proofreading the text.

Oiva W. Saarinen, Ph.D.
Professor Emeritus
Department of Geography
Laurentian University
Sudbury, Ontario, Canada

CHAPTER 1

The Unfolding of the Natural Landscape

The City of Greater Sudbury has within its boundaries some of the most complex geological features found anywhere in the world. Specifically, it is home to one of geology's greatest enigmas—the Sudbury Structure of the Precambrian Shield.[1] Any explanation of this unique formation's origin must take into account its setting in time and space. While the Sudbury Structure represents a localized feature, its origins cannot be divorced from broader spatial associations linked to Northern Ontario, the three structural provinces of the Precambrian Shield, and regional influences such as tectonic zones and fault lines. These murky associations occurred within the framework of other complicated events stretching back for eons. Some two million years ago, the geologic setting changed as the last Great Ice Age began. The advances and retreats associated with this glacial period dramatically altered Sudbury's physical appearance. Following the retreat of the last ice sheet about ten thousand years ago, the area underwent climatic and vegetative transitions, processes that have continued to the present day.

The area's geologic, glacial, climatic, and vegetative history has had a profound effect on Sudbury's economic raison d'être, distribution of population, topographical setting, and environmental appearance. While geographers generally reject the principle that such "environmental determinism" should be considered the dominant factor shaping urban development, the case can arguably be made that Sudbury serves as an exception to this rule.[2]

The Creation of the North American Continent

Since the formation of the earth some 4600 Ma (megaannum, or million years) ago, Northern Ontario has been part of a geological odyssey of epic proportions. Over time, the molten earth cooled, and processes came into play that reshaped the appearance and nature of the Sudbury area. These processes included plate tectonics, continental drift, and the creation of continental blocks. At various times, the site that became Sudbury was even a maritime/tropical environment. The cooling of the earth resulted in the formation of huge floating slabs of rocks on the surface, known as plates. While hard to conceptualize today, Northern Ontario was part of no less than four major continents: Arctica, Nena/Columbia, Rodinia, and Pangea (fig. 1.1). Starting 2 500 Ma ago, the original North American continent known as Arctica was created by smaller plates meshing together. The lower margins of Arctica later collided with another plate 1 900 Ma ago to create a continent known as Nena/Columbia.[3] Over 1 000 Ma ago, Nena/Columbia stretched, broke, and grew into a larger supercontinent called Rodinia. Rodinia was subsequently torn apart around 550 Ma ago. Then, 410 Ma ago, Rodinia gradually became part of yet another

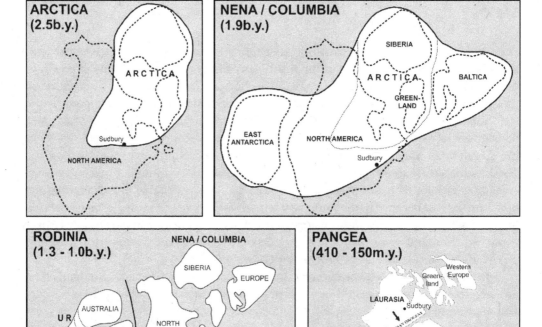

FIGURE 1.1   Early Continents. Adapted from John J.W. Rogers, "A History of Continents in the Past Three Billion Years," *Journal of Geology* 104 (1996): 93, 95, 99, and 101. Reproduced with permission of the University of Chicago Press.

supercontinent known as Pangea. It was around this time that Africa and South America careened into eastern North America. Pangea began to break up 225 Ma ago, eventually fragmenting into the continents we know today.

The Geology of Northern Ontario

These events provide the backdrop for the contemporary geological setting of Northern Ontario. This part of the province is dominated by the Precambrian Shield, a huge land mass composed of old bedrock that dips toward Hudson Bay to the north and Lake Ontario to the south (fig. 1.2a). The Shield is geologically important as it contains locally mineralized strips of rock running east to west known as "greenstone belts," often as much as 100 kilometres wide.[4] In the northern and southern reaches of Ontario, the Shield is covered by younger sedimentary rocks that were deposited when shallow seas covered these areas. The Shield rocks belong to the Precambrian Era, a time span that began 4 600 Ma ago and lasted until 543 Ma ago.

As illustrated in figure 1.2b, the rocks of Northern Ontario can be divided into three structural provinces: Superior, Southern, and Grenville. All were subjected, at one time or another, to a lengthy period of mountain-building (or orogeny). On this map, the Sudbury Structure appears as part of a zone situated between the Superior and Southern provinces. The Superior Province, containing 450 significant mineral deposits, features rocks that were affected by a lengthy mountain-building and faulting phase some 2 500 Ma ago. This was followed by a long period of erosion, which lowered the rock some eight kilometres. The eroded material was moved southward by rivers and deposited in shallow water to form a wide belt of sedimentary rocks 480 km long and up to 70 km wide and 12 km thick in the Sudbury area.[5] These banded rocks can be seen north of Highway 17 East, near Coniston. Elsewhere, they form the prominent ridges of the La Cloche Mountains made famous by the Group of Seven.

The Southern Province, which contains the bulk of the City of Greater Sudbury, extends northeast from Whitefish Falls to Lake Temiskaming. It consists of rocks that were formed some 1 600 to 1 900 Ma ago. Within this zone, a high mountain range comparable to the Himalayas today was formed in the Killarney area. Further south, the region underwent the Grenville mountain-building phase about 1 000 to 1 300 Ma ago, the last major geological event in the Sudbury region. The contact between the Grenville zone and Southern Province is known as the Grenville Front. It constitutes one of the most visible elements of the geologic skyline near Wahnapitae along Highway 17 East and Highway 69 South, around two kilometres south of Richard Lake. Other tectonic events and fault activities took place during this interval, indicating that the area may have represented the margin of a gigantic rift zone between converging crustal plates, one that extended into the mantle.[6] It can thus be said that the Sudbury area has been shaped by geological forces of global magnitude.[7] A better understanding of these forces has only developed since

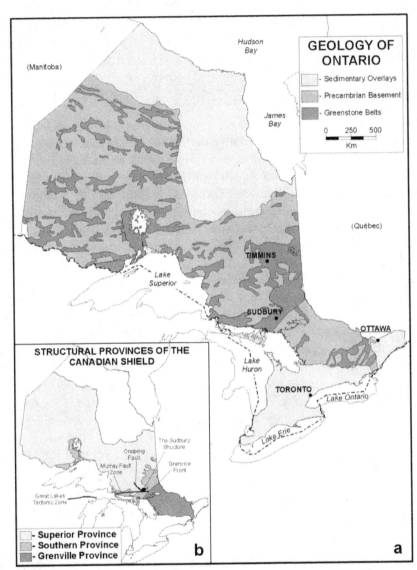

FIGURE 1.2 Geological Elements of Ontario, from J. A. Robertson and K.D. Card, *Geology and Scenery: North Shore of Lake Huron Region* (Toronto: Ontario Ministry of Natural Resources), 8. Copyright: Queen's Printer for Ontario, 1972. Used with permission.

the1960s, when experts advanced new theories that led to considerable debate and controversy. A consequence was the involvement of new researchers, including NASA scientists. Table 1.1 gives a more detailed accounting of the geological history of the Sudbury region. As per geological convention, the time scale runs from the present to the past.

TABLE 1.1   Geological time scale for the Sudbury Structure

Era/Period	Major geologic events in the Sudbury area	Millions of years (Ma) Ago
Cenozoic		
Quaternary	Continental glaciation (Pleistocene)	.01–1.8
Tertiary	Deformation of East range	
	Wanapitei Lake meteorite impact	37
Mesozoic	No Mesozoic or Paleozoic sedimentary deposition	
(65–250 Ma)	in Sudbury Area	
Paleozoic	Formation of Manitoulin Island	
(251–542 Ma)	Long period of erosion (Lipalean Interval)	
	Mountainous shield reduced to low peneplain	
Precambrian	Grenville Orogeny and the Grenville Front	
Proterozoic		
(543–2 500 Ma)	Deformation of the Sudbury Structure	1 000–1 300
	Deposition of Chelmsford and Onwatin	
	Formations in the Sudbury Basin	1 650–1 850
	Sudbury Event. Shatter Cones.	1 850
	Penokean Orogeny	1 700–1 900
	La Cloche Range uplifted and comparable to Himalayas in height	1 600–1 900
	Formation of quartzite ridges of La Cloche Range	2 200–2 450
	Large volumes of magma emplaced in the Sudbury area to form base of Huronian Supergroup	2 219–2 450
	Long erosional interval (Eparchean)	
	Wearing down of Algoma Mountains	
	Abrupt change from granitic to sedimentary rocks	
Archean (2 500–c. 4 600 Ma)	Kenoran Orogeny and formation of Algoman Mountains	2 500
	Formation of Levack Gneiss Complex, forming the basement of the Sudbury Igneous Complex (SIC), upon which the Sudbury Structure was later superimposed	2 654–2 711
	Rocks containing iron are injected northwest of Lake Wanapitei (Moose Mountain).	
	East–west trending Greenstone Mineralized Belts	2 600+

Source: Anthony J. Naldrett, "From Impact to Riches: Evolution of Geological Understanding as Seen at Sudbury, Canada," *GSA Today*, February 2003, 4–9; D.H. Rousell, W. Meyer, and S.A. Prevec, "Bedrock Geology and Mineral Deposits," in *The Physical Environment of the City of Greater Sudbury*, Special Vol. 6, *Ontario Geological Survey* (Sudbury: Ontario Geological Survey, 2002), 46; J.A. Robertson and K.D. Card, *Geology and Scenery of Lake Huron Region*, 5, and from other sources.

The Sudbury Event

The Sudbury Event brought the region onto the world stage of geology. This term has been used to describe a phenomenon of gigantic proportions that occurred 1 850 Ma ago: the collision of two worlds and a violent release of energy that formed the Sudbury Structure, which led to crustal intrusions of vast mineral deposits of nickel, copper, and precious metals.[8] The event "ushered in one of the most violent periods of explosive volcanic activity ever recorded in the rocks of the earth's crust."[9] It was unique in that it constituted the first demonstrated case of impact-triggered volcanism, a process long proposed to be the origin of lunar features.[10] The extraterrestrial impact came in the form of a huge meteorite that, in a microsecond, created 31 000 cubic kilometres of impact melt (six times the volume of Lakes Huron and Ontario combined, and 70 per cent more than the melt at Chicxulub, Mexico).[11] It transformed the local rocks into the Sudbury Structure, a huge geological feature encompassing some 15 000 square kilometres of land.[12] Of the 182 known impact structures on the surface of the earth, it is the third largest by diameter (~260 km) and likely the fourth oldest.[13]

This momentous event remained in the shadows of scientific discovery until the publication of Bell's geological map in 1891, at which time the Sudbury area became the focal point of attention for international geology. Bell, who undertook a detailed mapping of the area between 1888 and 1890, was the first to portray the existence of an oval geological basin with elevated northern and southern ranges.[14] Coleman later showed, in 1905 and 1913, that the basin was continuous, and connected to a larger body of surrounding igneous rock which was later known by several names, such as the Nickel-Bearing Eruptive, the Sudbury Nickel Irruptive, or Nickel Irruptive. More recently, it has become customary to call the surrounding belt the Sudbury Igneous Complex (SIC).[15] The brilliance of Coleman's maps at the time is shown by the fact that his placement of the rock types differs very little from present-day maps. Meanwhile, Barlow established that a close association existed between the nickel-copper ore deposits and the lowest sublayer of the SIC known as norite.[16] Most geologists today favour the idea that the ores were introduced as crustal intrusions of magma, possibly modified by a later meteor impact.

Since the discovery of the ores in 1883, geologists have paid a great deal of attention not only to the origin of the SIC itself but also to a plethora of associated features. These include the actual geographical extent of the Sudbury Structure, how the unusual rocks known as Sudbury Breccia were formed, and the shape of the ore body underlying the structure. It was thought until the 1960s that these issues could be explained by the theory that the complex had a volcanic (endogenic) origin. In two highly controversial landmark papers however, Dietz concluded in 1962 and 1964 that a meteorite impact was the only plausible origin for the Sudbury crater (astrobleme).[17] He correctly predicted that shatter cones (conical fracture surfaces with fanning striae up to three metres in length), a trademark of astroblemes, would be found in rock formations surrounding the basin. Since then, arguments both for and against this view have been the subject of numerous studies.

During the twenty-fourth annual meeting of the Geological Association of Canada held in Sudbury in 1971, a consensus emerged that support for the meteorite impact theory had grown; most geologists went from believing a meteorite impact was possible to probable.[18] While evidence for the newer theory is now almost overwhelming, there continues to be support for the belief in some form of volcanic involvement.[19] This debate underscores the fact that the Sudbury area continues to be one of the world's greatest geological enigmas; indeed, one author refers to this area as being the "Gordian Knot of the Canadian Shield."[20] Rousell and Card have cautioned against the complete acceptance of all aspects of the meteorite impact theory, noting that there has been a "tendency to ignore or force-fit certain aspects of Sudbury geology which are not in exact harmony with the impact model."[21] Before dealing with these extraterrestrial and volcanic theories in more detail, I present a brief review of the main features associated with the Sudbury Structure, the SIC, Sudbury Basin, and the outer footwall zone.

Components of the Sudbury Structure

The Sudbury Structure consists of three major components: (1) the SIC, (2) the inner Sudbury Basin (often referred to by locals as the Valley), and (3) an outer "footwall" zone of shatter-coned and fragmented rocks.[22] These spatial features are illustrated in figure 1.3a. The Sudbury Structure rests on a great dome of granitic rocks associated with the Superior Province. The structure can be thought of as a giant bathtub, with the SIC forming the basin, the sedimentary rocks of the Whitewater Group filling its inner lining, and the broken material called breccia covering its outer lining.

The SIC has an elliptical shape 60 by 30 kilometres in size, and a surface width of approximately 2 to 5 kilometres. Its rims are known as the North, East, and South Ranges. The complex is layered, consisting of a lower portion known as norite, and an upper one called micropegmatite. The base norite layer contains the majority of the ores associated with the Sudbury area. Its oval distribution is reflected on the surface through the circular pattern of minesites encompassing Falconbridge on the west, Garson and Creighton on the south, and Levack on the northeast. The micropegmatite layer on a geological map is easy to ascertain. Being less resistant to erosion than the norite zone, it has a lower elevation and is more deeply scoured; thus, it serves as a natural repository for many lakes, such as Fairbank, Whitewater, Whitson, Capreol, Joe, Nelson, Moose, and Windy.

The Sudbury Basin within the SIC is occupied by rocks of the Whitewater Group. This grouping, named after a lake, is approximately 2 900 metres in thickness and consists of four formations which are, in ascending order, called the Onaping at the base, Vermilion, Onwatin, and Chelmsford at the top. They play host to a variety of small mineral occurrences, whose origins are different from those of the SIC. The age of the Whitewater rocks are thought to approximate that of the Sudbury Event. Each of the formations associated with the Whitewater Group has added to the unique character of the Sudbury Structure.

The basal Onaping formation has a thickness of some 1 400 metres, and consists of fractured rock that was deposited very rapidly. Theories of how this formation came into being are contentious, and critical to unravelling the origins of the Sudbury Structure. While some continue to believe that the structure represents the product of explosive volcanism,

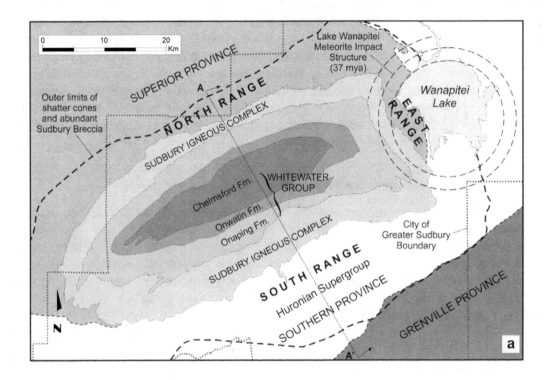

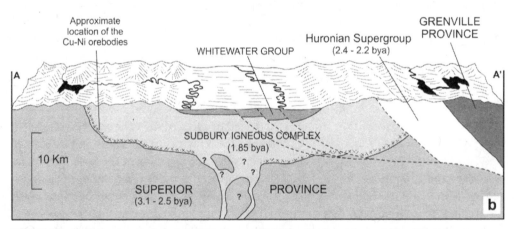

FIGURE 1.3   Sudbury Igneous Complex (SIC). From Léo Larivière, Cartographer and Map Librarian, Laurentian University.

others suggest that it is the remnant "fall-back" of broken and melted rock blown out of the crater by a meteorite impact, with perhaps some impact-induced volcanism.[23]

The Vermilion, a narrow formation with an average thickness of only 13.5 metres and limited surface exposure, hosts the zinc, copper, and lead deposits associated with some of the earliest minesites such as the Vermilion and Errington. The Onwatin Formation is estimated to be anywhere from 600 to 1 400 metres thick, and consists of black carboniferous slate-like material originally deposited in a deep basin of stagnant water. The presence of carboniferous material is possible evidence that life existed there at one time. There was some early interest in these veins as a source of fuel; however, this hope faded as the material did not burn well, due to its high ash content.[24] The Chelmsford Formation, approximately 850 metres thick, consists of beds of sandstones laid down by water currents. It underlies much of the flat agricultural land found in the Valley. This formation was folded after the Sudbury Event to form synclines and anticlines that appear as bedrock ridges and lowlands scattered throughout the Valley.

The outer footwall zone of the Sudbury Structure consists of Sudbury Breccia, a fragmented rock characterized by deformational features that resulted from the Sudbury Event. The footwall can be found 35 to 80 kilometres north and east of the SIC, and is about 9 to 40 kilometres wide to the south, where it abuts the Grenville Front.[25] This belt is of economic significance as it hosts nickel, copper, and other mineral deposits. The occurrence of these fragments set in rock at the outer margin of the Sudbury Structure suggests that the diameter of the original impact crater was much larger, perhaps in the order of 200 kilometres. Numerous shatter cones can be found in this zone as far as seventeen kilometres from the SIC. The ring exhibits shock features commonly associated with other impact sites, such as those found at the Vredefort ring in South Africa, and the Riess structure near Nordlingen, Bavaria in Germany.[26]

Origin of the Sudbury Structure

There are two basic theories to explain the origin of the Sudbury Structure and its ore deposits: a volcanic model and a meteorite impact model.

Volcanic Model

Geologists who adhere to the volcanic model, postulated since the late 1800s, consider the Sudbury Structure to be either an intrusive sill, a ring dike with the norite and micropegmatite layers as separate intrusions, or a laccolith, which is a funnel-shaped intrusion (fig. 1.4).[27] At the time, experts believed that the only source that could produce the energy to form the Sudbury breccias had to be volcanic in origin. Between 1890 and the First World War, this model was based on the sill theory. The theory, supported by Bell (1891) and Barlow (1904 and 1907), was simple in conception: magma had gradually worked its way up through the earth's crust via a fault or fissure, and spread out horizontally near the surface between the Whitewater Group of rocks and the Precambrian basement, where it then cooled, crystallized, and differentiated under the influence of gravity in situ to form the norite and micropegmatite layers. While Coleman was basically in agreement with this concept in 1905, he considered the SIC to be a sheet that had been folded into a lopolith (saucer/spoon-shaped) form. Since these concepts could not account for the fragmented character of the surrounding rock, and the amount of micropegmatite associated with the SIC, it was eventually replaced by other theories.

In 1926, Phemister endorsed the proposal Knight had made in 1917 that the Sudbury Basin was a caldera, cast doubt on the principle of magmatic separation in situ, and introduced the idea that the norite and micropegmatite layers were the result of two separate intrusions. This led Burrows and Rickaby (1929) to suggest that the elliptical outline of the SIC might reflect the former presence of a highly explosive ring of volcanoes around a sedimentary basin. Their idea was later pursued by Thomson and Williams in 1954, who posited that debris from this volcanic activity spewed through fissures as "glowing avalanches," settling in the lower-lying areas to form the Onaping Formation. This formation was then covered with the shales of the Onwatin Formation and the sandstones of the Chelmsford Formation, derived over eons of time from the erosion of volcanic fragments and subsequently deposited under water.[28] The major criticism of this concept centred on the fact that a volcanic eruption would not only have discharged breccia debris toward the centre of the basin but into the surrounding area as well; thus, the ring-dike theorists had difficulty explaining the absence of conglomerates outside the basin. In an effort to account for this anomaly, Thomson and Williams resurrected the caldera collapse theory. In 1959, they suggested that the rapid discharge of more than 1 250 cubic kilometres of avalanche debris caused an immediate dropping of a great block in the basin, forming a deep sink of perhaps one thousand metres. They postulated that magma intrusions then formed along the ring faults. Over time, the volcanic residue outside of the Sudbury Basin eroded, while the Onaping formation remained preserved within the basin itself. While the theory apparently gave the *coup de grâce* to the myth of the Sudbury lopolith, as late as 1978 Card still remained cautious, stating that it was too early to deduce the original shape of the SIC.[29]

In the final analysis, the majority of geologists today claim that the volcanic models fail to account for the tremendous source of heat required to produce the widespread fragmentation and shock effects found in the Sudbury area. As well, they assert that the volcanic proponents are focused mainly on the SIC, without accounting for its unusual chemistry, and have left the larger context of the Sudbury Structure for others to define.[30]

Meteorite Impact Theory

Interest in the geology of the Sudbury area was given a dramatic boost in 1964, when Dietz (prior to even visiting the area) made public his astonishing proposition that the Sudbury Structure was an astrobleme whose formation was initiated by a large meteorite, about four kilometres in size, in Middle Precambrian time. He suggested that a meteorite struck the earth at Sudbury, exploded, and excavated a shallow crater about 50 kilometres across and 3.5 kilometres deep (fig. 1.4). Shockwaves spread out and caused severe breakage in the rocks, giving rise to the Sudbury Breccia and forming shatter cones in the footwall rocks surrounding the outer margin of the SIC for distances of ten to fifteen

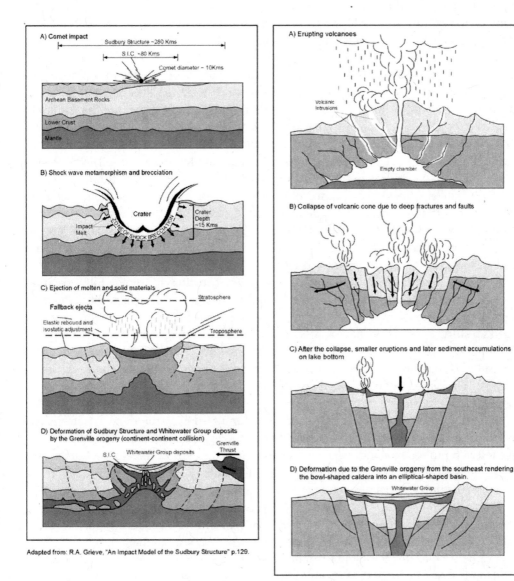

FIGURE 1.4 Volcanic and Meteorite Impact Models. Adapted from Grieve, "An Impact Model of the Sudbury Structure," in Lightfoot and Naldrett, *Proceedings of the Sudbury-Noril'sk Symposium*, 129; Peredery and Morrison, "Discussion of the Origin of the Sudbury Structure," 505–6; Stephenson et al., *A Guide to the Golden Age*, 1–6. Reproduced with permission.

kilometres. Dietz interpreted the SIC as an impact melt sheet, and the nickel, copper, and precious metal ores as extraterrestrial.[31] While many aspects of his thesis are now widely accepted, his theory regarding the origin of the ores has few adherents. Until recently, most researchers supported the view that the ores of the SIC were emplaced by magma from the lower crust immediately following the impact; however, there has been increasing support for the concept that the SIC was produced not by internal melting processes, but rather as an impact melt during the event itself.[32]

Newer studies give credence to the meteorite impact theory, citing, for instance, evidence of numerous shatter cones in the outer footwall rocks, and the presence of shock metamorphism within the Sudbury Basin. Newer assessments suggest that the meteorite was ten kilometres in diameter (roughly the size of Mount Everest), and may have penetrated the earth to a depth of ten to fifteen kilometres.[33] Estimates of the diameter of the Sudbury crater have also grown from 50 to 280 kilometres or more. Dietz's theory gained additional credence with the discovery that the Onaping Formation represented in part a "fallback breccia" composed of material explosively excavated from the crater by the impact that was immediately re-deposited. Assertions that the rocks and structural features associated with the Sudbury Structure were consistent with an origin of meteorite impact also increased his theory's credibility.[34] Support for this interpretation comes from researchers who have concluded that the impact event, the emplacement of the SIC, and the deposition of the Onaping Formation constituted "geologically instantaneous" events.[35] Others have been more definitive, stating, for example, that "the only improbable or rare event in this scenario is the large-scale impact—clearly the largest in Earth's history for which we have a well-studied record. The magnitude of the ore deposits in themselves demands one rare event or phenomenon."[36] In the same vein, Peredery and Morrison have asserted, "until it is demonstrated unequivocally that shock metamorphism can be caused by volcanic activity, the shock features will continue to be the criteria for identifying meteorite impact sites."[37]

During the 1980s and 1990s, more proof accumulated in favour of an extraterrestrial impact. Established in 1984 to initiate deep earth studies, Lithoprobe, the largest geoscientific research project ever undertaken in Canada, yielded in 1990 a three-dimensional view suggesting that the SIC is continuous beneath the Sudbury Basin, and that there are no structures or bodies at that depth that would be expected for an endogenous feeder system of magmatism.[38] Other evidence came in 1994, when geologists discovered fullerenes (or buckyballs, named after Buckminster Fuller, the inventor of the geodesic dome) in carbonaceous material in the breccias associated with the Onaping Formation. Fullerenes are pure carbon molecules containing helium arranged like a "soccer ball cage." The current thinking is that they originated from a meteorite previously associated with a red giant or carbon star nearing the end of its life.[39] In a 2005 study, Addison et al. concluded that ejected material associated with the Sudbury impact event could be found in an underground layer 650 kilometres to the northwest of Sudbury near Thunder Bay, and 875 kilometres to the west near Hibbing, Minnesota.[40] There has even been the suggestion that material from Sudbury could have been deposited as far east as southern Greenland.[41] In 2009, two researchers made the controversial proposition that Sudbury's continental shelf location and its submergence under, or nearness to, an existing ocean at the time of impact (figure 1.1) set into motion a giant tsunami that altered the ocean's ecosystem, thereby marking the beginning of life processes on our planet.[42] This concept of the Sudbury Event having an oceanic impact will undoubtedly spur further studies in this direction.

Card has remained cautionary with respect to the impact theory, stating that the Sudbury Structure must also be viewed as an integral part of its regional setting in space and time. He points out that both the location and shape of the Sudbury Structure are closely related to tectonic features, including the junction of three structural provinces and two main fault systems, all lying within a possibly major Precambrian rift system. Thus, the SIC is only one of a series of similar intrusions that could have occurred in this rift system.[43] While in general agreement with the meteorite impact theory, Rousell too has concluded that the meteorite impact was only one event in a complicated geologic setting that simply "accentuated on-going ore forming processes and magmatism in Sudbury."[44] Whatever the origin, Speers has suggested that the formation of the SIC was extremely rapid and "constituted one of the most violent periods of explosive volcanic activity ever recorded in the rocks of the earth's crust."[45]

Post-Sudbury Events: Mountains, Erosion, and the Wanapitei Crater

During the interval between the Sudbury Event and the end of the Precambrian Era, a complex series of crustal activities took place. There was significant mountain-building activity in the Grenville Province south and east of Sudbury on at least five different occasions. There were other disturbances as well, resulting in a landscape with northeast-trending structures. Rocks of the Whitewater Group in the centre of the Sudbury Basin were folded. The original shape of the Sudbury Structure became tilted and more oval in response to thrusts from the south, including the final closure associated with the Grenville mountain-building phase.

Following the Precambrian Era, a long period of erosion ensued. During this lengthy interval, which lasted until the start of continental glaciation, the mountainous terrain of the Precambrian Shield and the Sudbury Structure was reduced in elevation to its present level. Unlike the Manitoulin Island area, the Sudbury area was not affected by the deposition of sedimentary rocks during the Paleozoic or Mesozoic eras. The uniqueness of the Sudbury Structure was given an added dimension in 1972, when geologists revealed that another meteorite had hit Lake Wanapitei 37 Ma ago.[46] The original crater was estimated to be 8.5 kilometres across. While the meteorite had some influence on the shape of the Eastern Range of the SIC, its overall impact remains uncertain. If the Sudbury Event can be interpreted as the result of a meteorite, then the Sudbury-Wanapitei crater relationship must be considered one of the rarest combinations of impact events recorded on earth in the last three billion years.

Laurentide Ice Sheet

While the Sudbury Event was the most important one that shaped the local physical environment, the influence of the last ice age cannot be understated. Twenty thousand years ago, the Sudbury area was covered by an ice sheet one to two kilometres thick. Part of a larger continental glacier, this ice sheet covered all of Ontario and extended into the northern United States (figure 1.5). This, however, was not the first time that Sudbury was glaciated. Great ice sheets covered parts of Ontario on several occasions during the Precambrian era some 2.4 billion years ago. The most recent ice advance occurred in the Quaternary Period, the youngest period of the earth's history. The Quaternary began about 1.8 Ma ago, when the world cooled and large parts of North America and Europe were intermittently covered by continental ice sheets. The Pleistocene Epoch of the Quaternary (or the so-called Great Ice Age) began some one million years ago, during which time continental-scale ice sheets expanded and retreated. The Wisconsin Episode took place from 115 000 to 10 000 years ago. In the ensuing period, the Sudbury area was deglaciated and huge lakes formed that covered the entire area northward from Lake Huron and Georgian Bay up to the north rim of the SIC. These glacial lakes eventually retreated to form the present-day Great Lakes.

The earliest evidence related to local glaciation dates back to 1894, when abandoned shoreline features were found along the Canadian Pacific Railway (CPR) mainline near Cartier. In 1901, interest in glacial deposits was piqued by the discovery of placer gold deposits along the Vermilion River. While sporadic studies were undertaken later, it was not until the late 1960s that Quaternary geologists, including A.N. Boissonneau, G.J. Burwasser, P.J. Barnett, and A.F. Bajc began detailed mapping and analysis on a regional scale. Since then, a more comprehensive view of the region's glacial history has emerged.[47]

Glacial Advance and Retreat

The last glacial advance left a spectacular record of erosion and deposition. Evidence of its erosional effects is everywhere. By means of downward pressure and the assistance of rock fragments carried by the ice, the forward movement of the ice sheet sculptured the landscape through abrasion, from tiny etchings on bedrock surfaces to large, streamlined bedrock forms. The Sudbury area is distinctive in that it has some of the best sculptured forms seen in Ontario. Among the abrasional impacts are polished rock surfaces and striae (long linear scratches) that determined the direction of the ice's movement. Powerful subglacial meltwater discharges likewise had a major effect in creating bedrock-sculpted forms.[48]

The trend of striae and grooves to the south and southwest, paralleling the oval shape of the Sudbury Structure, indicates that the local topography influenced the movement of glacial ice. One striking example of this effect can be found at Bailey's Corner southwest of the Sudbury Airport, where the glacially sculptured bedrock is smooth and polished on the advance (northeast) side, and quarried and plucked on the downflow (southwest) side. The flattening impact of the ice, however, is best viewed from a distance. While the topography of the Sudbury area appears rugged at ground level, the skyline effect known as peneplanation on the south and north rims of the SIC becomes evident when viewed from any height, such as from the Parker Building on the Laurentian University campus or on the Southwest Bypass. It has been estimated that the lowering effect of glacial advance was in the order of tens of metres.[49]

The advancing ice sheet was also responsible for depositing debris by direct placement at its base, or along its frontal margins during a temporary stop phase. The direct lodgement process involved the deposition of unsorted sediment known as till or ground moraine. While absent or deeply buried in the Valley, till deposits can readily be found elsewhere, generally in thicknesses of less than one metre over bedrock. Not very productive for agricultural purposes, these areas were nonetheless favoured by Finnish settlers, who acquired farms in places such as Wanup, Long Lake, Whitefish, and Beaver Lake. While these thinly mantled areas are frequently associated with bedrock exposures, an offsetting factor has been their ability to produce blueberry patches. These poor till deposits form a sharp contrast to the more extensive till plains found just north of Toronto, where they are deeper and devoid of rock features.

A different form of till deposit known as an end or recessional moraine was laid down north and east of the Valley, where the retreating ice sheet came to a temporary halt, and a large accumulation of flow till developed along its frontal margins. This resulted in the formation of a moraine ridge some five to ten metres high. This ice-marginal feature is

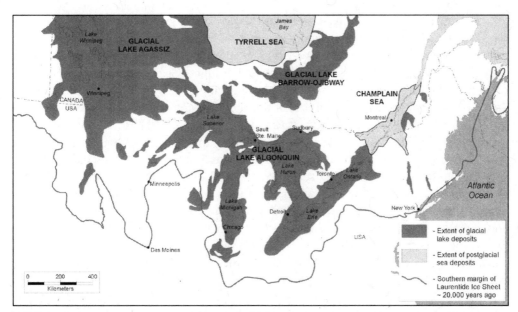

FIGURE 1.5 Extent of the Laurentide Ice Sheet and Postglacial Lakes. From J.T. Teller, "Proglacial Lakes Along the Southern Margin of Laurentide Ice Sheet" in *North America and Adjacent Oceans During the Last Glaciation*, DNAG Volume K-3, ed. W.F. Ruddiman and H.E. Wright (Boulder, CO: Geological Society of America, 1987), 39–69.

part of a larger Northeastern Ontario formation that runs from Batchawana Bay on Lake Superior through Cartier and Capreol on the North Range to the southern end of Lake Temiskaming. The large boulders (known as erratics) that were carried by the ice sheet to the Sudbury area from other areas attests to the former presence of continental glaciers. These erratics litter the surface of the region and are particularly noticeable in the Onaping Falls area.

Stratified Deposits and Postglacial Lakes

Within the Valley and along the northern and eastern rim of the SIC, glacial deposits of a different kind can be found. Unlike the unsorted till deposits discussed above, these consist of sediments reflecting the influence of meltwater as the ice sheet started to retreat around 10 500 years ago. As the deglaciation process occurred during a relatively short period of time, approximately five hundred years, many deposits did not fully develop. These stratified materials, known as glaciofluvial deposits, vary according to their method of deposition; that is, whether they were laid down on, within, or under the glacier (known as ice-contact deposits), in contrast to those laid down beyond the ice margin (referred to as outwash deposits). The Valley acted like a huge trap that captured the incoming sediments carried by glacial meltwater streams such as the Onaping River, Sandcherry Creek, and the Nelson, Rapid, and Vermilion Rivers. Finer-grained sediments were dispersed and deposited in the Sudbury area under Glacial Lake Algonquin, the precursor to the present-day Great Lakes (figure 1.6).

Stratified sediments formed under or at the edge of the ice sheet constitute an important feature of the Quaternary deposits found in the area. Many of these deposits are readily observable because they now serve as locations for companies providing sand and gravel, mine backfill, and crushed and screened products. Several types of these layered deposits exist. The largest consists of a delta running southwest of Wanapitei Lake to Falconbridge and Garson, forming part of the Sudbury Airport Glaciofluvial System. Laid down along the receding ice margin, it is a major landscape element. The northern part of the system was laid down in an ice-walled channel that may have attained a maximum width of six kilometres. The sand and gravel delta was formed by meltwater discharge that flowed from the ice sheet into a glacial lake. It created a large plateau that now serves as the site of the Sudbury Airport and several sand and gravel quarries. The plateau's edge can readily be seen on the west side of the airport as a sharp terrace. Up to fifty-five metres of sorted material has been discovered above the bedrock in the vicinity of the airport.

Associated with this feature are a complex of twenty-five kettle lakes and an esker that runs southwest from Bolands Bay toward the Falconbridge and Garson townsites. This complex was created when large blocks of ice became separated from the main ice sheet. As these blocks were later covered and insulated by sands and gravels, they remained long after the main ice sheet left. When they eventually melted, they left large holes (kettles) now occupied by ponds.[50] With their deep terraces and depths sometimes in excess of forty-five metres, these kettles represent some of the most distinctive landforms in the area. Other ice-marginal deposits, known as outwash deposits and deltas, occur within all of the major south and east flowing river systems associated with the North and East Ranges. A number of these deposits have as many as five or six terraces linked to the declining levels of Glacial Lake Algonquin. The Dowling delta which occupies some eight to ten square kilometres is an excellent example of this feature; others include the Capreol and Suez deltas, situated northeast of Hanmer. Another system of eskers and outwash deposits extend southwest from the centre of Levack Township to beyond Windy Lake.

The retreat of the last ice sheet occurred in phases, controlled by the location of the receding ice margin, and the opening of outlets in the North Bay area.[51] The melting ice sheet resulted in the formation of Glacial Lake Algonquin, which, at the time, extended beyond the boundaries of today's Lake Huron and Georgian Bay. The Sudbury

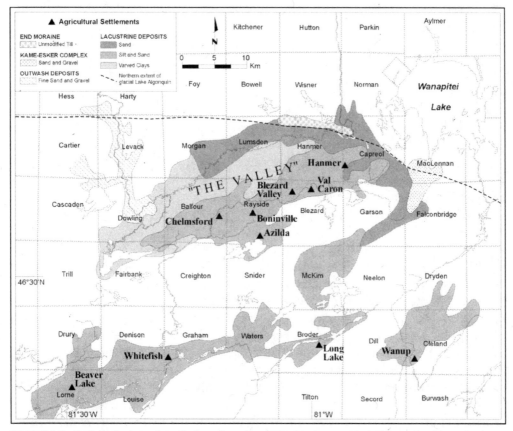

FIGURE 1.6 Surficial Geology of the Sudbury Area. From O.W. Saarinen and W.A. Tanos, "The Physical Environment of the Sudbury Area and Its Influence on Urban Development," in *The Physical Environment of the City of Greater Sudbury*, Ontario Geological Survey, Special Vol. 6, ed. D.H. Rousell and K.J. Jansons (Sudbury, ON: Ontario Geological Survey, 2002), 8. See also A.N. Boissonneau, *Algoma, Sudbury, Timiskaming and Nipissing Surficial Geology*, Map No. S465 (Toronto, ON: Ontario Department of Lands and Forests, 1965).

Basin remained totally under water until 9 500 years ago. Following the uncovering of an outlet in the Mattawa area, Lake Algonquin disappeared, leaving in its wake the stratified plain that currently occupies the centre of the Valley.[52] The nature of these deposits is closely related to the topography of the Valley, which slopes gently in elevation to the southwest from about 300 to 260 metres over a distance of about 39 kilometres. The lake left sandy deposits behind in the shallower areas associated with Chelmsford and Hanmer, and laid down clays in the deeper areas to the south and southeast. These sediments are up to 125 metres thick. The presence of these glaciolacustrine deposits, along with the groundwater resources (now known as the Valley East Well System) encouraged an extensive agricultural economy to flourish among the francophone population in the Valley throughout the past century.[53] Outside of the Valley, smaller pockets of lacustrine clays were deposited in several areas, such as those found southeast of the Falconbridge townsite, in parts of the former city of Sudbury, and in the low-lying stretches of land adjacent to Long, McFarlane, and Richard Lakes.

Postglacial Landscape

While the main elements of the Sudbury landscape were fashioned by geological and glacial forces, other influences subsequently came into play which modified the existing landscape. New drainage patterns and lake systems developed. Glacial and postglacial deposits were transformed into soils of varying characters. Climate impacted the soil to create vegetative communities. Over thousands of years, these elements combined with one another to create a distinctive landscape that provided the physical and economic setting for settlement of the area by Aboriginal cultures.

Waterway System

A waterway system consisting of rivers, lakes, and swampy zones emerged in the postglacial era as an integral aspect of the natural environment. Widely used as a mode of transport by the Aboriginal population, explorers, surveyors, and the first white settlers, parts of the system assumed an important economic role, serving as the highway for log drives in the forest industry, electrical power generation for the mining industry and local municipalities, and a source of water for companies, businesses, and homes. After the First World War, rivers and lakes became attractive for residential and cottage purposes, a trend that has continued to the present day. More recently, the system has become an important asset to the quality of life for the local citizenry, not only for its scenic and ecological value but also for outdoor recreational uses such as swimming, fishing, ice skating, and boating. Indeed, the use of these waterways has developed so much that Sudbury is increasingly viewed as a "city of lakes" rather than a "city of rocks."

Due to the prevailing south-facing topography, the Sudbury area lies entirely within the Georgian Bay (Lake Huron) watershed. The regional drainage pattern is dominated by two main river systems: the French and Spanish (figure 1.7). Geologically controlled by faults, the flow of water reflects a northeast–southwest orientation. Examples of this phenomenon are visible in Vermilion and Long Lakes. Within the main watershed, twenty-five sub-watershed units exist.[54] The French River system drains the eastern part of the city through the Wanapitei subsystem. The Wanapitei River, with its main channel some 210 kilometres long, flows from the north into Lake Wanapitei and from the south to the French River. The Spanish River system drains the rest of the municipality via the

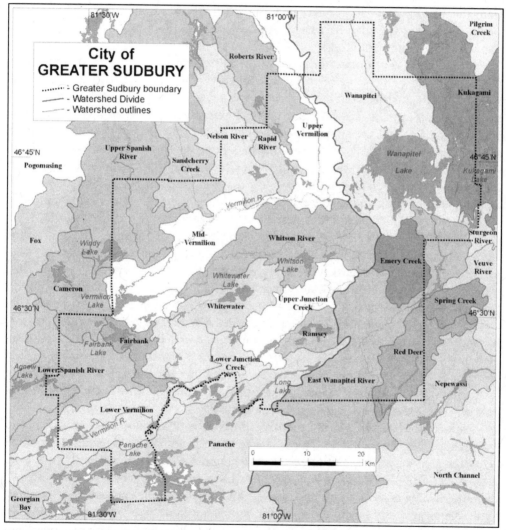

FIGURE 1.7   Local Watersheds. From Rousell and Jansons, eds., *The Physical Environment of the City of Greater Sudbury, Ontario Geological Survey*, Special Vol. 6 (Sudbury, ON: Ontario Geological Survey, 2002), Map 3.

Vermilion. Five subdivisions of this system exist: Vermilion River, Onaping River, Whitson River, Junction Creek, and Levey Creek.

The Vermilion River, the largest in the area, has a length of approximately 246 kilometres. It enters the basin near Capreol, and proceeds from there westward to the confluence of the Onaping River before broadening to form Vermilion Lake. Along the Vermilion River's course, alluvial deposits have formed intricate meanders (winding shapes) that contrast sharply with the more rugged river systems found outside of the basin. Both Joe and Nelson Lakes flow into this system. The river eventually flows into Kusk Lake, and from there to Espanola and the confluence with the Spanish River that flows into the North Channel of Georgian Bay. The Onaping subsystem drains the northwest sector of the Basin, including Windy Lake.

The physical control the north rim of the SIC exerted on this river has resulted in powerful stream gradients before it meets with the Vermilion River. The north rim provides the setting for the A.Y. Jackson Lookout and a scenic view of High Falls as it cascades fifty-five metres into the Valley. Whitson River drains much of the central valley as well as Whitson and Garson Lakes. Junction and Nolin Creeks, because of their close association with residential settlement, have been modified more by subdivision activity and highway construction than by geology and topography. Levey Creek connects Whitewater Lake with the Vermilion River. As many of the drainage areas in the Valley and downtown Sudbury are low-lying and highly flood-prone, damage from flooding has been a major occurrence in Greater Sudbury on at least forty-six occasions between 1890 and 1979.[55]

The wide profusion of lakes within the municipality is significant in that it gives the city a natural environment unrivalled elsewhere. In fact, Greater Sudbury has one of the highest concentrations of urban lakes of any city in Canada.[56] There are over 330 lakes greater than 10 hectares, and 47 lakes over 100 hectares in size within its boundaries. Of the city's total surface area, 12.2 per cent is covered by water and another 4.3 per cent by wetlands, for a total of 601 square kilometres.[57] Lake Wanapitei (13 254 hectares in size), has the distinction of being the largest city-contained lake in the world. Lake Ramsey, situated in the heart of the municipality, offers numerous opportunities for recreational and sporting activities, including summer and winter fishing. The attractiveness of regional lakes for residential living is shown by the fact that in 2003, approximately 7 000 people, or 4 per cent of the city's population, lived on a lake.

Numerous wetlands, wholly or partly covered with cattails, algae, and wet and spongy soils, dot the rural landscape. These areas are an important element in Sudbury's ecosystem, serving as excellent filters for absorbing metals and sulphur. In the shallower areas, small depressions with low water tables have accumulated inorganic and organic detritus, yielding bogs, swamps, and marshes often less than one metre deep. Larger wetlands, featuring grassy types of vegetation, have emerged in many channels linking the dispersed lake system of the region. These sites all have considerable aesthetic appeal and add to the diversity of the environment.

Climate and Changing Ecosystems

In the postglacial period, changing climates and ecosystems emerged along with differing forms of wildlife and vegetation. Temperatures then were about ten degrees cooler than modern conditions.[58] Gradually the climate came to feature a high degree of seasonality, with some of the steepest north–south temperature gradients in Canada.[59] Located in the heart of the North American land mass, the climate evolved into relatively long and severe winters and short temperate summers.[60] The climate of the Sudbury area is thus comparable to that found in southern Sweden, which is also buffeted by arctic and tropical air masses. Compared to the Prairie provinces, there is some modification of the climate due to the Great Lakes in the west and south, and Hudson Bay to the north.[61] Another aspect of the climate is the presence of four distinct seasons along with differing temperatures and precipitation types.[62]

As evidenced by the flow pattern of superstack emissions, these air masses generate prevailing wind directions from the north and southwest. Precipitation in the form of rain and snow is uniformly distributed throughout the year. While the annual frost-free period is in excess of 110 days, the growing season is considerably longer, lasting six months, beginning around April 25 and ending about October 25.[63] The Sudbury area, therefore, has one of the longest growing periods in Northern Ontario. Since the region is located along some of the major storm tracks of North America, the passage of high and low pressure systems produces wide variations in the day-to-day weather. This is especially true in summer, when warm, humid air masses from the south alternate with cooler and drier air from the north, resulting in two or three days of fine weather followed by warmer and more humid weather, often with changeable winds and rain.

While the local climate fails to reflect the superlatives of many other parts of Canada, it does have the distinction of recording Canada's eighth largest tornado on August 20, 1970, an event that left 5 dead, 200 injured, 750 homeless, and caused some $10 million in damage (figure 1.8).[64] Many local residents would be surprised to learn that the well-known *Places Rated Almanac* has given the Sudbury area the dubious honour of the lowest ranking in a bad-weather list made up of 354 American and Canadian metropolitan centres.[65] In the past few decades, however, the Sudbury climate has moderated, taking on some of the characteristics associated with Southern Ontario; some climate experts have estimated that Sudbury's average temperature since the 1970s has warmed by one degree Celsius. Some comfort for future generations can be garnered from the fact that by the next century, global warming is expected to give the Sudbury area the same weather that the Cleveland area now experiences.[66]

The climatic warming that brought about the retreat of the Laurentide Ice Sheet was followed by a lengthy period of revegetation. As the climate warmed, the boreal forest, consisting of spruce, poplar, and later, jack pine, white birch, and alder, was replaced around 7 700 years ago by more mixed species associated with the Great Lakes–St. Lawrence Forest zone such as white pine, hemlock, and beech.[67] This forest complex has persisted to the present. The white pine stumps now found in many locales hint at the huge size of the pine forests in parts of the region.[68] In contrast to areas found along Georgian Bay and parts of Southern Ontario, Aboriginal agriculture had little impact on the forest cover. The first surveyors in the Sudbury area found a surface cover that showed evidence of previous forest fires, followed by a later successional regrowth of birches and poplars. During the period of early settlement, the vegetative cover of the area was sufficient to provide material for logging, railway construction, and roast beds for the mining industry. Later, sulphur dioxide emissions from smelters had a great influence on the forest cover. Wildfires caused by railway engines, and fires deliberately set by prospectors to reveal the rock cover, also affected the forest. One positive outcome of these events was the flourishing of blueberries. In many areas, fair-sized trees survived the fires, as evidenced by the naming of the first Sudbury Catholic parish as St. Anne of the Pines. This alteration of the landscape that took place throughout the first half of the twentieth century continued after the Second World War, but in a new fashion when reclamation programs were initiated, first by the mining companies, and later by the Regional Municipality of Sudbury.

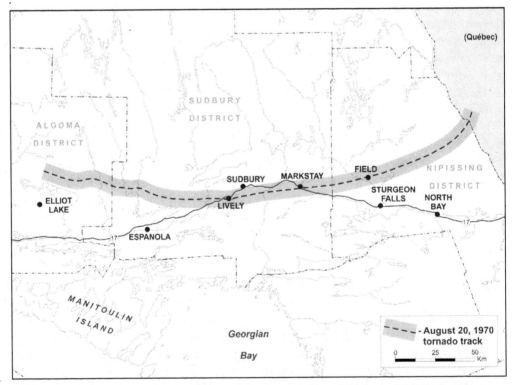

FIGURE 1.8   The Path of the Sudbury Tornado (August 20, 1970). From G.W. Gee and B.F. Findlay, *The Sudbury Tornado—August 20, 1970*, Technical Memoranda 764 (Ottawa, ON: Atmospheric Environment Service, Department of the Environment 1972), 17.

Given the short period that has passed in geological terms since the retreat of the ice sheet, there has been little time for the development of mature soils. Other factors that have influenced soil development include the type of parent material, the climate, and the nature of the vegetative cover. The dominant types of soil include those formed from (1) till, (2) lacustrine materials (loams-clays), (3) fine sand outwash (sand), (4) coarse and medium outwash (sands and gravels), and (4) organic soils (see figure 1.6). Soils derived from till are typical of the agricultural areas found in the scattered stretches of land running west to east from Beaver Lake to Wanup. Clays and loams can be found in the western part of the Valley. While these are good soils, they are often poorly drained with a high organic content. In the middle and eastern parts of the Valley, better quality loam soils prevail.

Since the parent materials were derived from Precambrian Shield bedrock, the soils are acidic with a low nutrient status. The cool, humid climate has likewise encouraged the formation of soils known as podsols. In these soils, elements such as calcium, magnesium, iron, and aluminum have been leached from the upper layers and deposited below. The presence of a mixed conifer-deciduous forest cover has enhanced the process of podsolization. In some areas other soil types exist, such as the drained sandy soils that have accumulated in pockets of exposed rock knobs and ridges, in saturated soils, and those that have formed on rock outcrops where the organic matter shows less decomposition.[69] The soils of the area have been influenced by human factors as well, such as logging (which has promoted soil erosion), and smelting activities that have increased copper and nickel levels.

Compared to Southern Ontario, the soil resource base is limited. According to the Canadian Land Inventory, the former Regional Municipality of Sudbury had 41 902 hectares of land in the 1970s designated as having class 2, 3, and 4 soils.[70] These designations mean the soils have moderate to severe limitations that restrict the range of crops and require farmers to use conservation practices. The limitations of these soils include low fertility, high acidity (or low pH), and low water-holding capacity. Of the 41 902-hectare total, 55 per cent was associated with the Valley, and most of the remainder with the former municipality of Walden. While the soils in the Valley are concentrated, those found in the outer districts are dispersed. With the addition of Dill and Cleland Townships to the new municipality in 2001, there was some augmentation of the soil resource base.

Fauna in the area includes fish, marsh birds and waterfowl, small and large mammals, woodland birds, and amphibians and reptiles. The numerous lakes in the municipality support an abundance of aquatic life and fish species such as perch, lake trout, brook trout, rainbow trout, yellow pickerel (walleye), northern pike, yellow perch, smallmouth and largemouth bass, and whitefish. Their presence has served as a major tourist and recreation attraction, a phenomenon that started with Lake Panache as a destination for American cottagers and tourist camps after the First World War. While the fish habitats of some lakes have been impacted by the smelting operations of the mining companies, recent efforts to restore them to their original state have met with some success.

The many rivers and the presence of large wetlands support numerous species of marsh birds and waterfowl. While the presence of waterfowl in the area has historically been considered only fair to poor, recent changes due to the re-greening of the environment and projects by Ducks Unlimited have improved the situation. Ducks, geese, teal, herons, gulls, loons, sandpipers, and terns are now common. The mixed environment of forests, lakes, and wetlands provide habitat for many small mammals, woodland birds, and amphibians and reptiles. At least forty-seven mammal species are known to exist locally. Among them are herbivores (white-tailed deer, moose, rabbits, and hares), omnivores (bears, raccoons, striped skunks, red foxes), carnivores (minks, otters, martens, weasels, fishers), rodents (beavers, muskrats, chipmunks, squirrels, mice and rats), as well as bats, moles, and shrews. Open woodlands have attracted ruffed grouse (partridge), and other types of woodland birds, including hawks, owls, swallows, woodpeckers, flycatchers, nuthatches, whip-poor-wills, crows, starlings, and sparrows. While amphibians and reptiles are more limited, frogs, toads, turtles, and small snakes can be found dispersed throughout the wetlands.[71] In recent years, there has been evidence of new species migrating from Southern Ontario.

This natural environment sets the framework for the chapters to follow. It will be shown, time and time again, that an appreciation of this natural environment is vital to understanding the how and why of Greater Sudbury's economic and socio-cultural development since permanent settlement began in late 1800s.

CHAPTER 2

The Aboriginal/Colonial Frontier

After the retreat of the continental glacier, the Sudbury area gradually acquired more favourable climatic conditions, and large game and plant communities began to emerge, making it possible for a fledgling Aboriginal culture to evolve. Unlike the areas situated further to the west, south, and east at locations such as Sault Ste. Marie, Manitoulin Island, the French River, and North Bay, the Sudbury region remained a fringe zone for Aboriginal settlement. There was little to attract serious interest of any kind. In the 1600s, Europeans invaded the area and began the process of exploration and mapping. It was through this process that the geographical territory of what became Sudbury began to be revealed in cartographic form. This colonial phase was marked by a gigantic struggle involving both the British and French empires that lasted for centuries. The period also witnessed the ascendency of the Hudson's Bay Company, based on the coinciding rise of the fur trade. With the signing in 1850 of the Robinson-Huron Treaty, the development of the reserve system, and the creation of two Indian reserves in the Sudbury area, the Aboriginal/colonial period came to an end.

Early Aboriginal History

Due to the paucity of archeological evidence resulting from climatic wear, the breakdown of artifacts in acidic soils, and large areas without survey data, an understanding of Aboriginal cultural history must be placed within the broader geographical context of tribal life around the drainage basin of the upper Great Lakes region.[1] Prior to the arrival of Europeans in the seventeenth century, local tribes had, for thousands of years, left their imprint on the land. While evidence is fragmented and vague, archeological research provides some clues as to the nature and extent of this imprint. Unlike Southern Ontario, the northern region is characterized by a high degree of similarity, a feature which allows researchers to draw general interpretations from limited archeological data.[2]

Archeologists have recognized four phases associated with the Aboriginal settlement of the Great Lakes region: the Paleoindian Period (ca.11 000–7 500 years ago); the Archaic Period (ca. 7 500–2 000 years ago); the Middle Woodland Period (ca. 2 000–1 000 years ago); and the Late Woodland Period (ca. 1 000 BP–European Contact).[3]

It is thought that the first peoples to occupy Northern Ontario came from the western plains. Associated with the late Paleoindian Plano culture, they laid the groundwork for all subsequent settlement.[4] There is evidence of the existence of this culture in the Manitoulin Island, Killarney, and southern Georgian Bay districts, where people used fine-grained white quartzite quarries to fashion stone tools. Archeologists have discovered stone tools that were made ca. 9 500 years ago at Sheguiandah on Manitoulin Island, and at George Lake near Killarney.[5] These sites stood on the ancient postglacial shorelines of the present-day Great Lakes. None, however, have yet to be found in the immediate Sudbury area. Throughout its 4 000 years of existence, the Plano culture survived by hunting large game animals in the existing environment; many of their sites were associated with caribou crossings. Aquatic resources, such as fish and beaver, were also important to their sustenance. As the continental ice sheets retreated, the Plano culture moved north.

In the ensuing Archaic period, people known as Shield Archaic and Laurentian Archaic spread throughout Northeastern Ontario, including the northern shore of Georgian Bay, and closer to the Sudbury area in the Spanish River watershed. Aboriginals adapted to the geographical setting by fishing in the summer and hunting and trapping in the winter. Following the same exploitative pattern adopted by their Plano culture ancestors, they were well adapted to the Great Lakes and boreal forest environments. They began to use copper and the bow and arrow. Groundstone-grooved gouges were used to build dugout boats. Climatic amelioration brought with it more extensive forests. In some regions, massive forest fires resulted in new plant growth that favoured an increase in moose population over that of caribou. Sustained by hunting, fishing, and gathering, the Shield Archaic people became more forest-oriented, and spread in small bands to almost every remote lake and river system in Northeastern Ontario.[6] Because of the presence of water in the summer and snow in the winter, these early groupings must have possessed watercraft and snowshoes.

The Middle Woodland period that began around 2 000 years ago was marked by the first appearance of ceramic pottery vessels and the development of wide trading patterns among the Algonquian language group of Indians, involving the exchange of copper tools, furs, and later, pottery vessels for corn and fish nets from the Hurons and Ottawas in Southern Ontario. Water bodies such as Lake Panache, Lake Wanapitei, the French River, and the Spanish-Vermilion Rivers served as routes for these trading patterns. During this period, two archeological traditions emerged based on the use of differing pottery styles: the Laurel and the Point Peninsula, at sites in the Sudbury area and along the North Shore of Lake Huron. Some locations indicate that moose and beaver served as important food animals. Wild rice gathering may have also been important. Burial mounds from this period exist at the Killarney Speigel site, and along Georgian Bay. The artifacts made of flint, shell, copper, and other materials associated with this period came from across a wide region, extending from Ohio to Lake Superior.

The beginning of the Late Woodland Period brought with it a more temperate climate, and a marked shift in the population of the upper Great Lakes that saw Ojibway, Cree, Algonquian, and Ottawa tribes established in the region. Forest fires forced various bands to shift from one location to another in an erratic fashion, with a homogenizing effect upon the language and the culture.[7] Another outcome of this nomadic shifting and the different seasons was the construction of temporary housing types, such as wigwams. Algonquian hunters did not practice cultivation, so they exchanged furs for agricultural products from the Iroquois situated further south. As it was difficult for men to obtain marriageable women within their own tribes due to close blood or clan relations, they sought women from outside areas. It has been postulated that this high degree of female mobility gave the Algonquians of Northern Ontario a cultural and linguistic homogeneity that today permits Natives of northern Saskatchewan to converse with Natives in Labrador. Dawson summarized the lifestyle at that time: "The situation indicated by the archeological record suggests that the people had an ability to exploit a broad variety of resources over an extended area. They developed capability to live in an unstable environment and survive on any food resource that was abundant at any given time, a response necessitated by an environment with sparse, unevenly distributed and seasonally unreliable resources."[8]

The massive attack by the Iroquois against the Hurons in 1648 was accompanied by the appearance of Iroquois war parties throughout Northeastern Ontario. According to one account in 1660, the impact of these wars was felt locally:

History records say that in the 1600s an Iroquois war party came into our area and in their chain of destruction wiped out a local link. The story is that when reaching the junction of the Vermillion [sic] and Onaping Rivers they heard of a tribe of Ojibowa [sic] camped north of the Onaping River. They made a small detour, because the Iroquois were travelling south at the time, up High Falls and then north again to the vicinity of Moose Lake where the Ojibowa had a fairly large encampment. . . . It is believed that there were only a small band of Northern Iroquois "braves" who in the short period of four hours killed possibly 200 to 300 men, women and children.[9]

After the fall of Huronia, the Algonquians came back to the area. In 1671, Father André and other French missionaries discovered that the Nipissings had returned to their homes around the lake of that name, and along the French River. Local Ojibways returned to their traditional hunting grounds, which stretched from the watershed of the Wanapitei to the drainage system of the Vermilion River, and from the Lake Panache and Tyson Lake areas northward to the height of the land. Due to cultural forces introduced after the arrival of European explorers, missionaries, and traders, Aboriginal tribes were forced to adopt new ways of life and had to deal with situations such as the devastating outbreaks of smallpox in 1670 and 1763 that depopulated parts of Northern Ontario.

The year 1763 was historic for two other reasons. Under the Treaty of Paris, France signed away its Canadian empire to Britain; a Royal Proclamation of the same year also marked the first time that Britain gave legal recognition to Aboriginal rights. Following the American Revolution, and the signing of the second Treaty of Paris (1783) and Jay's Treaty (1794), which ceded the Ohio valley and the post at Michilimackinac to the Americans, many Ojibways returned to the North Shore of Lake Huron. By this time, the area had become part of the newly created Province of Canada. Locally the importance of the fur trade increased, and tribes in the area became, in one way or another, dependent upon this activity. In turn, this brought about a heavy reliance on the use of the regional waterways. The geographical implications of these factors have been succinctly summarized by Buchanan:

> The proliferation of aboriginal sites to the south and south-west of Sudbury indicate, at least, that the French and Pickerel Rivers and the north channel of Georgian Bay were in constant use during post Nipissing times, and in all probability, for thousands of years before. Since the Wanapitei River empties directly in the French system and the Vermilion/Spanish into the North Channel, it would seem plausible that these highly navigable waters have also been utilized for a similar period of time.[10]

In the Sudbury area, three prehistoric sites have been found which bear testimony to Buchanan's conclusion. The Svensk site, situated where the Vermilion River leaves Kusk Lake near Highway 549, has revealed one quartzite point, a quartzite scraper, and a number of quartzite flakes dating from the early Archaic period. At the Post Creek site, located on the northeastern shoreline of Lake Wanapitei on Wanapitei Indian Reserve No. 11, archeologists discovered two pottery shards typical of Algonquian ware from the north shore of Lake Huron. It is thought that the site represented a fishing station during the late spring or early fall. The Erkki Saarinen site, situated on the eastern shore of Red Deer Lake in Cleland Township, also contained an exquisite ground slate gouge, a flake knife, and several chert flakes typical of the Laurentian Archaic era. It is thought that the gouge represents a trade relation with contemporaneous neighbours to the south via the Wanapitei/ French River system and Georgian Bay.[11] There are a number of other sites (more than twenty) reported across the Sudbury region, most located along waterways, including Kelly Lake and Lake Ramsey. According to Julig, the paucity of sites does not necessary mean that the Sudbury area served only as a peripheral outpost for prehistoric Indian activity; rather, it may simply be a result of the lack of adequate research in this area.[12]

The War of 1812 against the United States stimulated trade on Lake Huron, and commerce continued after the signing of the Treaty of Ghent in 1814. By 1830, the Ojibway had been forced to cede much of their territory in Southern Ontario; as well, the Ottawa were expected to migrate after the boundary disputes between the Americans and British in the upper Great Lakes were resolved in 1828. To accommodate these tribes, plans were made to create a "refuge" on Manitoulin Island, beyond the limit of colonial settlements, where a general reserve would be founded. To this end, the government of Upper Canada signed the Bond Head treaty in 1836 with the Ojibway and Ottawa tribes, who held the existing rights to Manitoulin Island, establishing Manitowaning as a self-sufficient Indian administrative centre for the entire upper Great Lakes region. By this time, a group of Ottawa families had settled in Wikwemikong. In following years, the white population along the North Shore of Georgian Bay increased, and this pressure brought about a new treaty in 1862 whereby all of the territory west of the Wikwemikong peninsula was surrendered to the Crown.

European Exploration and Mapping

In the seventeenth century, the Aboriginal landscape was transformed by the arrival of Europeans. Physical geography dictated the pace and nature of the early European exploration of Northern Ontario, and the first appearance of the Sudbury area on world maps. There were two main entryways into this area: Hudson Bay and the St. Lawrence River. The northern entry was initiated by the British and the southeastern entry by the French. The region became a frontier zone that served as the site of a titanic struggle between these two European powers (figure 2.1). From the first day that outposts were established by the British in the Hudson Bay area, and the French along the St. Lawrence River, these two nations sought to extend their holdings into their rival's territories using three classic weapons: trade with the Natives, diplomacy, and war.

For nearly two centuries the conflict raged, ending only in 1763, when the last of the wars for empire ended with Britain's victory and France being ousted from eastern North America. While the Sudbury area, lying between Hudson Bay and the Great Lakes, was never strategic and remained peripheral in a continental context, the region was impacted by British and French explorers, Recollet and Jesuit missionaries, coureurs de bois, and local bands who occupied traditional areas. In 1610, Henry Hudson sailed through the strait and bay now carrying his name during his doomed search for the Northwest Passage, until he reached one of the most southern harbours at the "bottom of the bay." Mutineers, however, eventually assumed control of his ship (known as the *Discovery*) and Hudson himself was cast adrift in a small boat, never to be seen again. The remaining crew returned to England in 1611. His death notwithstanding, Hudson made a considerable contribution to the mapping of the area.[13]

Despite its imperfections, this map proved to be useful to spur on the further exploration of Northern Ontario. According to one historian, "probably no voyage since that of Christopher Columbus had so stirred the minds of men, kindled their enthusiasm and fired their imagination as the 1610 voyage of Hudson."[14] So great was the optimism that, in July of 1612, Britain's government granted a royal charter to 288 merchants known as the Governor and Company of the Merchants of London, Discoverers of the North-West Passage. The charter gave these merchants a monopoly over any passage in the regions of Hudson Strait and westward. Their hopes eventually waned as the voyage of Thomas Button on the *Resolution* in 1612–13 proved beyond a doubt that the western shore of Hudson Bay was a barrier of unknown width between the Atlantic and Asia. In 1625, Henry Biggs produced the first map on which the name Hudsons bay [*sic*] appears.[15] Later voyages by Luke Foxe and Thomas James in 1631–32 contributed to the further mapping of Hudson and James Bays; between them they completed the map of the southern shore of the bay.[16] When the publication of Henricus Hondius's map of 1636 affirmed that there

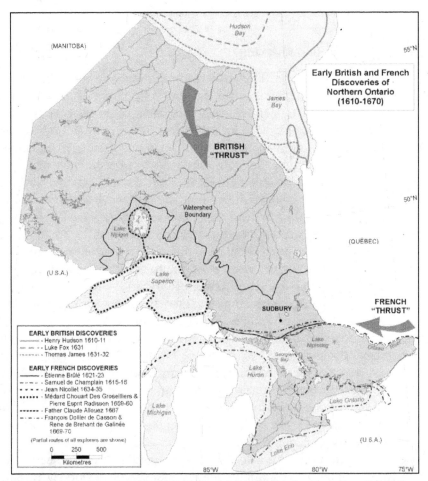

FIGURE 2.1   Northern Ontario as a British and French Exploration Zone. From D.G.G. Kerr, ed., *A Historical Atlas of Canada* (Toronto, ON: Thomas Nelson & Sons, 1960), 14, 16, and 20; and R.C. Harris, ed., *Historical Atlas of Canada*, vol. 1, plates 36 and 58. Reprinted with permission of the University of Toronto Press.

was no westward lead out of Hudson Bay, this paved the way for a shift in strategic attention by the British to the interior, closer to the Sudbury area.

In the meantime, the French started to explore Northern Ontario using the Great
Lakes system rather than Hudson Bay as the basis for their mapping. The expansion of the
French from Quebec into the Great Lakes area began with the epic journeys of Étienne
Brûlé and Samuel de Champlain. Brûlé arrived in the New World in 1608 with Samuel
Champlain, Royal Geographer to Henry IV, who in 1610 sent Brûlé into the wilderness to
live with the Huron Indians to "observe the country, its rivers and communications, find
out about minerals and other resources; if possible, make your way westward to the great
lake [Huron]."[17] He was the first incarnation of what came to be known as the coureur de
bois. Unfortunately, Brûlé left no written records of his journeys; consequently, many of
his achievements cannot be fully verified, and must be inferred from indirect accounts.[18]
As well, his unsavoury reputation caused others to minimize his exploits. By going into
Huron country, he was the first European to ascend the Ottawa as far as the Mattawa River.
He may also have been the first white person to stand on the shores of Lake Nipissing,
descend the French River, and discover Georgian Bay on Lake Huron.[19]

In 1615, Champlain embarked on his epic journey with Brûlé. On July 26, the two
reached the Nipissing tribe of Indians on the shore of Lake Nipissing, at the site of the
present city of North Bay, which Champlain regarded as being "very pleasant."[20] From
here they canoed to Georgian Bay on *Mer Douce* ("the sweet sea," Champlain and Brûlé's
name for Lake Huron because they were not yet aware of the other Great Lakes) by
means of the French River, where they met up with Ottawa Indians from Manitoulin
Island. This may have been the closest distance that any French explorer came to the
Sudbury area. Champlain returned to Quebec in the following year. Later, in 1620–21,
Brûlé reported that he saw the rapids at Sault Ste. Marie and discovered Lake Superior. It
is possible that he visited all of the Great Lakes except Michigan, and was the first white
man to do so.

Unlike Brûlé, Samuel de Champlain left numerous accounts of his travels issued in
1604, 1613, and 1619, which were brought together in a single volume accompanied
by his later journals and his map of 1632.[21] This map is outstanding because of its early
representation of the Great Lakes (figure 2.2). For the first time, all except Lakes Michigan
and Erie are present. The huge lake to the west (*Grand lac*) is Lake Superior. *Mer Douce*
combines Georgian Bay and Lake Huron. A masterpiece, it was long unequalled in seven-
teenth-century cartography.[22] Using Huronia as a base, the Jesuit missionaries who had
replaced the Recollets in Huronia after 1625 added to the knowledge of the north by following Indian trade routes into this unfamiliar territory. The travels and explorations of
the Jesuit missionaries in Ontario between 1610 and 1791 are fully documented in the
famous *Relations* documents.[23] By the 1640s, they had reached the strategic point of Sault
Ste. Marie, where Indian tribes came from all directions to trade for the whitefish caught
by the Chippewa. By the end of the decade, Indian informants had given the Jesuits a
rough idea of the size and relationship of all of the Great Lakes. Around 1638, Jean Nicolet

made his way through Lake Huron to Sault Ste. Marie. From 1654 onward, Pierre Esprit Radisson, Sieur des Groseilliers, and Médard Chouat were among the coureurs de bois who ventured throughout Northern Ontario, and may even have reached Hudson Bay. These northward journeys, undertaken with the conviction that the true passage back to Europe was by way of Hudson Bay rather than the St. Lawrence system, initiated a period of hostility between the British and the French at the "bottom of the bay."[24] Other French explorers also passed through the upper Great Lakes region in the late 1600s.

As a result of these explorations, the cartography of the Great Lakes region began to unfold with greater clarity. From the beginning of the seventeenth century until the fall of New France in 1763, it was the French who excelled in outlining the geography of the Great Lakes landscape.[25] Nicolas Sanson d'Abbeville's maps of 1650 and 1656, which brought together for the first time all five of the Great Lakes, represented the first major improvement on Champlain's map of 1632. They were soon superceded by Francesco Bressani's Huronia map of 1657, derived from records made prior to 1648. It can be inferred from the

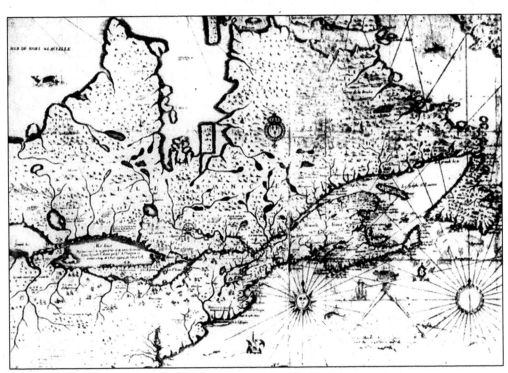

FIGURE 2.2 Champlain's Map of New France (1632). From Trudel, *Atlas de la Nouvelle France: An Atlas of New France* (Quebec City: Les Presses de l'Université Laval, 1968), 87; reprinted with the permission of Les Presses de L'Université Laval. The western half of this map, the last and greatest drawn by Champlain, served as the prototype for later European maps for nearly a century. There is a growing, but still confused portrait of the Great Lakes. While Lakes Superior (*Grand lac*) and Huron (*Mer Douce*) are shown, Lakes Erie and Michigan are still missing.

map, by the correct placement of the height of land to the north, the inclusion of tepees, and the three rivers which appear to be the present-day Whitefish, Spanish, and Serpent, that missionaries and traders had visited the Sudbury area prior to 1848.[26] Galinée's map of 1670 was important in that it revealed a more accurate route of the waterways linking Georgian Bay with the Ottawa River. The Jesuit map of 1672 followed shortly thereafter; its delineation of Lake Superior and the northern parts of Lake Huron were not surpassed until the detailed surveys of the nineteenth century. Jacques Nicholas Bellin's map of 1755 depicts the state of mapping as it pertained to the Great Lakes area just before the end of the French period (figure 2.3). The fringe position of the Sudbury area throughout this period is depicted on these maps by the use of vague terms, such as Indian "Algonquin" or "Outaouacs" territory, and phrases indicating a lack of territorial knowledge such as "toute cette coste n'est pas connue." Even as late as 1813, maps continued to refer to the area as being part of "Chippewa Hunting Country."[27] The Sudbury area, with its terra incognita status, had to bide its time before it became known to the world.

Hudson's Bay Company and the Fur Trade

With the rise of the fur trade, the Sudbury area emerged as part of a wide economic system dominated by two great players: the Hudson's Bay Company (HBC) controlled from England, and the North West Company based in Montreal. Chartered in 1670, the Hudson's Bay Company was formed as a fur trading organization planning to reach the interior of North America by means of Hudson Bay. HBC's realm eventually encompassed the drainage basin of Hudson Bay known as Rupert's Land; it could be said that the company turned this territory, including Northern Ontario, into a "company town writ large."[28] In the early years, the English and French engaged in a series of naval and land battles for control of Hudson and James Bays. At the same time, another battle was raging for the loyalty of the Indians. In 1713, by signing the Treaty of Utrecht, France acknowledged England's claim to Hudson Bay. For the next sixty years, the HBC erected posts at the mouth of the major rivers flowing into the bay. The company, however, was reluctant to penetrate the interior. The French took advantage of this situation, and began to compete effectively for the Indian trade. After the Treaty of Paris (1763), the company's French rivals were replaced by a more formidable opponent, the Montreal-based North West Company (NWC). It evolved as a major force in the fur trade from the 1780s to 1821, at which time it merged with the Hudson's Bay Company. Throughout this period, the French River assumed a new strategic position and, according to one source, "from 1770 to 1820 the French river became as busy as highway 401 is today."[29] In the area north of Lake Huron and the French River, the two companies established several posts. The NWC had a post at Lake Nipissing known as La Ronde; further west, the HBC erected posts at La Cloche Island, Little Current, Mississaugie, Green Lake, and Whitefish Lake; the last area had an estimated population of 1 100 Aboriginals in 1856.[30] In addition to Whitefish Lake, other HBC posts and a store were established in the Sudbury area (table 2.1).

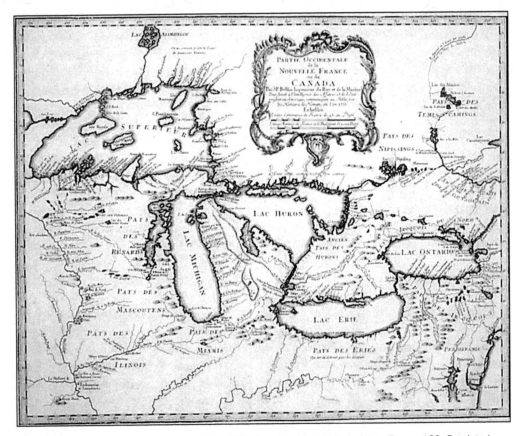

FIGURE 2.3 Jacques Nicholas Bellin's Map (1755). From Trudel, *An Atlas of New France*, 120. Reprinted with the permission of Les Presses de L'Université Laval. This map is a magnificent rendering of the Great Lakes landscape in the middle of the eighteenth century. It also shows the strategic French ports guarding the fur trading routes and the Indian settlements in the region just prior to the French and Indian war.

TABLE 2.1 Hudson's Bay Company posts and stores in the Sudbury Area (1822–1900)

Wahnapitae Post	1822–23 and 1879–91
Whitefish Lake Post (1) at Whitefish Lake	1824–87
Whitefish Lake Post (2) at McNaughtonville	1887–96
Larchwood Post	1885–92
Sudbury Store	1886–1900

Source: D. McKechnie and G.R. Stock, *Report to the Tourist and Convention Committee on the Historic Sites in the Sudbury District*, Appendix 111 (Sudbury, ON: Chamber of Commerce, n.d.).

The Whitefish Lake post was established in 1824. The key post in the Sudbury district, its main purpose was to prevent independent traders from Michigan, Wisconsin, and what is now Southern Ontario from gaining a foothold in the area north of the French River. All trade in furs south of the height of land was controlled by the officer in charge of this post. It was reasonably successful in achieving this aim. Originally located on the west shore of Whitefish Lake, and adjacent to the portage route to Wakemi Lake, the post later moved to Naughton in 1887 near the recently constructed Algoma Branch of the Canadian Pacific Railroad (CPR). With the advent of the lumber industry and the development of mining, fur returns steadily declined, and in 1896 the post was closed.[31] The HBC also operated a dry goods store near the corner of present-day Larch and Elgin Streets in downtown Sudbury; this site was nominally the headquarters of the district. It superceded Fort La Cloche on La Cloche Island, which had become redundant due to increased competition in the interior. The Sudbury store remained in operation for fourteen years (from 1886 to 1900). During this interval, members of the Whitefish Lake Reserve used the Junction Creek watercourse to paddle to Sudbury to trade with locals for goods.

The HBC established smaller satellite outposts to protect the trade of the Whitefish Lake Post at Larchwood (Vermilion) and Onaping adjacent to the CPR line, and on Wahnapitae Indian Reserve No. 11.[32] Located adjacent to a small waterfall just southwest of the junction of Onaping and Vermilion Rivers, the Larchwood Post enabled the HBC to control the trade of those Indians who trapped along the Onaping River south of the height of land. The first phase of the Wahnapitae store lasted only one season (1822–23). Its second phase lasted from 1879 to 1891. This post was strategically located, as it controlled all Indian trade using the upper Wanapitae and Sturgeon Rivers. This enabled the HBC to close down the Nipissing Lake Post in 1879 and ship all furs from the upper Sturgeon River to Fort La Cloche by way of the Wahnapitae Post. With respect to the latter, a rare picture of the old HBC post on the north shore of Lake Wanapitae and along the west shoreline of Post Creek can be found in Burwash's report to the Ontario Bureau of Mines in 1896.[33] Aside from the Whitefish Lake post, no white settlement occurred, as the "rock-bound, unfriendly shield discouraged the penetration of settlers.[34] The land above Georgian Bay, offering possibilities for logging and mining, awaited further exploration.

Geopolitical Setting

Another aspect of Sudbury's peripheral location was its fluctuating political ownership (figure 2.4). Prior to 1608, the area constituted the northern fringe of what was termed "Indian Territory." From the founding of Quebec in 1608, France placed its firm stamp on the history of the continent, and all of Ontario south of the Hudson Bay drainage divide became part of an enlarged part of New France. The French presence was then weakened by the Treaty of Utrecht (1713), which ceded the Hudson Bay region to Britain. Following the Quebec Act (1774), the Sudbury area became part of the western extension of an

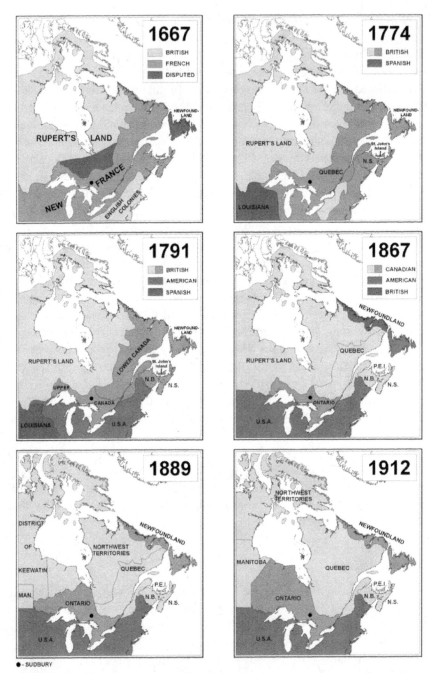

FIGURE 2.4 Sudbury's Geopolitical Situation (1667–1912). Adapted from Norman L. Nicholson, *The Boundaries of Canada, Its Provinces and Territories*, Geographical Branch Memoir 2 (Ottawa: Department of Mines and Technical Surveys, 1964), 45, and R.L. Gentilcore et al., *Historical Atlas of Canada*, vol. 2, plate 21. Reprinted with permission of the University of Toronto Press.

enlarged Province of Quebec. Ontario north of the Great Lakes watershed remained part of Rupert's Land. The influx of United Empire Loyalists after the American Revolution brought political pressures to bear on the British Parliament that resulted in the Province of Quebec being divided into Upper and Lower Canada respectively by the Constitutional Act (1791). Following aborted rebellions, both Upper and Lower Canada were merged into a reunified entity called the Province of Canada. Within this union, the Sudbury area became part of Canada West, although government officials continued to use the terms of Upper and Lower Canada in the Legislative Assembly. This situation remained until Ontario became a province of Canada in 1867. The boundaries of Ontario were gradually pushed north to Hudson Bay in 1889 and 1912, completing the province's expansion to its current borders. Even though it was largely uninhabited, the Sudbury area thus experienced considerable political change, at varying times being part of New France, the Province of Quebec, Canada West/Upper Canada, and finally, Ontario.

Robinson-Huron Treaty No. 61 (1850)

In 1845, the Province of Canada commissioned a study known as the Rawson Report of the conditions of Indian life in the Huron-Superior area. While the study noted that Aboriginal life was difficult at the time, it nonetheless concluded that settlement in Upper Canada had reached such a stage that the hunting grounds on the shores of Lakes Huron and Superior should be opened to white colonists. The discovery of minerals provided yet another impetus for allowing white men to enter the area. Thus, in 1850, the chiefs of all the bands in the area were called to a grand council at Sault Ste. Marie, where most agreed to cede their lands in return for presents, a cash payment, a yearly annuity, and land reserves. This agreement was known as the Robinson-Huron Treaty No. 61.[35] The treaties were signed by seventeen Ojibway Chiefs representing 1 422 Indians who inhabited the eastern and northern shores of Lake Huron. As a result of this treaty, the province established two reserves in the Sudbury area: Whitefish Lake Indian Reserve No. 6 and Wahnapitae Indian Reserve No. 11.

CHAPTER 3

Drawing Lines on the Map

By the 1840s, most of the desirable land south of the Precambrian Shield had been occupied, forcing settlement northward. To facilitate this process, the provincial government took various measures such as constructing colonization roads, granting mining claims, selling timber tracts, and issuing free land grants for agricultural purposes. These measures, in turn, forced the province to develop a surveying system using Salter's baseline and a principal meridian as the foundation for the establishment of timber berths and settlement on a township basis. The province then initiated a process of administrative reform that transformed the north from its *terra incognita* status into official districts. The creation of these political boundaries gave the North Shore of Lake Huron and the Sudbury area a more defined geography to the rest of the world. The publication of the first geological map of the Sudbury area by R. Bell in 1891 provided yet another spur for settlement and economic development.

Opening of the North

The emergence of the north from its frontier status began in 1853 when the government passed the Public Lands Act, providing colonization roads to open unsettled areas of the province.[1] Several routes were started that year, serving the fringe areas between Lake Simcoe and the Ottawa River, and some even went as far north as Parry Sound. Northern extensions of this policy took place in 1857 and 1865, when surveys were launched for a line known as the (Great) Northern Road intended to link the North Shore area with Parry Sound.[2] Parts of this road, which would have run south of Lake Panache, were never completed. The Colonization Road Program was transferred from the Bureau of Agriculture to the Department of Crown Lands in 1862, when it became the responsibility of the Colonization Roads Branch. In 1900, the program was moved to the Department of Public Works. It remained active until 1947, when the government repealed the legislation governing colonization roads.

The North Shore of Lake Huron was likewise the focus of considerable mining activity; between 1846 and 1856, the government issued seventy-eight location tickets for mining tracts.[3] Among these was the Wallace Cobalt Mine situated near the mouth of the Whitefish River, west of Whitefish Falls.[4] Lumbering also spilled over to the Shield, as evidenced by the erection of Waddell's Sawmill on the north shore of Collins Inlet.[5]

In line with the policy of British administrations in Canada to survey land before it was opened for settlement, the province established a method for drawing lines on the map. The first step in this process, the alienation of Aboriginal claims by the Crown, had already been completed with the signing of the Robinson-Huron Treaty. The next phase came in the form of crown surveys by surveyors, who could be called the *sine qua non* of land settlement. These surveys consisted of two processes: establishing the basic grid of meridians and parallels, and then the surveying of townships, lots, and concessions for transference to settlers. Only when these surveys were completed could administrators apply the necessary nomenclature to the townships for referencing purposes.

Early Surveys

The first major exploration of the upper Great Lakes area occurred in 1788, when Captain Gother Mann undertook a military survey in case of future hostilities with the United States. "A great solitude, little known or frequented except by some Indians," was Mann's description of the Georgian Bay region of the time.[6] This "great solitude" was later scrutinized through Captain Henry Bayfield's surveys (1817–21), John Bigsby's scientific tour (1819–20), David Thompson's surveys (1821), and Lieutenant John Carthew's Navy exploration (1834).[7] Other surveys in Northern Ontario were made in conjunction with the mining claims issued in the 1840s. The 1847 and 1848 explorations of Alexander Murray were the first to make reference to extensive valleys of fertile lands that could be found behind the rocky hills skirting the northerly shore of Lake Huron; his travels up the Spanish River caused him to mention in passing a "third tributary" near Whitefish Lake—an early reference to the Vermilion River. His 1848 survey likewise noted an area of abnormal magnetic attraction that provided the first hint of the Creighton Mine ore body.

Murray's exploration and mapping in 1856 was significant, as it provided the first detailed map of the Sudbury area (fig. 3.1). Given the importance of waterways as the major transportation network at the time, the main lines on the map focused on the French, Wahnapitae [sic], White Fish [sic], and Spanish watersheds that ran through the area. On the east, the relationship between Lake Wanapitei, the Wanapitei River, and the French River is clearly outlined. Murray shows Richard, McFarlane, and Long Lakes as being linked to Round Lake, Lake Panache, and the Whitefish River.[8] His sketch of the Whitefish River was based on information provided by Provincial Land Surveyor Salter, who had explored that part of the territory the year previous. The linking of the Onaping, Vermilion, and Whitson Rivers to the Spanish River system awaited further exploration, as did the nature of the Junction Creek–Lake Ramsey–Kelly Lake system.

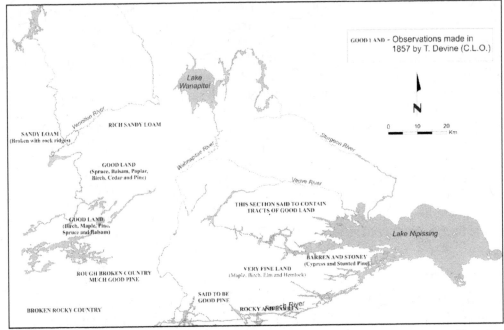

FIGURE 3.1 Murray's Pioneering Map of the North Shore and Lake Huron. From *Geological Survey of Canada, Report of Progress for the Years 1853–54–55–56*, Appendix No. 52 (Toronto, ON: Legislative Assembly, 1857), Map 7. Land descriptions from Salter's map of 1857 have been added.

Salter's Baseline and Principal Meridian

Salter had been instructed in 1855 by Joseph Cauchon, Commissioner of Crown Lands, to ascertain the position and extent of lands northwest of Lake Huron, and to determine their suitability for settlement (biography 1).[9] He started this task with a coastal exploration from Sault Ste. Marie to Lake Nipissing. Reporting that there was much land available for settlement, he was directed to draw base, meridian, and range lines from Lake Nipissing to Lake Superior on which to project townships. In the summer of 1856, he explored farther inland and produced the required baseline from a point on the Sturgeon River, near its mouth, as far west to Whitefish Lake and the Spanish River, a distance of 144 kilometres (fig. 3.2).[10] From Whitefish Lake he surveyed a meridian line some twenty kilometres north, continuing twenty-nine kilometres farther the following year. South of the baseline, his meridian passed though the eastern side of Round Lake; to the north, the same line passed through Whitefish Lake. Other surveyors were directed to draw meridian lines southerly from the baseline to Lake Huron twenty-nine kilometres apart in order to determine the agricultural and lumbering capabilities of the terrain. In 1883, Salter's meridian line became the west boundary of Waters and Snider Townships. Ladell makes the enormity of this surveying feat clear:

The running of a "baseline" was particularly rigorous. This was a line upon which future surveys would be based, and no deviation from the preselected compass course was tolerated. Thus, in 1856 a provincial land surveyor by the name of Albert Salter was instructed to start from what is now Sturgeon Falls and run a baseline due west until he reached the northeastern shore of Lake Superior, some 160 miles (257 km) away. He and his men were also required to run "meridians" at stated intervals and of stated lengths. These were north–south lines upon which future townships would be "erected," to use the terminology then in vogue. It took Salter and his survey parties two seasons to finish the work. Today, Salter's Line is still used by land surveyors as a reference line.[11]

Biography 1

ALBERT PELLEW SALTER
A Surveyor Who Left His Mark (1816–1874)

Albert Salter deserves recognition for his pioneering role in opening up the North Shore of Lake Huron and the Sudbury area for exploration and development. He was but one of a hardy and remarkable group of surveyors who braved the rigours of wild Northern Ontario in order to lay out grid lines and maps, thereby giving the world its first real glimpse of this part of the Precambrian Shield. Their task was not easy, as they had to assess the natural features of the terrain and its waters, scan the heavens to determine latitude and longitude, locate and fix regional boundaries by means of baselines and meridians, and determine precise dimensions, directions, and relative positions.

Salter was born in Teighnmouth, Devonshire, England in 1816.[a] He came to Canada in 1834, and temporarily resided at Plympton in the County of Lambton. He later fought in the Rebellion of 1837. After getting married in 1839 he worked as a teacher, during which time he obtained a degree in land surveying and civil engineering, graduating in 1844. The Salter story as it applies to Northern Ontario started in 1846, when he wrote to the Commissioner of Crown Lands expressing his interest in surveying the northern shores of Lake Superior. This proposal bore fruit, and in 1847–49 he was hired to work on a survey of mining locations in the Sault Ste. Marie vicinity. During this period, he laid out the first township in Northern Ontario. In 1859, he moved to Chatham. When Joseph Cauchon became Commissioner of Crown Lands in 1855, he hired Salter to undertake an exploratory survey of the North Shore of Lake Huron, which had acquired new importance because of the 1850 Robinson-Huron Treaty. By June 1855, Salter had already conducted an examination of the country around Sault Ste. Marie and the Bruce Mines. In the next six months he worked his way eastward, past the lower reaches of the Thessalon, Mississagi, Blind, Serpent, Spanish, and French Rivers. In his final report, he suggested that the region presented a very rugged and barren appearance, and went on to affirm its vast timber resources, rivers with magnificent waterpower, and mineral potential. Salter ended his report with a recommendation that the

American system of survey be used to open the region. His work formed the basis for how townships were laid out in Northern Ontario and many in the Canadian Prairies as well.

In the summer of 1856, he returned to the region and laid down what later became known as Salter's Line, the first baseline in Northern Ontario. The line represented the initial step toward the placement of several tiers of townships along the North Shore. As his Principal Meridian, he laid down a north–south line that passed though the eastern side of Round Lake. While running his meridian north of Whitefish Lake, Salter's attention was drawn to an outcrop of minerals so unusual that he made a point of mentioning them to Alexander Murray of the Geological Survey. Salter's find was ignored for the next few decades. In 1857, he completed his baseline, and for the next few years laid out townships and colonization roads around Sault Ste. Marie.[b] He passed away in 1874.

[a]Much of the information for this biography has been taken from the Association of Ontario Land Surveyors' *Annual Report* of 1915, and John H. Ladell's *They Left Their Mark: Surveyors and Their Role in the Settlement of Ontario* (Toronto, ON: Dundurn Press, 1993), 159–62.
[b]Salter's supervision of the survey of the North Shore of Lake Huron between 1857 and 1859 is recorded in: Commissioner of Crown Lands of the Province of Ontario, *Report for the Year 1872* (Toronto, ON: Hunter, Rose & Co., 1873), 44–52.

Timber Berths

It soon became clear that the passage of the Free Grants and Homesteads Act of 1868, which had marked the watershed of agrarian influence in land policy in Southern Ontario, was not having its intended settlement effects on the Precambrian Shield.[12] First there was the question of bad timing, as other areas and countries were beckoning settlers. Second, the lands made available for free grants were often unsuitable for agricultural purposes. The provincial interest in the north thus turned to lumbering. Supported by the Crown Timber Act of 1849, which confirmed the principle of Crown ownership, the province began to view the forests as an important source of revenue. As well, timber sales were spurred by dreams of creating an "Empire Ontario," to counter the eastern influence of CPR-based incursions from Montreal.[13] While the first signs of interest in timber were evident as early as the 1850s, and several land licences granted, little lumbering actually took place. This was due to the availability of pine from other sources, such as the Ottawa Valley and northern Michigan. By 1859, the first inland timber berths, relying on Salter's baselines, were being depicted on maps. While they were not very detailed, these maps gave the northern parts of the province a higher degree of geographical visibility than they previously possessed. In the 1860s, lumbermen showed interest in timber limits along the French, Whitefish, Spanish, and Wanapitei Rivers. After Confederation, the newly created Province of Ontario, desirous of alternate sources of revenue, began considering applications for timber limits on the North Shore.[14] In view of this growing interest, the Commissioner of Crown Lands recommended, in 1871, that certain lands remaining unsold and unlocated in the districts be offered for sale as timber limits in berths not exceeding twenty square miles (52 square kilometres).[15] Beginning in 1872, the province conducted a series of auctions involving the sale of 8 000 square kilometres of North Shore limits, including territory now found in Dieppe, Graham, Louise, and Waters townships. Each limit consisted of a projected township depicted at the time only as a number (fig. 3.2). Henceforth, whenever the provincial treasurer required additional revenue, the Commissioner of Crown Lands would auction off another batch of timber limits. As the issuing of these timber berths involved new surveys, by 1877 the Crown Lands Department had acquired considerable information on the timber, soils, and topography of the region that hitherto had been lacking.

Laying Down the Six-mile Square Township Grid

When the rough country of the Shield began to be surveyed, some surveyors began to express grave doubts that the large townships provided by the 1 000- and 2 400-acre section systems used in Southern Ontario were suitable for administering areas where forestry and mining were the major industries. The successful use of the six-mile township in the United States was noted, and in 1859 a system that was almost identical

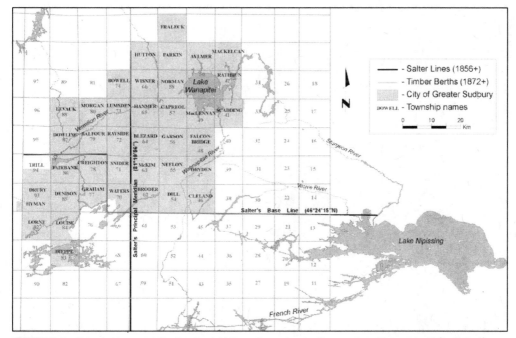

FIGURE 3.2 Salter's Lines and Timber Berths. From Crown Lands Department, *Reduction of Mr. Salter's Plan of the North Shore of Lake Huron* (Toronto, ON: Crown Lands Department, 1857) and Hallsworth, "'A Good Paying Business,'" Map B, 149.

was adopted for the country north of Lake Huron (fig. 3.3). The adoption of this system, firmly advocated by Salter, was momentous in that it was used to lay out over 560 townships in Northern Ontario. Each township measured six miles square, embracing an area of thirty-six square miles. As with the American model, no specific road allowances were surveyed, but 5 per cent of the acreage was reserved for roads, which were run when required. Lots were patented by sections and quarter sections. This was called

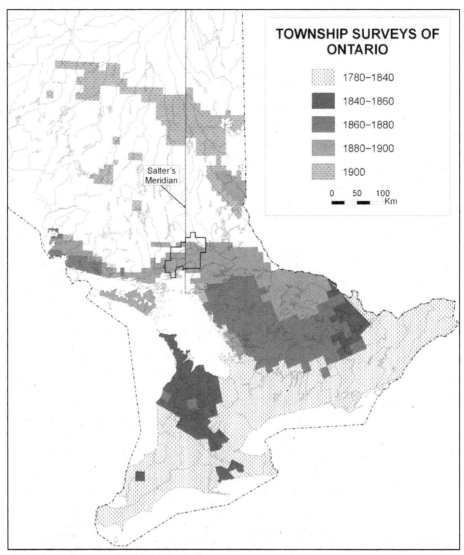

FIGURE 3.3 Township Surveys of Ontario (1780+). From Ladell, *They Left Their Mark*, 254. Reproduced with permission.

the 640-acre sectional township. In 1874, a modification was made, providing it with the familiar system of six concessions running east and west, each containing twelve lots. Instead of dividing the square mile section into 160-acre quarter sections, it was divided into two 320-acre lots. In Northern Ontario, therefore, the "standard lot" became one of 320 acres. This system was used from 1874 to 1906 for all townships completed by Ontario Land Surveyors. When the province opened up the Great Clay Belt, this system was abandoned in favour of the 1800-acre section township. It was used from 1906 until 1935, when all township surveys ceased.[16]

Table 3.1 provides a listing of the townships currently found in the City of Greater Sudbury, including the original surveyor, year of survey, and the origin of the township name. Dryden was the first township surveyed in 1882; the last surveys occurred in 1899 in Bowell and Fraleck Townships. The surveying process occurred in two main stages: between 1883 and 1887, there were twenty-two townships laid out, and from 1890 to 1894 another ten townships were formed. An important aspect associated with the surveying of these townships is the detailed information shown on the survey maps and reports published annually in the *Reports of the Commissioner of Crown Lands of the Province of Ontario*. These reports clearly showed, for instance, that much of the Sudbury area had previously been ravaged by fire. Bolger, reporting on McKim Township which he surveyed in 1883, gives us a glimpse of Sudbury almost from the moment of the settlement's birth. He stated that the "whole of the township is completely divested of timber, the bush fires having made a clean sweep of every bit of vegetation."[17] A review of other township survey notes from Dryden on the east to Lorne on the west provides additional support for the observation that there was extensive burning of the Sudbury area prior to the arrival of the CPR and the mining industry.

Shaping the Boundaries of the Sudbury District

Until the late 1700s, the northern part of the province served merely as an administrative *terra incognita* referred to on maps as a "Great Tract of Wilderness" or "Chippewa Hunting Country."[18] When Upper Canada's first lieutenant-governor, John Graves Simcoe, issued a proclamation in 1792 that divided the province into nineteen counties, Northern Ontario received its first official designation as an undetermined extension of Kent County.[19] The real process of administrative organization in the north started in 1853, when the Legislative Assembly of the Province of Canada passed An Act to Make Better Provision for the Administration of Justice in the Unorganized tracts of Country in Upper Canada. This act laid out roads, provided for the well-being and protection of those connected with lumbering and mining, deterred people from inciting the Aboriginal population, and created Provisional Judicial Districts.[20] This measure was followed, in 1857, by the passage of An Act to Provide for the Better Administration of Justice in the Unorganized Tracts of Country Within the Limits of the Province, permitting the formation and division of temporary judicial districts, each with their own magistrates, constables, jails, clerks, bailiffs,

TABLE 3.1   Townships in the City of Greater Sudbury by surveyor, date of survey, and origin of name

Township	Surveyor and Date of Survey	Origin of Township Name
Aylmer	W.B. Ford, 1898	Lord Aylmer, Soldier
Balfour	J. Degurse, 1884	William Douglas Balfour, MPP Essex South
Blezard	F. Bolger, 1885	Thomas Blezard, MPP Peterborough East
Bowell	C. Fairchild, 1899	Sir Mackenzie Bowell, Prime Minister
Broder	W.R. Burke, 1886	Alex Broder, MPP Dundas
Capreol	J. Laird, 1893	Frederick C. Capreol, Railway Promoter
Cleland	E.J. Rainboth, 1890	Charles Cleland, MPP Grey North
Creighton-	J. McAree, 1884	David Creighton, MPP Grey North
Davies		Tom Davies, Chairman, Regional Municipality of Sudbury (1997)
Denison	W. Burke, 1886	George T. Denison, Police Magistrate
Dieppe	D. Beatty, 1890	Dieppe, France
Dill	I. Bowman, 1886	Jacob W. Dill, MPP Muskoka and Parry Sound
Dowling	W. Burke, 1884	John F. Dowling, MPP Renfrew South
Drury	W. Johnson, 1884	Charles Drury, Ontario Premier
Dryden	T. Bolger, 1882	John Dryden, MPP Ontario South
Fairbank	F. Bolger, 1884	J.H. Fairbank, MPP Lambton East
Falconbridge	E. Stewart, 1892	Sir W.G. Falconbridge, Chief Justice
Fraleck	T.J. Patten/H.R. McEvoy, 1898/99	Fraleck, Mining Engineer
Garson	J. Degurse, 1887	William Garson, MPP Lincoln
Graham	W. Johnson, 1883	George P. Graham, MPP Brockville or Peter Graham, MPP Lambton East
Hanmer	J. Tiernan, 1894	Gilbert Hanmer, Farmer from Brant County
Hutton	W. Galbraith, 1898	Maurice Hutton, Professor
Levack	C. Bowman, 1885	Mary Levack, Sir Oliver Mowat's Mother
Lorne	E. Stewart, 1884	Marquis of Lorne, Governor General
Louise	E. Stewart, 1884	Marchioness of Lorne
Lumsden	J. Laird, 1887	Alex Lumsden, MPP Ottawa
Mackelcan	G.L. Brown, 1898	F. McKelcan, QC
MacLennan	E. Stewart, 1892	James MacLennan, Supreme Court Judge
McKim	F. Bolger, 1883	Robert McKim, MPP Wellington North
Morgan	J. Degurse, 1886	William Morgan, MPP Norfolk South.
Neelon	J. McAree, 1883	Sylvester Neelon, MPP Lincoln or James Neelon from St. Catharines
Norman	E.J. Rainboth, 1893	Robert A. Norman, MPP Hastings East
Parkin	H.R. McEvoy, 1899	Dr. George R. Parkin, Principal, Upper Canada College
Rathbun	D. Beatty, 1894	Edward W. Rathbun, MPP Hastings East
Rayside	J. Bowman, 1884	James Rayside, MPP Glengarry
Scadding	W. Davis & A. Griffin, 1892	Henry Scadding, Historian
Snider	I. Bowman, 1883	Elias Snider, MPP Waterloo North
Trill	W. Johnson, 1885	Unknown
Waters	W. Burke, 1883	John Watters, MPP Middlesex West
Wisner	J. Robertson, 1894	W.S. Wisner, Manufacturer

Source: Nick and Helma Mika, *Places in Ontario: Their Name Origins and History*, Parts 1–3 (Belleville, ON: Mika Publishing, 1977, 1981 and 1983); Alan Rayburn, *Place Names of Ontario* (Toronto, ON: University of Toronto Press, 1997); and information provided by Ministry of Natural Resources, Survey Records, OSG and the Information Resource Management Branch.
Note: In December of 1997, the name of the Township of Creighton was changed by the Legislative Assembly of Ontario to that of the Township of Creighton-Davies.

modes of trail, and a Registrar of Deeds.[21] In 1858, the Sudbury area then had its terri-tory split between the Temporary Judicial Districts of Nipissing North and Algoma East (fig. 3.4). According to the proclamation, the line dividing the two districts ran north from the mouth of the French River until it reached the southern limit of Rupert's Land.[22] This line later constituted the western boundary of McKim Township. Thus, citizens in McKim Township and those to the east later found their judicial and administrative ser-vices centred in North Bay, whereas citizens who resided in Snider or Waters Township and farther west had to travel to Sault Ste. Marie for the same purposes. The effect of this partitioning on the local setting was evident in the name chosen for Sudbury's sec-ond hospital in 1894: the Algoma and Nipissing Hospital. The *Sudbury Journal* recognized this reality by proclaiming itself to be devoted to the mining interests of Nipissing and Algoma Districts.[23]

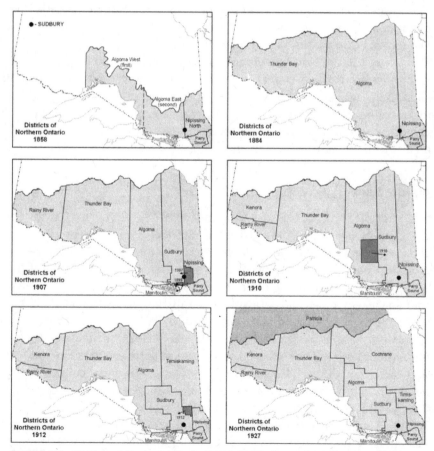

FIGURE 3.4 Districts of Northern Ontario (1858–1927). From John A. Farquhar, "The Historical Evolution of the Territorial District Boundaries in Northern Ontario with Reference to their Suitability at Present" (B.A. thesis, Department of Geography, Laurentian University, 1972).

In the meantime, the newly created province of Ontario made other adjustments to the district boundaries. First, they extended the boundaries of Algoma and Nipissing Districts northward to the Albany River in 1879; and second, the territorial limits of Algoma District were reshaped when Thunder Bay District was created in 1884. This geographical division lasted until 1899, when a portion of Algoma District became part of the Provisional District of Manitoulin. In 1907, the growing influence of the Town of Sudbury led the province to pass An Act to Create the Provisional Judicial District of Sudbury. By this act, parts of the District of Algoma and Nipissing were separated to create the Provisional District of Sudbury, with the Town of Sudbury serving as the district town.[24] In 1910, the Chapleau area, constituting some 8 000 square kilometres, became part of the District of Sudbury. This addition came about because of the importance of the CPR line between Chapleau and Sudbury. To compensate for the loss of its northern territory due to the creation of the Temiskaming Territorial District in 1912, the provincial government awarded a sizeable section of the District of Nipissing to Sudbury in the same year, marking the final stage in the formation of the District of Sudbury. In 1927, the shaping of the district boundaries throughout Northern Ontario was finally complete.

While the districts in Northern Ontario did not constitute municipalities, as was the case for counties in Southern Ontario, they nonetheless served a variety of administrative and judicial uses pertaining to the province. However, the railway network began to diminish in importance starting in 1927; as a result, these district boundaries have now lost much of their previous significance. Today, the traditional map of Northern Ontario based on district boundaries reflects an anachronistic throwback of a bygone era.

Mapping the Sudbury Structure (1891–1905)

After the Geological Survey of Canada was founded in 1842, one of its main areas of investigation was the North Shore of Lake Huron.[25] Although some lake and river systems in the area had been traversed by Alexander Murray and others, no comprehensive geological map of the region existed when prospectors discovered ore around Sudbury during the mid-1880s. To remedy this situation for prospectors working in the area, the Geological Survey of Canada sent R. Bell and his assistant, A.E. Barlow, to begin a program of regional mapping in 1888. Their resulting map, published in 1891 at a scale of four miles to an inch, was the first geological map of the area (fig. 3.5). On this map, the geologists recognized the existence of a central basin bounded by a mineralized zone later known as the North and South Ranges.[26] While this map did not reveal the entire continuity of the ore belt around the basin, it hinted at the enormity of the mineral belt lying north of the village of Sudbury. In 1904, Barlow followed up on Bell's work, publishing two maps depicting the South Range and introducing what would later be termed the Sudbury Nickel Irruptive.[27] These geologic maps proved to be of inestimable value in directing exploration and promoting settlement in the Sudbury area for many years, and were among the

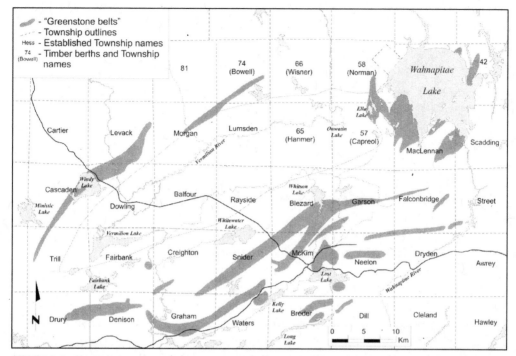

FIGURE 3.5 The First Geological Map of the Sudbury Area (1891). From R. Bell, "Report on the Sudbury Mining District (1888–1890)," in *Geological Survey of Canada Annual Report*, 1891, New Series, 5, Part 1, 5F–95F.

survey's most important contributions to the mining industry of Ontario throughout the twentieth century.[28]

Following the establishment of its Bureau of Mines in 1891, Ontario became more actively involved in the geology of the Sudbury area. In 1902, A.P. Coleman commenced mapping here and his subsequent map, published in 1905, showed for the first time that the ore zone was continuous, forming a closed, oval-shaped basin (now known as the Sudbury Igneous Complex) that was fifty-six kilometres long and twenty-seven kilometres wide.[29] Coleman's 1905 map, which represented the culmination of all the mapping work that had taken place since the 1890s, made the geological enormity of the entire Sudbury Structure clear to the rest of the world.

All of these developments—the first surveys, the establishment of Salter's Baseline and Principal Meridian, the delimitation of the timber berths, the laying down of the six-mile square township lines, and the mapping of the Sudbury Structure—were historically significant as they laid the geographic foundation for what was to become the "nickel capital of the world."

CHAPTER 4

Forging of a Local Monopoly

From Prospectors and Speculators to the
International Nickel Company (1883–1902)

The discovery of nickel-copper ore at Sudbury in 1883 was the impetus for the strangest boom in Canadian mining history. There was no "rush" at the start, such as occurred in the Klondike or at Cobalt. Only a handful of dreamers expected more from Sudbury than the echo of train whistles and smell of sawdust. The work of prospecting proceeded slowly, and when the speculators arrived, it became clear that they were not in a financial position to undertake the actual process of production. The widespread nature of the ore deposits and the problems associated with their refining and marketing, compared to gold or silver, dictated the need for large mining corporations with technical expertise and deep pockets. Thus it was not until the 1890s that any real development of the mining industry was possible. Within a short period of time, Sudbury's dominance in global nickel production came to be matched by the local supremacy of one company: the International Nickel Company. How this dominance came about is one of the great stories in Canadian mining history.

Prospectors and Speculators

The discovery of minerals in the Sudbury area is the stuff of legend. While it was suspected as early as 1856 through findings made by A.P. Salter and later by Alexander Murray that minerals likely existed in the area, their assessments were lost in government reports and did not attract much attention.[1] The discovery of minerals was a byproduct of the construction of the CPR mainline as it wound westward. In 1883, clearing parties in advance of the railway noticed a peculiar rusty or "gossan" look to some of the rocks northwest of Sudbury. According to early accounts, it was a blacksmith named Thomas Flanagan who made the discovery in August of that year. Whatever the merit of this assertion, it paved the way for bringing the mineral potential of the Sudbury area to international attention. Flanagan's site has since been commemorated as a historical site by the Province of Ontario.

The find attracted the attention of John Loughrin from Mattawa, who applied for permission to purchase the rights to 310 acres (at the statutory rate of one dollar an acre) on Lot 11, Concession 5, in the Township of McKim on February 25, 1884. This was the first deposit of nickel/copper ore found in the Sudbury area. On October 1, 1884, Loughrin, along with Thomas and William Murray and Henry Abbott, acquired a mining patent for this historical property, later known as the Murray mine. Most of the other discoveries in 1884 were limited to within eight kilometres or so of the original find alongside the CPR mainline. A second ore body was patented by Rinaldo McConnell, Joseph Riopelle, and John Metcalf in the southern half of Lot 1, Concession 4, and part of Lot 2, in the Township of Snider. This property, which became the site of the Canadian Copper Company's (CCC) Nos. 4 and 6 Mines, was also the location of the Clarabelle Pit. A chance discovery of the Worthington ore body in Lot 21, Concession 2 in the Township of Drury was made by Francis Charles Crean in the same year during the construction of the Sault Ste. Marie Branch of the CPR. Other finds in 1884 included the Elsie, located by Francis Crean and patented by Henry Totten, the Frood Mine by Thomas Frood and A.J. Cockburn, the Frood Extension by A.J. Cockburn, the Lady McDonald named after Sir John A. Macdonald's wife, and CCC's No. 2 Mine.

In the spring of 1885, prospectors fanned out from the railway and made more discoveries. The scene became chaotic as other prospectors made new finds, such as the Creighton Mine by Henry Ranger, Copper Cliff by Thomas Frood (biography 2), the Mount Nickel and Stobie Mines by James Stobie, Crean Hill by Francis Crean, and the Evans Mine by F.J. Eyre. While these prospectors legally held the rights to potentially unlimited wealth, the reality of their economic situation was somewhat different. Aeneas McCharles, the prospector who discovered the North Star mine, was not in the region long before he learned a hard truth: "The Sudbury district is not a poor man's camp. A few big companies are going to make all the money there is in mining there. It takes large capital to work nickel mines, and if a prospector happens to find a good body of ore, the only thing he can do with it, is to try and sell it."[2]

While some temporary work took place at Crean Mine in 1885, the first property that had an active mine is usually credited to the Copper Cliff (Buttes) Mine, opened in May of 1886. It was the first to make ore shipments, and also served as the site for the area's first smelter in 1888. In its early stages, it was an open pit operation, establishing the pattern for mines in those days; it later became an underground operation. The Copper Cliff Mine operated until 1905.[3] Other mines opened for production in 1886 included the Stobie and Evans Mines. It remained until 1889, however, before James Stobie and Rinaldo McConnell made the first discovery of ore at Levack on the North Range.

According to Main, the arrival of the speculator made it more difficult for prospectors to prosper. As the true scope of the region's mineral wealth became evident, speculators arrived on the scene in large numbers. Since the land had already been surveyed, the provisions of the Mining Act made it easy for speculators to apply for mining patents without any exploration of the land. Speculators would often apply for a patent first and then send out prospectors to determine whether or not any mineral wealth existed. Even if a prospector made a discovery on unpatented land, it was easy for a speculator to "jump" his claim by applying for the land as soon as he had any knowledge of that discovery. Other provisions in the act favoured the speculator over the prospector, notably the need for a minimum but costly purchase of eighty acres (thirty-two hectares). Thus the prospector was forced to give way to the speculator. While a few prospectors, such as McConnell and Stobie, derived considerable wealth from their finds, many got little for their years in the wilderness.[4] Bray, however, has a different perspective on the relationship that existed between speculators and prospectors. He is of the opinion that few of the so-called prospectors were actually "prospectors" in the conventional sense, most coming to the area for other reasons and then turning to prospecting when the situation arose. So, in a sense, they were also speculators. As well, the so-called speculators were generally small-time operators who made very little out of their speculation.[5]

The geographical location and distribution of many early minesites in the Sudbury area from 1884 to 1917 and their associated ownership patterns, shown in table 4.1 and figure 4.1, give some indication of the widespread nature of the ore deposits and the fluidity of the corporate structure that existed at the time. Many of these sites went on to become legendary in Canadian mining history.

Local Dominance of the Canadian Copper Company

One of the interesting aspects of the Sudbury area's mining history concerns the ease with which this economic wealth transferred from prospectors and speculators to a single company with a monopoly. How did it come to be that, five years after the discovery of minerals in the area, the CCC remained the only enterprise in the field? According to Main, there were several reasons for this development. First, the sheer size of the deposits encouraged only those producers that could achieve cost benefits from large-scale mining operations. Second, fewer producers were encouraged by the Canadian political atmosphere and existing legislation favouring a concentration of production. Third, the mineral deposits in the area were concentrated in the hands of speculators who were unable or unwilling to assume the financial risks of developing the properties; their main concern was to secure someone else to purchase their property at a good profit. Fourth, English and French companies showed little interest in the area, leaving the door wide open for Americans. Finally, there was simply circumstance. S.J. Ritchie and the CCC were not lured by the mineral prospects of the area; rather, they were making a desperate attempt to salvage their fortunes after floundering in unsuccessful efforts to exploit the iron ore fields of Hasting county in Central Ontario.[6]

It was this history that caused Ritchie to move into the Sudbury area in 1885 and purchase the holdings of Metcalf and McAllister (biography 3). In January 1886, the

Biography 2

THOMAS FROOD
Prospector and member of the "Big Four" (1843–1916)

Thomas Frood was one member of the group known as the "Big Four" prospectors (the others being Rinaldo McConnell, Francis Crean, and James Stobie) who arrived in Sudbury seeking their fortune through prospecting.[a] Under the permissive terms of the Mining Act of 1869, men such as Frood were able to purchase mineral lands in blocks of 80, 160, or 320 acres at the nominal price of one dollar per acre, without having seen the land in question, or having established the existence of mineral deposits on the property being patented. Like many prospectors in the Sudbury area, he hailed from the Ottawa Valley. Born in McNab Township, Upper Canada in 1843, he later moved to Hamilton where he became a teacher and married Mary Biggar in 1865. They had two daughters. After teaching school for a time, he was lured by the wilds of Northern Ontario. Following the CPR into Sudbury in 1883, he first worked as a timekeeper. Despite lacking any expertise in the field other than the ability to identify gossan (the tell-tale rust stain on outcroppings), he became a part-time prospector.

In November 1883 he was successful in finding copper deposits in a railway cut west of Sudbury; this property, however, was patented by others and it later became the Murray Mine. On May 18, 1884 he and another prospector, James Cockburn, located a vein of ore on Lots 6 and 7, Concession 6, McKim Township. An ownership dispute then ensued. Lot 6 was allocated to Cockburn and Lot 7 to Frood, who made application for the southern half. Cockburn transferred his claim to two speculators from Pembroke: J.H. Metcalf and W.B. McAllister. Frood then formed a syndicate for his property that included Peter Campbell and R.J. Tough. This property, which became the famous Frood Mine, was sold to the CCC for $30 000. Of this sum, Frood received $12 500. Frood discovered other promising properties in 1885. He located deposits on the north half of Lot 12, Concession 2 in McKim Township that became the site of the Copper Cliff mine and CCC's smelting works. In association with Metcalf and McAllister, he began development work on a find in Lot 1, Concession 3, Township of Snider that became known as the McAllister Mine (later the Lady Macdonald). For his efforts in finding the Copper Cliff and McAllister Mines, he received the paltry sum of a few hundred dollars. Extremely bitter, he hoped that "in the coming changes of relations of Capital and Labor, there will ... be more encouragement given to individual effort and less crushing by capitalists."[b]

Lacking capital to invest and with a family to support, Frood felt that he never received the full value of the properties he had discovered. Conditions at the time, which demanded the purchase of at least eighty acres of land, favoured the speculator over the prospector. In evidence given before the Mineral Resources Commission of Ontario in 1888, Frood stated that he was forced to give away 95 per cent of his mineral land to secure the other 5 per cent.[c] Thus it was inevitable that the efforts of the early prospectors gave way to giant speculators who had extensive holdings in the range of 5 000 acres. In turn, given the risk and capital required to develop ore deposits in the north with its severe winters, many speculators opted to sell their properties for a quick profit to the larger mining corporations, especially the CCC.

In 1886, Frood's wife passed away and three years later he married Christine McKay. He continued to prospect, but made no strikes of any magnitude. In 1899 he moved to Wallace Mine, a community at the mouth of the Whitefish River on Lake Huron's North Shore. When it was clear that his luck in prospecting had come to an end, he turned his attention to becoming a booster of the "New Ontario." He felt that there was good agricultural potential on the North Shore and encouraged the province to divert population northward. For a number of years, he worked as provincial woods ranger. He loved the wilderness and claimed, after a canoe trip from the Whitefish River to Vermilion Lake, that the "Vermilion was the Windermere of Canada."[d] He passed away on May 5, 1916.

[a]For a review of Thomas Frood's life, see Matthew Bray, "Thomas Frood," in *Dictionary of Canadian Biography*, vol. 14 (1911 to 1920), ed. Ramsay Cook and Jean Hamelin (Toronto, ON: University of Toronto Press, 1998), 378.
[b]Ibid.
[c]Cited in Mineral Resources Commission, *Report of the Royal Commission on the Mineral Resources of Ontario and Measures for their Development* (Toronto, ON: Legislative Assembly, 1890), 307.
[d]"Thomas Frood and His Faith in the New Ontario," *Sudbury Star*, April 8, 1978, 3.

TABLE 4.1   Early mine sites in the Sudbury area (1884–1917)

Mine	Township Location	Ownership	
Big Levack (Coleman)	Levack	(1889)	James Stobie
		(1913)	Mond Nickel
		(1929)	Inco
Blezard	Blezard	(1888)	C. Ducharme & E. Hillman
		(1889)	Dominion Mineral
		(1903)	Great Lakes & Copper
		(1913)	Mond Nickel
		(1929)	Inco
Cameron	Blezard	(1892)	Robert McBride & Dominion Mineral
Chicago/Inez/Travers	Drury	(1889)	Benjamin Boyer
		(1890)	Drury Nickel
		(1897)	American Nickel Works
		(1902)	Inco
Clarabelle (Open) Pit/Copper Copper Cliff No. 4 and 6/ McAllister, or Lady McDonald, or No. 5 Lady Violet	Snider	(1884)	Rinaldo McConnell Joseph Riopelle & John Metcalf
		(1889)	Canadian Copper
		(1902)	Inco
Copper Cliff North Mine Or No. 2	McKim	(1884)	Thomas Frood
		(1886)	Canadian Copper
		(1902)	Inco
Copper Cliff South or No. 1	Snider	(1885)	J.H. Metcalf W.B. McAllister
		(1886)	Canadian Copper
		(1918)	Inco

continued

TABLE 4.1 (continued)

Mine	Township Location	Ownership	
Crean (Evans)/Copper Cliff South	Snider	(1885)	F.J. Eyre
		(1886)	Canadian Copper
		(1918)	Inco
Crean Hill	Denison	(1885)	Francis C. Crean
		(1904)	Canadian Copper
		(1918)	Inco
Creighton	Creighton	(1887)	Canadian Copper
		(1902)	Inco
Elsie/Murray	McKim	(1884)	Henry Totten
		(1900)	Lake Superior Power
		(1912)	British America Nickel
		(1924)	Inco
Falconbridge	Falconbridge	(1916)	E.J. Longyear
		(1928)	Falconbridge Nickel
Frood/Frood Extension	McKim	(1884)	Thomas Frood & A.J. Cockburn
		(1886)	Canadian Copper
		(1911)	Mond Nickel
		(1929)	Inco
Garson	Garson	(1891)	J.T. Cryderman
		(1899)	Mond Nickel
		(1929)	Inco
Gertrude	Creighton-Davies	(1892)	William McVittie & George Jackson
		(1899)	Lake Superior Power
		(1913)	British American Nickel
		(1925)	Inco
Kirkwood	Garson	(1892)	William McVittie & George Jackson
		(1900)	Canadian Copper
		(1913)	Mond Nickel
		(1931)	Inco
Levack	Levack	(1889)	James Stobie & R. McConnell
		(1913)	Mond Nickel
		(1929)	Inco
Little Stobie	Blezard	(1885)	James Stobie
		(1901)	Mond Nickel
		(1929)	Inco
Long Lake Gold	Timber Berth 69	(1908)	Canadian Exploration
MacLennan	MacLennan	(1907)	Abraham Boland
		(1909)	Canadian Copper
		(1918)	Inco
McIntyre	Denison	(1888)	D.L. Lockerby
Moose Mountain	Hutton	(1907)	National Steel Corporation
Mount Nickel	Blezard	(1885)	James Stobie
		(1899)	Great Lakes Copper
		(1915)	T. Travers, T. Smith & J.A. Holmes

Murray Mine	McKim	(1884)	Thomas & William Murray, Henry Abbott & James Loughrin
		(1889)	H.H. Vivian
		(1894)	J.R. Booth
		(1912)	British America Nickel
		(1924)	Inco
North Star/McCharles	Snider	(1898)	Aeneas McCharles
		(1903)	Mond Nickel
		(1929)	Inco
Sheppard/Beatrice/ Davis	Blezard	(1887)	O.B. Sheppard
		(1927)	Noranda
		(1947)	Blezard Nickel Syndicate
		(1949)	Sudbury Shepherd Nickel
		(1951)	Inco
Stobie	Blezard	(1885)	James Stobie
		(1886)	Canadian Copper
		(1902)	Inco
Sultana	Trill	(1891)	S.E. Miller
Tam O'Shanter	Snider	(1893)	T. Baycroft
		(1913)	Canadian Copper
		(1920)	Inco
Totten	Drury	(1885)	Francis Crean
		(1935)	Inco
Trillabelle/Gillespie	Trill	(1891)	Ralph Gillespie
		(1892)	W.O. Washburn
		(1931)	Nickel Metals
		(1931)	Inco
Vermilion	Denison	(1887)	Henry Ranger
		(1888)	Vermilion Mining
		(1896)	Canadian Copper
		(1902)	Inco
Victor	MacLennan	(1890)	J.K. Leslie & G.S. MacDonald
		(1907)	Dominion Nickel-Copper
		(1917)	British America Nickel
		(1926)	Inco
Victoria (Mond)	Denison	(1886)	R. McConnell
		(1899)	Mond Nickel
		(1931)	Inco
Whistle	Norman	(1897)	I. Whistle & A. Belfeuille
		(1900)	R. McConnell
		(1909)	Dominion Nickel-Copper
		(1913)	British America Nickel
		(1927)	Inco
Worthington	Drury	(1884)	F.C. Crean
		(1890)	Dominion Mineral
		(1913)	Mond Nickel
		(1929)	Inco

Sources: Royal Ontario Nickel Commission, *Report of the Royal Ontario Nickel Commission*, 30–94, and Stephenson et al., *A Guide to the Golden Age: Mining in Sudbury, 1886–1977*, 50–70 and Appendix.

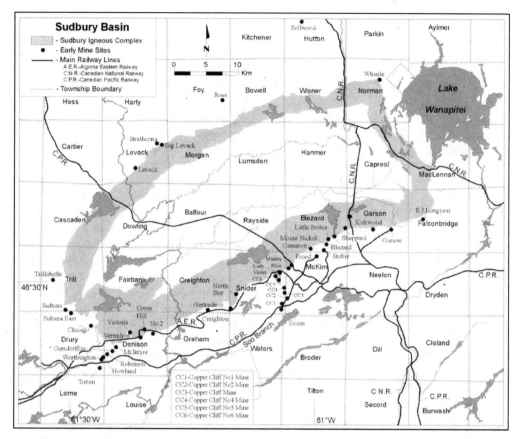

FIGURE 4.1 Early Mine Sites in the Sudbury Area (1884–1917). From the Royal Ontario Nickel Commission, *Report of the Royal Ontario Nickel Commission*, 1917, 104.

Canadian Copper Company (CCC) was formed in Cleveland, Ohio. In the same year, the company acquired the deposits held by the McConnell group and entered into a contract with an American smelting firm, the Orford Copper Company. As can be inferred from the company names, the owners assumed at the time that copper was the only game in town. The 1886 discovery of nickel in the ores, however, changed the economic situation. In the words of S.J. Ritchie, "We found we had a great nickel deposit instead of a great copper deposit, or, to be more correct, we had a great nickel and copper deposit."[7] The finding prompted the CCC to consolidate its position, a strategy made easier by the fact that speculators charged a much higher price for the rich copper-nickel ores rather than marginal copper ore, which had the effect of discouraging other companies from entering the field. When its monopoly control over the Sudbury mineral deposits was temporarily challenged by the Vermilion Mining Company in 1888, which had discovered gold, nickel, and copper at its Vermilion deposit, the CCC moved swiftly to acquire a controlling interest in the company and closed down its operations. Meanwhile, fuelled

Biography 3

SAMUEL J. RITCHIE
"Father" of the Canadian Copper Company (1838–1908)

In the annals of Sudbury, Samuel J. Ritchie is best known as the founder of the CCC, the precursor to Inco, founded in 1902.[a] In the broader context, however, he symbolized how Sudbury's destiny from the start was dominated by the American military-industrial complex and the global arms race. He was born in Boston Township, Ohio, in 1838. He married Sophronia Hale in 1865 and had two sons and a daughter. Ritchie was an irrepressible entrepreneur, who in the early 1880s had invested in some iron deposits near Trenton in Southern Ontario and gained control over a small railroad that he renamed the Central Ontario Railway. When the ore deposits proved worthless, he looked for other sources of revenue for his railway. To this end, he acquired a charter to extend his small railway from Trenton to the CPR mainline at Callander. After learning about the presence of copper samples taken from Sudbury, he directed his attention in 1885 to the copper deposits in the area. He subsequently formed the CCC in 1886. With Ritchie as president, the company quickly gained ownership of some 12 100 hectares of land, including the Copper Cliff, Creighton, Evans, Frood, and Stobie mining claims. He started open-pit mining in May 1886 at the Copper Cliff site.

Since shipping green ore with 7 per cent copper content to smelters and refineries abroad was too expensive, Ritchie decided in December 1888 to smelt the ores in Copper Cliff to reduce their export bulk by half. Following the advice of his engineers, a smelting system involving roast heaps and a furnace was introduced, one that eventually wrought extensive devastation to the surrounding area. When it was discovered that the ore also contained nickel, Ritchie was faced with the twin dilemmas of acquiring nickel refining technology and fostering new markets for the metal. The first problem was solved in 1891, when the Orford Copper Company of New Jersey entered into a contract to refine any matte (a molten mixture of copper and nickel produced during the smelting phase) that the Navy Department purchased from the CCC. After the Orford Company solved the problem of nickel refining with its "Orford process," the relationship between Orford and the CCC was cemented. When later experiments proved that a nickel-steel alloy was seven times more resistant to armour-piercing shells than the regular steel plating then in use, the American Congress voted to develop the new alloy as a military necessity. The CCC was awarded a lucrative Navy supply contract, and the Orford Copper Company was given an exclusive mining contract with the CCC.

The success of the Navy tests changed Sudbury forever. Local mining operations received a tremendous boost from contracts that effectively closed the American market to all but Ritchie and the CCC. Nickel mining in the Sudbury area became a one-company show, much to the chagrin of other speculators and Canadian promoters. The Province of Ontario was likewise dismayed, as it had sold the Sudbury deposits for a pittance, and had not even provided for the enactment of royalties to tap into the wealth of these mineral reserves.

Ritchie did not fare well either. Suspicious of his speculative forays and his favouring of the Central Ontario Railway, the directors of the CCC had Ritchie removed from the company's presidency in 1887, and in March 1891 forced him out of the CCC entirely. Distrusted by the CPR and the federal government as well, his only support came from Sudbury. During a personal visit to Sudbury on June 18, 1891, the *Sudbury Journal* reported that township Reeve Stephen Fournier told Ritchie, "Sudbury owes to you what it is today."[b] By disposing of Ritchie, however, the CCC lost a supporter and gained an ardent foe. Over the next fifteen years, Ritchie launched a series of lawsuits against the company. In 1897 he became part of a Canadian syndicate opposed to the CCC. He argued that the CCC had hindered the development of the Sudbury area by suppressing its rivals and requiring the Sudbury ores to be refined in the United States. The battle between Ritchie and the CCC carried on in the federal and Ontario legislatures and the newspapers. In the end, after threatening to close operations in Sudbury, Canadian Copper triumphed. Ritchie did gain some measure of satisfaction in 1902, when the CCC withdrew all the suits it had against him in the courts of Ohio and Ontario. The International Nickel Company was formed just two weeks later. The 1902 settlement allowed Richie to retire to palatial surroundings in Akron, Ohio. Ritchie passed away in 1908.

---

[a]Ritchie's life is chronicled in Matt Bray, "Samuel J. Ritchie," in *Dictionary of Canadian Biography*, vol. 13 (1901 to 1910), ed. Ramsay Cook and Jean Hamelin (Toronto, ON: University of Toronto Press, 1994), 873–76. Also see "Ritchie: A Tower in Early Mining," *Sudbury Star*, December 1, 1979, 3.
[b]Cited in Gilbert A. Stelter, "Origins of a Company Town: Sudbury in the Nineteenth Century," *Laurentian University Review* 3, no. 3 (1971): 17.

by coke imported from Cleveland via Algoma Mills, Canadian Copper erected two smelting furnaces in 1888 and 1889.[8] Within three years of its formation, therefore, Canadian Copper had succeeded in gaining complete domination of the Sudbury mining field.[9] Due to the huge size of the company, the enormity of its deposits and the lessons learned from previous experience mining in the Great Lakes area, Canadian Copper from the outset managed to introduce more sophisticated methods of mining and processing. The diamond drill, for example, was extensively used from the beginning, as were compressed air drills.[10]

From Canadian Copper Company to International Nickel Company (1889–1902)

The year 1889 looms large in the history of the CCC and the fortunes of the Sudbury district. In this year, two other enterprises appeared on the local scene to compete with the CCC: the Montreal-based Dominion Mineral Company, and H.H. Vivian & Company, each with its own smelter. The latter was the first English company to foray into the Sudbury mining field. Both proved to have short lives due to poor market conditions and internal management issues.

When researchers in France and England discovered in 1888 and 1889 that the addition of nickel to steel made the resulting alloy an excellent material for plate armour, the

American government was quick to recognize its strategic implications. The United States Navy ran a series of studies to assess its long-term consequences for military purposes. Following the publication of a glowing report by the US Navy Department regarding the use of armour plating in 1890, mining interest in the Sudbury district grew exponentially.[11] This sealed the fate of the Sudbury area as a pivotal resource for the emerging American colossus. Canadian Copper's economic position was strengthened in 1891 when it was awarded a contract to supply the American government with matte, from which nickel could be refined for use in plate armour experiments.

The CCC promoted the US Navy test results to spur demand for its nickel deposits. Interest was likewise aroused by newspaper accounts of evidence taken before the Royal Commission on the Mineral Resources of Ontario (1888) implying that enormous profits could be expected from the future exploitation of local deposits.[12] Prospectors swarmed into the area, and speculators bought up all the available Crown land where nickel was thought to exist. While ten companies were formed in the first six months of 1891, these were ultimately not successful because they had acquired small deposits in marginal areas.

Despite its near-monopolization of the mining sector, Canadian Copper faced a major technological problem with respect to smelting its ore into a matte suitable for blast furnace processing methods. To resolve this problem, the company hired Edward Peters, Jr., a metallurgist, who introduced the infamous German process of heap-roasting to the area on August 25, 1888.[13] From this day onward, Sudbury's image became intrinsically associated with clouds of sulphurous acid gas and environmental degradation. Although the process was crude, it was inexpensive and effective in eliminating excess sulphur; for this reason, heap-roasting became the prevailing smelting process throughout the district.

Until 1891, the CCC's main markets were in England and Europe. When the company started to penetrate the American market, however, it discovered that Sudbury matte was the kind that could only be handled efficiently by refineries experienced in treating it. To rectify this situation, the CCC signed a far-reaching agreement with the Orford Copper Company in New Jersey, which had been incorporated in 1887 for the purpose of refining copper ore. After some experimentation, the Orford process for extracting nickel was also developed between 1890 and 1892.[14] Given the contractual links between the Orford Copper Company, the CCC, and the US Navy Department, the stage was set for the establishment of a powerful triumvirate of interests within the emerging US military-industrial complex. This link enabled Canadian Copper to maintain its monopoly in the Sudbury area. Although Canadian Copper had secured domination in the Sudbury area from 1890–95, the position of S.J. Ritchie was weakened by his split with other directors. After briefly using Ritchie's influence with the US Navy to curry favour in Washington in the fall of 1890, the CCC would have nothing to do with him; he lost his influence and was removed from the company's board. The dismissal of Ritchie and the company's failure to refine its ores in Canada had strong repercussions for Canadian Copper. Even after rival threats had disappeared, for the next twenty-five years the company faced concerted attacks in Canada for its failure to consider any Canadian nickel refineries.

Since the US market for nickel was still small in comparison to Europe, Orford Copper turned its attention to that market, and in so doing, brought about a clash with Le Nickel, a company the Rothschilds formed in France in 1882. Le Nickel dominated the nickel refining business in Europe, and had complete control of the purchase of nickel ore from New Caledonia. After gaining a significant share of the European market, Orford signed an agreement in 1896 with other producers there that ensured the company an ongoing presence on the continent. Two years later, the outbreak of the Spanish-American War began an armament boom that resulted in increased shipments of nickel from Sudbury to the United States. As Thompson and Beasley make clear, the significance of the war was enormous for Canadian Copper, Orford, and the entire future of the nickel industry in Canada:

> On April 24, 1898, Spain declared war on the United States. On December 10, 1898, the war was over. Between the two dates were two events that changed the weapons of the world. The events were the Battle of Manila Bay and the Battle of Santiago de Cuba. In these two battles, the American warships clad in nickel steel armor destroyed almost the entire Spanish fleet without the loss of a single vessel.... Immediately, there was a quickened economic and political interest in nickel.[15]

While both the Vivian and Dominion Mineral companies ceased operations in 1894 and 1895 respectively, the rising demand for nickel encouraged the entrance of other Canadian, British, and American capitalists into the Sudbury region. The most important result of this new surge of interest was the formation of the Mond Company in 1900.

Meanwhile, the importance of nickel for plate armour manufacturers brought the steel industry into the picture. The steel manufacturers, who possessed greater financial resources than the nickel producers, decided to take control of the nickel industry. Led by the Orford Company and the United States Steel Corporation, a new syndicate known as the International Nickel Company was formed in 1902 that combined the refining works of the Orford Company with the mining works of CCC. This consolidation of the refining and mining interests in the nickel industry placed the future destiny of the Sudbury area firmly under the control of US steel producers and set the stage for an even wider global monopoly.

CHAPTER 5

Sudbury (1883–1939)

The interval between 1883 and the outbreak of the Second World War witnessed numerous changes in the nickel belt area, among them the creation of the Town/City of Sudbury. The arrival of the CPR in 1883 was the first step to shedding the region's previous isolation. As Sanford Fleming noted, "no civilized man, so far as known, had ever passed from the valley of the upper Ottawa through the intervening wilderness to Lake Superior."[1] As late as the 1880s, transformation was barely noticeable. There was little evidence that any settlement of significance would emerge from this lonely dot on the transcontinental mainline. Only the foolhardy could envisage that by the Second World War, Sudbury would emerge as the largest urban centre in Northeastern Ontario with more than 30 000 inhabitants. Locally, Sudbury was rivalled only by Inco's company town showcase of Copper Cliff with its almost 4 000 residents.

Arrival of the Canadian Pacific Railroad

A condition of British Columbia's entry into Confederation in 1871 was the completion of a transcontinental railway from the Atlantic to Pacific Ocean. Preliminary surveys undertaken along the North Shore in the 1870s were followed by the incorporation of the CPR in 1881. One of its first acts was to acquire the Canada Central Railway and other lines that gave the company a connection between Pembroke and Montreal. It was originally intended that the main trunk line would stretch westward from Sudbury via Sault Ste. Marie, with Callander serving as the regional headquarters of the CPR and McNaughton as a major depot on the Sault line. Late in 1882, a crucial decision was made to build the railway further northward, bypassing Sault Ste. Marie in favour of a line running through the Shield to the top of Lake Superior.[2] The Sault Branch, however, was to remain part of the mainline, providing connections with the American Midwest. For administrative purposes, the northern route linking Fort William to Callander west of North Bay was known as the Lake Superior East Section. At the time, Callander served as the western junction of the Canada Central Railway. From 1882 to 1885, the Sudbury Junction at the intersection of the Sault Branch line and the CPR mainline became a hub of construction activity. The size of the transient labour force varied, estimated to be as low as 1 500 and as high as 3 500.[3]

Construction of the mainline from Callander westward began early in 1883. By February, a tote road (one built for hauling supplies) had been completed to a junction site north of Lake Ramsey. This was the same tote road made famous by Florence R. Howey, who became Sudbury's "first lady."[4] The site was given the name of Sudbury by James Worthington, in honour of his wife's birthplace in England. Worthington had not intended to use this unimportant spot on the map for such a designation, but the station up the line expected to be the real centre of the area had already been named for Magistrate Andrew McNaughton.[5] The Sudbury site was selected by accident. The original course of the line as planned in 1881 was to be as follows:

> From Callander the line follows the course of the Vase river to the Forks of the same, thence by the north shore of Lake Nipissing and across the Sturgeon River immediately below the falls, thence in a north-westerly direction along the course of the Veuve river, and by the North Branch of the same to near the Wahnapitae River in Township 47 [Dryden] and crossing the latter river at the township line between Townships 47 and 55 [Neelon] thence in a south-westerly direction by the northerly side of Long Lake, to near the west line of Township 62 [Broder] thence westerly through Township 70, [Waters], thence south-westerly crossing Vermillion [*sic*] River in Township 77 [Graham] and continuing in the same course to the left bank of the Spanish River near the big bend, thence by the left bank of the last mentioned river, and crossing the same near the south line of township 99, [Nairn] thence still following a south-westerly course near to the right bank of the Spanish River until it reaches the shore of Lake Huron.[6]

While the intention had been to locate the railway "south of a lake of irregular shape lying in an east–west position, about four miles long by one mile at its greatest width," a CPR Chief Engineer named William Allen Ramsey inadvertently ran the line farther north. Known then as "Lost Lake," it was subsequently renamed in error as Lake Ramsay. In 1929, the spelling was changed to Ramsey to reflect the proper name of the engineer.[7] Had the original intention held, the junction of the main and branch lines would likely have been near the north shore of Long Lake, in either Broder or Waters Township.

On November 28, 1883, Sudbury's connection to the outer world was established with the arrival of the first supply train from the east. Regular freight and passenger service to and from this direction started late in the following year, bringing Sudbury firmly into Montreal's hinterland. The appearance of the first transcontinental train on June 28,

1886, and the CPR's launch of telegraph operations in the same year marked the incorporation of Sudbury as part of Canada's east–west communication network. Sudbury's regional connection was then strengthened by the completion of the Sault Branch in 1887. These advances reduced Sudbury's fringe position and enhanced the quality of life of its local residents.

A CPR Townsite

According to one account, the distinction of being Sudbury's first resident in the CPR townsite perhaps goes to Jules Collin, a railway cook. Another possibility was Joseph Boulay, a timber-cutter under contract to the CPR.[8] Following the survey of McKim Township in 1883, the company started the process of expanding the community into a railway townsite. In its first year, the village consisted of a haphazard collection of primitive buildings owned by the CPR, several boarding houses for managerial and clerical staff, a store, a company hospital (General Hospital) on Elm Street, and a house for the company doctors, Struthers and Arthur. The CPR also made formal application for ownership of 770 hectares of land on Lots 5, 6, and 7 in Concessions III and IV of McKim Township.[9] While this request was denied, the CPR managed on September 17, 1884 to purchase 192 hectares of "Mining Lands" from the Crown under the General Mining Act of 1869 for the nominal sum of $470.00 (fig. 5.1). Within its jurisdiction, the CPR had total control, a power sanctioned by the Public Works Act in 1882.[10]

Another power broker in the fledgling community was the Catholic Church. As part of the territory of the Vicariate of Northern Canada established by Pope Pius IX in 1874, its first bishop, John Francis Jamot, had established Jesuit missions and chaplains along the CPR mainline under the jurisdiction of the Diocese of Peterborough (after 1904, the Diocese of Sault Ste. Marie). The Jesuit presence in Sudbury dates from 1883, when three Jesuits were assigned to provide the canonical requisite for a religious community.[11] Cognizant of the sizeable Roman Catholic population in the Sudbury area, the diocese took steps in 1884 to acquire 121 hectares of land from the Crown northeast of CPR's property. The Jesuit Fathers thus became the first private property owners in the Sudbury camp.

These land acquisitions were significant because they dictated from the outset a geographical pattern that left an indelible mark on Sudbury: sprawling population clusters interspersed by rocky terrain. These land purchases were followed by both the CPR and the Jesuits surveying and dividing the land into lots for residential and other purposes. The need for land subdivision was dictated by the influx of people attracted to the bourgeoning mining industry. By the end of 1886, a plan of subdivision had been prepared for the CPR. Completed on December 29, 1886 and registered on August 16, 1887, the plan included 66-foot-wide (one chain) streets, 50-by-120-foot lots, and 20-foot lanes. The wide swath of land accorded to the CPR mainline and its Algoma Branch, and the winding presence of Nolin and Junction Creeks gave early evidence of the look that dominated the community for decades (fig. 5.2).[12]

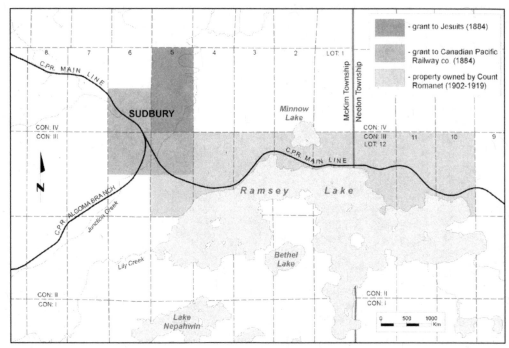

FIGURE 5.1   Early Land Grants and Count Romanet Properties (1884–ca. 1919). From Sudbury Land Registry Office, *First Registration Books*.

In accord with its standard practice, the CPR named many of the north–south streets after Governors General, and east–west streets after trees. This plan was followed, on October 31, 1890, by the registration of Father Cote's subdivision on behalf of the Jesuits, encompassing the broad area found north and west of Nolin and Junction Creeks respectively (fig. 5.3). The Jesuit tract was dismembered into a west and east section because of the CPR's acquisition of land for its Stobie Branch, which ran diagonally across the entire property.

Under the authoritarian eye of James Worthington, the townsite was run as a fiefdom of the CPR. Due to the Jesuit's strategic acquisition of its own land and the transformation of Sainte Anne's mission into the higher status of a parish, however, the Roman Catholics strengthened themselves as a political power in the community. The other major force was the CCC. Formed in 1886, its offices remained in the townsite until 1890, when they moved to Copper Cliff. Thus, the CPR, the Roman Catholic Church, and the CCC emerged as the trinity that dominated early Sudbury. On March 1, 1885, the Public Works Act for the Sudbury region was rescinded, and in 1886 the Township of McKim came into being.[13]

By 1885, Sudbury had a population of around 1 500. Following a decision by the CPR to move its construction camp to Biscotasing and to keep North Bay as its regional divisional point, the population dropped to 300 virtually overnight. With only lumbering

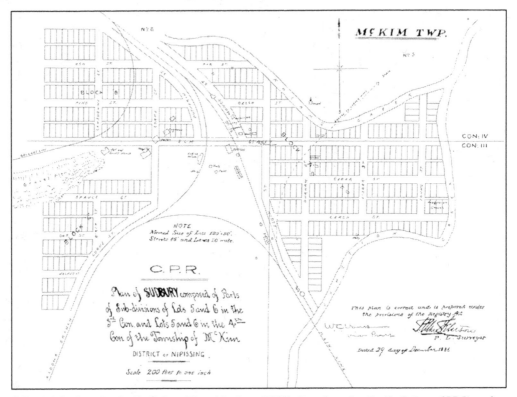

FIGURE 5.2 Canadian Pacific Railway Plan of Sudbury (1886). From Canadian Pacific Railway, *CPR Plan of Sudbury* (Montreal: Canadian Pacific Railway, December 29, 1886).

offering local employment, it appeared that Sudbury was well on the way to becoming a ghost town like many other CPR construction centres. After the formation of the CCC in 1886 and the beginning of mining in Copper Cliff the same year, the setting began to show new promise. Unlike the mining boom towns of Dawson City or Cobalt, there was no rapid influx of people; rather, the population grew slowly to about 1 000. According to one account, life in Sudbury at the time was primitive, and "cows wandered hungrily over the town."[14] Nevertheless, by the turn of the 1890s, the physical nature of the townsite had significantly transformed. Mindful of the power of the Protestant churches, the CPR granted the Anglicans, Methodists, and Presbyterians prime properties. A fledgling school system emerged. The first public school (McKim No. 1) opened in 1884 in the rectory of St. Anne's Church; in the following year it was moved to a CPR log building in the downtown area. At the time, it served all students in the townsite. In 1887, another public school was built on the corner of Elm and Lorne streets.

The CPR structures around the rail yard were soon complemented by the construction of new accommodations. The latter sites included legendary hotels and watering

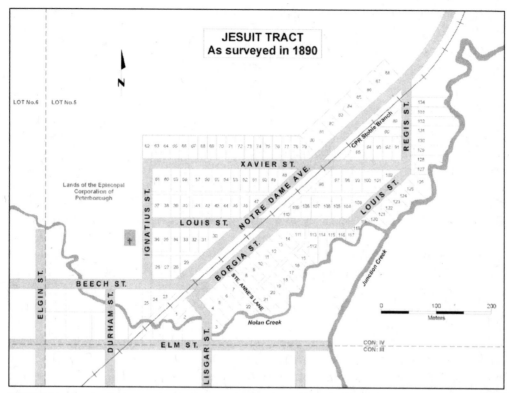

FIGURE 5.3 Rev. A. Cote's Jesuit Tract (1890). Sudbury Land Registry Office, *Registered Plan 1 S of Part of the Village of Sudbury*, October 31, 1890.

holes such as the Revere (later the New American and Hotel Coulson), Balmoral, Montreal House, and Queen's Hotels. By 1894 there were seven hotels. In these palaces, miners, lumberjacks, and prospectors could be found sharing their "vision of untold wealth" and where "land lots were bought and sold with intense rapidity."[15] The gala opening of the new, fully equipped White House on Elm Street in 1890 (replaced by the equally prestigious Nickel Range Hotel in 1915), which made all other hotels second class, indicated Sudbury's emergence from depot status.[16] In this year the first bank, known as the Ontario Bank (later the Bank of Montreal in 1906) also opened its doors in Sudbury. New stores (e.g., Hudson's Bay) appeared, all clustered around the Durham and Elm Street intersection. At this time, it was still a common sight to see "Indians from Whitefish Bay come in with their dog teams."[17] Already considered the centre of town, this intersection maintained its status as the business hub throughout the next century. Among those merchants whose names came to be associated with the commercial life of the community were John Frawley, Frank Cochrane, Stephen Fournier, James McCormick, Max Rothschild, and Thomas J. Ryan. The distinction of establishing Sudbury's first business goes to John

Frawley, who opened a general store in 1884. Despite these signs of urban maturity, however, Sudbury had many frontier characteristics. There were no sewers, and water was still drawn from the gravel pit (now Queen's Athletic Field). The streets, rutted by wagons, became quagmires in the rains, and cattle, pigs, and dogs roamed around at will.

From Village to Town

During the early 1890s, members of Sudbury's business and professional elite, fearful that a company town mentality was becoming fixed in the minds of its citizens and unsatisfied with Sudbury's setting as a mere CPR village, opted for change. The desire for a new status was a reflection of the boosterism exhibited in James Orr's *Sudbury Journal,* which first appeared on March 5, 1891. Shortly thereafter, a number of prominent citizens banded together in a "company of 100 associates" and petitioned the province for incorporation as a town. This petition appeared in the *Sudbury Journal* on January 14, 1892.[18] Despite the fact that the townsite did not appear to have the requisite 2 000 inhabitants required by the Municipal Act, the provincial legislature passed the bill incorporating the Town of Sudbury on April 14, 1892.[19] The act increased the size of the town to include Lots 4, 5, 6, and 7 in Concessions III and IV, an area encompassing 990 hectares, and established three wards (Ryan, Fournier, and McCormack). A council consisting of a mayor and three councillors from each ward was instituted (in 1899 the number of councillors, elected at large, was reduced to six). The new administrative structure made it possible to finance many needed improvements to local services. The incorporation of Sudbury resulted in the population of McKim Township being reduced to fewer than 700 persons, most living in Copper Cliff. The voters of Copper Cliff after 1892 thus controlled the municipal government of the Township of McKim.

Incorporation brought with it a major problem that plagued the Sudbury mining district until the creation of the regional government in 1973: sources of revenue largely limited to property taxation and licensing. As no mines or smelters were located within the town's boundaries, Sudbury received no taxes from the industry that gave the community its reason for existence. The assessment situation was exacerbated by the decision of the CCC in 1890 to move its company office from Sudbury to Copper Cliff. McKim Township suffered a similar fate when Copper Cliff was incorporated as a town in 1902. The removal of its largest concentration of population and tax base from the township jurisdiction left McKim impoverished. This situation was not unique, as it was an accepted practice by mining and forestry corporations throughout Northern Ontario to shirk regional responsibility for the well-being of area residents by locating their plants outside of existing municipal jurisdictions and forming their own company townsites. As Nelles has written, the province served as a willing accomplice to this corporate tactic until the latter part of the 1960s.[20] Another financial difficulty for the new town was the difficulty of getting adequate assessment from the CPR, the community's largest landowner.

With a limited tax base, the town found it difficult to provide adequate infrastructure for its residents. Given the high rate of infant mortality and the widespread prevalence of communicable diseases, one of its main concerns was public health. For this reason, the town government appointed a health board in accordance with the Public Health Act of 1884. By this time the CPR hospital had disappeared, and three other hospitals had been established in its stead. The town hired a chief of police and built a new jail that served as a common "gaol" for Nipissing, Algoma, and Manitoulin Districts. While the core of the city was serviced with sewers in 1895, effluent was still being dumped into Nolin and Junction Creeks that ran through the centre of the town. In the spring, when these watercourses flooded, matters got worse, often resulting in accidental drownings. This problem has lasted into the present, as evidenced by the fact that since 1920 there have been twenty-two documented deaths involving Junction Creek.[21] In 1896, waterworks and an electric light system were installed. The latter was unique: Sudbury was the first Ontario municipality to own and operate an electrical plant. Fire services were operated on a voluntary basis.

Schooling emerged as another pressing issue. The public school, located on the corner of Elm and Lorne Streets, was moved in 1906 to grounds where the Sudbury Arena now stands. Known as Central Public School, it incorporated high school classes until 1907, when they were shifted to Jubilee Hall in the Michaud Block at the corner of St. Anne's Road and Durham Street. In 1913, a second elementary school (Elm Street Public School) was erected on the corner of Elm and Regent Streets; College Street Public School followed in 1916. Meanwhile, an important event occurred in 1915 involving the division of the Sudbury Separate School Board into French and English sections. This separation was brought about by the increasing "nationalism" French Canadians showed and their reaction to Regulation 17, approved by the province in 1912, which forbade French language instruction. Because of the strongly worded assertions by local French-speaking politicians and religious leaders that their "cause was sacred," and with the tacit approval of many of the leading anglophone elite, Regulation 17 in the Sudbury area was rendered meaningless.

While high school courses were offered in 1908 at the Jubilee Hall, it was another year before Sudbury High School opened on a block of land purchased from the CPR; its curriculum expanded one year later to include the first mining courses in Canada. The Sudbury Mining and Technical School later opened its own doors in 1921; however, it did not become an independent school with its own principal until 1934. This decision resulted in a major rivalry between the two schools. A bilingual program was initiated at Sudbury High School in 1930 for francophone senior students; among the first of its kind in the province, it was the beginning of a program that lasted thirty-nine years.[22] These remained the only public high schools in the area until Capreol and Copper Cliff High Schools opened in 1935 and 1937, respectively.

On the business front, two developments of note included the appearance of the first issue of the *Sudbury Journal* on March 5, 1891, whose masthead proudly proclaimed it "devoted to the mining interests of Nipissing and Algoma Districts," and the incorporation of a Board of Trade in 1895. Despite these developments, Sudbury continued to serve only as a small "railway hub" and "commercial entrepôt."[23] The CPR's presence was enhanced in 1907 by the construction of a new station.

Up to the 1920s, the railway companies remained the largest providers of employment in the town; the CPR had more than 100 workers, and the CNR and AER each had more than fifty. The economic situation changed in 1905 following the town's acquisition of hydroelectricity from the Wahnapitae Power Company's plants situated at Coniston (1905–1915), Stinson (1925), and McVittie (1912). These plants provided the necessary power for the establishment of industrial operations and the construction of an electric street railway. New enterprises included the Sudbury Construction and Machinery Company in 1909 alongside the CPR's Algoma Branch, and the nearby Sudbury Brewing and Malting Company built in 1907 which advertised its famous Silver Foam beer. The latter company accompanied two other bottling operations, New Ontario Bottling Works and Taylor and Pringle. In 1909, the Sterling Bank, located on the northwest corner of Elm and Elgin Streets became the fourth banking institution in the town. Six years later, the bank constructed one of Sudbury's most interesting buildings, featuring classical decorated stonework that now houses the menswear store known as Fiorino's at the Port.[24]

In 1910, the Manitoba and Ontario Flour Mill Company erected Sudbury's largest mercantile building to date, a seven-storey brick mill with six massive concrete grain elevators. Another large operation was the Canada Creosote Company, a subsidiary of the CPR. In 1920–21, this company constructed the largest facility of its kind in the country on Lorne Street. Other companies created around this period included Co-operative Creamery, Liquid Air Company, Woollen Mills, and Northern Ski Factory. These businesses complemented other employers established earlier, such as the W.A. Evans and J.B. Laberge lumber stores, Sudbury Steam Laundry, and Cochrane-Dunlop Hardware.

Other developments included the start of *The Sudbury Daily Star* newspaper in 1910, the closure of the *Sudbury Journal* in 1917, and the opening of the Finnish newspaper *Vapaus* in 1917. The growing position of Sudbury vis-à-vis the surrounding mining belt encouraged the construction of more banks and the expansion of hotels. Large neon signs associated with the latter dominated much of the downtown's skyline, a feature that lasted well into the 1950s. Of great importance to the image of the town's core was the construction of the federal post office in 1915 at the corner of Elm and Durham Streets. It was a beautiful stone building, complete with a domed clock which faced the four compass directions; unfortunately it was reduced to rubble in 1959. While these business and industrial developments expanded Sudbury's economic base, the role of the CPR remained significant. A picnic sponsored by the CPR on Lake Ramsey in 1924, for instance, was attended by some 2 000 people.[25]

Between the turn of the century and the outbreak of the First World War, Sudbury's regional position was strengthened by communication and transportation advances that enabled the community to emerge as the "hub of the North" (see table 5.1 in appendix A).

The era of communications was heralded by the installation of the first telephone in Sudbury in 1899. In 1902, Bell Telephone had acquired its own franchise with only 61 subscribers, and by 1911 long distance service had been extended to North Bay and Sault Ste. Marie. On the transportation front, the year 1908 was momentous as both the CPR and Canadian Northern Railway (CNoR after 1918, and CN after 1960) completed direct railway links to Toronto, negating the need to travel south via North Bay. Thus began Sudbury's first steps toward supremacy in Northeastern Ontario and the severing of its ties with Montreal in favour of Toronto.

Sudbury attained another milestone in 1912 when the trunk highway to Sault Ste. Marie was initiated as part of the province's new northern road-building program, a development that enabled direct traffic to flow between North Bay and Sault St. Marie via Sudbury by the end of the First World War.[26] This was followed in 1913 by the completion of the Algoma Eastern Railway's (AER) 140-kilometre line (known as the Manitoulin & North Shore Railway until 1911), bringing Manitoulin Island into Sudbury's growing sphere of influence.[27] During the First World War, the AER expanded its operations to service the British America Nickel Corporation's smelter at Nickelton as well as International Nickel's roast yards at O'Donnell. Following the cessation of roasting at O'Donnell and the shutdown of Abitibi's Pulp and Paper Company's mill in Espanola in 1929, the railway passed into the hands of the CPR. The final regional advance in transportation occurred in 1915 when the Canadian Northern Railway completed its transcontinental mainline from Vancouver to Quebec via Capreol, giving the Sudbury area access to two trans-Canada railway connections.

Meanwhile, improvements were made to the local road network through the Colonization Roads program that expanded Sudbury's reach throughout the mining district. By 1910, most farming communities in the area had links to Sudbury by road or rail. The first bus service between Sudbury and Copper Cliff was initiated in this year. The opening of the Sudbury-Copper Cliff Suburban Electric Railway Company's streetcar system in 1915 was very important; it used power supplied by the privately owned Wahnapitae Power Company (acquired by the Hydro-electric Power Commission of Ontario in 1930). This connection between Sudbury and Copper Cliff ended any lingering pretensions that the latter may have had competing with Sudbury for local dominance.[28]

The growing population in the Sudbury area prompted changes at the provincial and federal levels that worked to Sudbury's advantage. Facing criticism that the old provincial electoral district of Nipissing was too large for political representation, the province divided the district of Nipissing into Nipissing East and Nipissing West for the 1902 election. As a result, Sudbury got its first provincial representative in cabinet, Frank Cochrane (see table 5.2 in appendix). A former mayor of the Town of Sudbury (1897, 1898, and 1902), the owner of Cochrane Hardware Limited, and a shareholder in the Wahnapitae Power Company, Cochrane became the provincial Minister of Lands, Forests and Mines from 1905 to 1911. Cochrane was also Sudbury's first federal Member of Parliament; he opted to move from provincial to federal politics in 1911. Following his victory at the national level, he was appointed Minister of Railways and Canals in the Borden government, a position he retained until 1917. A consummate politician, he carefully tended his northeastern fiefdom, which he called "Greater Ontario," and spread his government's largesse in the form of colonization roads, railway subsidies, educational aid, and settlement assistance. To reward his zeal, the Town of Cochrane was given his name.

Sudbury was proclaimed to be its own judicial district in 1907, so that people with legal business no longer had to go to North Bay. Thus, Sudbury became the base for various provincial governmental services, a role symbolized by the construction of imposing buildings such as a registry office and courthouse. In 1911 Charles McCrea, another Conservative, succeeded Cochrane in the province and continued to be re-elected until 1934. During his tenure, McCrea was a strong supporter of the Sudbury area and served both as Minister of Mines and Minister of Games and Fisheries; he was also influential in having the trunk road constructed between Sault Ste. Marie and Sudbury prior to the First World War.[29]

During this formative stage of community development, other features emerged that characterize Sudbury even today. Sudburians developed a passionate interest in all kinds of sports, including hockey, curling, lacrosse, baseball, rifle shooting, and bicycling, and enjoyed recreational activities such as camping, canoeing, fishing, and swimming on Lake Ramsey. The latter site was especially popular.

The Sudbury Boating Club (later Sudbury Canoe Club) and boathouse was established in 1902 and the club held Dominion Day regattas throughout the decade. The boathouse remained a fixture on the shore of Lake Ramsey until it was relocated to the foot of Elizabeth Street in 1999. Since fish could easily migrate to the lake from the Vermilion River system via Kelly Lake and Lily Creek, this activity attracted anglers until the 1920s, when Kelly Lake became polluted. Baseball had a huge following.[30] Curling and ice hockey were also in vogue following the erection of a rink on land located east of Durham and south of Larch Street purchased from the CPR in 1902. The site later evolved into the well-known Palace Rink, where activities expanded to include roller skating and dancing. It was during this period that Jimmy "Trump" Davidson started his famous career in Sudbury, one that was to span more than fifty years. Beginning as a trumpeter in the Canadian Legion band, he moved to Toronto in 1929 where he became a cornet soloist and one of the leading pioneers and promoters of jazz in the country, showcasing the delights of Dixieland music through his many CBC broadcasts.[31]

People in the late nineteenth century created a wide variety of secular organizations, especially lodges. By 1891, Sudbury boasted more than ten lodges, prompting the claim that "Sudbury had more lodges than any other Canadian town of its size."[32] These fraternal groupings, which attracted people along ethnic and religious lines, were later accompanied by ethnically based halls. In the downtown area, the Grand Opera House that opened in 1908–9 (later renamed the Grand Theatre) served as a major entertainment venue. Modelled after the Grand Opera House in Toronto, vaudeville, stage, and movie presentations were shown there until the Second World War.[33] All of this construction notwithstanding, many in the citizenry felt that the cultural needs of the municipality were not being met by the province. As noted by the *Sudbury Journal*, "During the decade, Sudburians were very conscious of the fact that 'the Ontario government [had] taken $3 000 000 out of this district in the past few years from mineral lands, and millions of dollars for timber, but not one percent [had] been spent here."[34] Finally, the seeds for the Sudbury area's well-known labour militancy were sown. While Sudbury's first recorded strikes in 1895 and 1899 failed, they were a harbinger of battles to come.[35]

The Francophone/Catholic Presence

While the British dominated Sudbury's population until the Second World War, there was one feature that distinguished the community's citizenry from most other large Ontario municipalities, that is, the presence of a strong francophone/Catholic sector. As shown in table 5.1, the French ethnic group consistently comprised more than one-third of Sudbury's population up to the Second World War period. This trend developed early

TABLE 5.3   Population by ethnic origin for Town/City of Sudbury (1901–1951)

	Year					
Origin	**1901**	**1911**	**1921**	**1931**	**1941**	**1951**
English	253	705	1 762	2 860	5 339	
Irish	540	837	1 279	2 117	4 262	
Scottish	342	661	1 241	1 737	3 355	
(Total British)	(1 139)	(2 218)	(4 331)	(6 790)	(13 379)	(15 502)
French	702	1 518	3 091	6 649	10 772	16 060
Czech/Slovak				145	533	431
Chinese	2	27	97	130	109	141
Netherlands	1	3	22	70	157	284
Finnish			206	1 374	1 241	1 478
German	38	52	58	277	516	845
Native	1	1	3	12	30	69
Italian	48	78	281	627	959	1 502
Jewish	73	87	129	192	239	172
Polish		14	50	393	815	1 127
Russian	1	6	42	71	122	143
Scandinavian	14	14	31	172	352	389
Ukrainian			13	761	1 617	2 571
Other/Multiple	8	132	267	855	1 362	1 696
TOTAL	2 027	4 150	8 621	18 518	32 203	42 410

Source: Statistics Canada, *Canadian Censuses* (1901–1951).
Note: For these censuses, a person's ethnic group was traced through the father.

in Sudbury's history due to the land grants given to the Jesuits in 1884 and the fact that many of the workers on the CPR mainline were francophones from Western Quebec or Eastern Ontario who decided to remain in the area. This development, in turn, encouraged the subdividing of lands acquired from the French-speaking Jesuit fathers. As Dennie states, "the French Canadians originally owned most of the land in a funnel-like northeast direction starting from Lot 5, Concession 3. This established the pattern of residency for French Canadians for the next eighty years."[36] This tract of land was dominated on the west by a highland referred to as Mount St. Joseph and by lower-lying lands to the east known as O'Connor Park and the Flour Mill.

With respect to religion, the figures pertaining to the Roman Catholic faith were even more significant, ranging between 50 and 60 per cent (table 5.2). This higher proportion was due to a strong Irish presence at the outset and the arrival of Italians after 1921. Unlike francophones, the other segments of the Roman Catholic population tended to congregate in other parts of the community, such as the Gatchell, the West End and the Donovan. Through the parish and its associated parochial institutions, holy orders, notably the Jesuit Fathers, the Grey Nuns of the Cross (Soeurs Grises de la Croix), known since 1964 as the Sisters of Charity of Ottawa (Soeurs de la Charité D'Ottawa), and the Sisters of St. Joseph emerged as powerful forces within these Roman Catholic neighbourhoods.

TABLE 5.4   Population by religion for Town/City of Sudbury (1901–1951)

	Year					
Religion	**1901**	**1911**	**1921**	**1931**	**1941**	**1951**
Anglican	277	677	1 160	1 901	3 094	3 210
Baptist	29	81	113	273	596	747
Greek Orthodox	7	3	78	276	539	716
Jewish	71	85	126	190	232	184
Lutheran	10	12	149	1 302	1 431	1 971
Mennonite	13	7				
Methodist	193	379	777	3		
Mormon					3	21
Pentecostal					4	76
Presbyterian	330	666	1 474	1 380	1 723	1 687
Roman Catholic	1 095	2 155	4 637	10 785	18 466	25 366
Salvation Army	6	3	18	31	69	
Ukrainian Catholic						1 152
United Church				2 184	5 685	6 880
Protestant		2	2		88	47
Christian	7	1	3			
Other/None	7	80	88	95	211	490
TOTAL	2 027	4 150	8 621	18 518	32 203	42 410

Source: Statistics Canada, *Canadian Censuses* (1901–1951).

Mount St. Joseph and the Flour Mill

Prior to the First World War, francophones began to move in increasing numbers to the Notre Dame area east of O'Connor Park. For several years after its founding in 1897, the J.B. Laberge Lumber mill and its construction division provided employment for local francophones.[37] Their presence increased after Manitoba and Ontario Flour Mill put up their building and grain silos in 1910, and the area began to be referred to as French Town or the French Quarter. Over time, the name Le Moulin à Fleur, or the Flour Mill gained favour. Despite the demise of the milling company in 1919, the use of the term "Flour Mill" has continued to the present day due to the visibility of the silos that still dominate the local skyline. There has been an ongoing debate regarding the future of these silos, one that has also lasted into to the present.[38]

The roots of this ethnic concentration were set in 1883 with the formation of St. Anne of the Pines (Sainte-Anne-des-Pins) parish by Father Nolin and the erection of a chapel and rectory. From this mother parish would eventually come all the other Catholic parishes in Sudbury. In 1884, a public school was temporarily incorporated into the church building. A separate school was then established in 1886 in a private home situated near St. Anne's and, two years later, a Catholic Separate Schools Commission was created. These facilities proved to be inadequate, and in 1887–89 a large church was completed on the site that included a school and parish hall.[39] The *Sudbury Journal* described it as "the principle ornament of our town, from an architectural point of view." Unfortunately, the church (but not the rectory) was destroyed by fire in 1894.[40] Undeterred by this tragedy, another smaller, yet still imposing church was constructed in the same year. In line with a motion passed by the Separate Schools Commission that allowed students to receive schooling exclusively in their mother tongue, a new school known as the Brown School was erected to replace the one that had formerly been in the church. In 1898, the Grey Nuns came to Sudbury to assume responsibility for the Catholic Schools.[41] In 1905, the parish's Jubilee Hall was opened across from the church on Beech Street and became the centre of activities for St. Anne's parish and often the wider community as well.

Due to the growth of Catholic families, St. Anne's church was enlarged in 1914–15, and numerous activities came to be associated with it. Many older parishioners belonged to either la Ligue du Sacré-Coeur or the Dames de Sainte Anne. Younger males often became altar boys, and many girls joined the Cercle Marguerite Bourgeoys. When language tensions and different attitudes held by Irish and French-Canadian members regarding the First World War conscription issue threatened the well-being of the congregation, an English-speaking parish named St. Joseph (consisting largely of Roman Catholic Irish parishioners) was created in 1915. While the French Canadians were not opposed to the creation of an English parish, they were upset by the decision of the Irish Bishop, Monseigneur David J. Scollard, to give the new parish land which they regarded as theirs.[42] Beginning in 1917, English-speaking services were temporarily held in the Jubilee Hall. In 1921, St. Joseph's rectory was erected on the western side of St. Anne's property. Between 1923 and 1928, a second impressive church known as St. Joseph's (later changed to Christ the King in 1935) was built on the same location.[43] These side-by-side Catholic churches have since continued to provide one the most enduring images associated with the downtown area.

Father Paré from St. Anne's in 1927 then acquired Brown School situated north of the church for use as an orphanage (Orphelinat d'Youville); run by the Grey Nuns, it opened in 1929. Meanwhile, the Catholic presence in the surrounding area had been strengthened by the construction of two separate schools on MacKenzie Street: St. Louis de Gonzague (formerly Central) in 1915 for French-speaking students, and St. Aloysius for English-speaking students in 1923. Two other separate schools were subsequently built in the Flour Mill area in 1928: Ste. Marie and Nolin.

The geographical influence of St. Anne's parish in promoting Catholicism and the French language in the surrounding area was profound in other ways. For example, the Jesuits early on realized that the facilities of the two existing hospitals, that is, the Sudbury Hospital run by Dr. W. Hart and later Dr. J.S. Goodfellow and the Algoma and Nipissing led by Dr. Struthers, were inadequate to meet the needs of the community. The latter facility, located at the corner of Elm and Durham Streets, was renamed the Sudbury General Hospital in 1895 and leased to St. Anne's parish in 1896. The parish priest, Father Lussier, called on the Grey Nuns to take charge of the institution under the direction of the Jesuit Fathers.[44] Renamed again as St. Joseph's Hospital, the building was seized by the town for unpaid taxes. This prompted the Grey Nuns to build a new hospital on Mount St. Joseph in 1898 on land donated by St. Anne's parish.[45] Locals described St. Joseph Hospital as another "beautiful" Catholic building, and it strengthened the French character of the neighbourhood. In 1911, the hospital opened the first training school for nurses in the area under the direction of Sister Elizabeth.[46] Known as the St. Elizabeth School of Nursing, its first students graduated in 1913; while it was registered with the province as early as 1921, it was 1931 before it became an approved school in Ontario.[47] Additional wings were added to the hospital in 1907, 1921, and 1929.[48] In 1921, the English-speaking order of the Sisters of St. Joseph arrived as a second teaching order, and in 1923 they opened a convent at the rear of St. Anne's church.[49]

A major event was the establishment of a Jesuit-run classical college for boys known as Sacred Heart College (Collège du Sacré Coeur) on Notre Dame Avenue. It was built on land owned by Father Caron that also supported a cemetery, sand pit, and a pool.[50] This project was undertaken by Father Lefebvre, who was responsible for St. Anne's parish from 1904 to 1914. The college, given the green light by the Jesuit Father General in Rome in 1912, came at a propitious moment, as the English-speaking Sudbury High School had opened in 1908. This opening concerned Franco-Sudburian elites, who thought that the public high school would attract many of the French-speaking students.[51] Sacred Heart opened its doors in the fall of 1913 to 110 students. In 1914 it was granted a charter by the Ontario Government to establish schools, colleges and universities, a role that was to have major significance fifty years later.[52] By 1917 it had gained a French-only configuration. It was the second institution of its kind in the province, and the only higher education facility in Northern Ontario.[53] In 1916 Sacred Heart became affiliated with the University of Ottawa; however, when disputes arose due to the differing educational approaches the Jesuits used in Sudbury compared to the Oblates (who controlled the University of Ottawa at the time), Sacred Heart signed a new agreement in 1927 with Laval University, one that lasted until 1956. The school developed linkages later on with the Universities of Toronto and Western Ontario for postgraduate studies.[54] From these beginnings, the college graduated thousands of Sudbury area men who went on to distinguish themselves as doctors, lawyers, teachers, and priests.

Another Catholic development in the area was the construction of St. Mary's Ukrainian Greek Catholic Church on Beech Street in 1928. Built with financial assistance from Inco in order to counter the growth of Communism and Bolshevism in the community, the church was eventually taken down as part of Sudbury's massive downtown urban renewal project of the 1960s.[55] Due to the increase in the number of French Canadians in the area north of Sacred Heart College in the late 1920s, a new parish known as St. Jean de Brébeuf was created in 1930. Construction of the church was started in the same year, again on land provided by the Jesuits.[56] It remained, however, until 1958 before the church was completed.[57]

The influence of the Jesuits was likewise exerted through the subdivision of their original land grant for residential purposes. This process started in 1890 and 1896, when the Jesuits subdivided parts of their lot east of St. Anne's Church. Between 1901 and 1925 the Jesuits opened another three subdivisions. Three more were opened by French Canadians who had originally acquired lands from the Jesuits.[58] Through this process, the Jesuits ensured that future settlement in the area would remain French Canadian in character. The success of this long-term objective can be measured by the fact that Sudbury managed to attain the position of being Ontario's second-most-important Franco-Ontario centre, surpassed only by Ottawa.[59] Another indication of Sudbury's position in Ontario came in 1925, when a group of three hundred French Canadians from Quebec headed by Henri Bourassa, editor of *Le Devoir*, included Sudbury in their tour of the province and were given a civic reception.[60] The final factor that strengthened the French presence in this area was the rise of associations that catered to the community. There were at least eight French fraternal, religious, or financial associations. These groupings formed the basis for the evolution of a small elite and a greater sense of solidarity within the French-Canadian community. While situated farther to the south, another important Catholic landmark deserves attention: the establishment of the Grotto of Notre Dame of Lourdes (Our Lady of Lourdes Grotto) in 1907 by Sudbury's famous French Count, Nicholas Julian Frédéric Romanet du Caillaud.[61]

A Protestant "Downtown"

While the Catholic faith was influential in shaping the community in the northern and northeastern reaches of the town, a Protestant-based religious zone emerged within a stone's throw of the Elm and Durham Street intersection, the core of the downtown. Protestant citizens built a number of imposing brick and stone church buildings mainly along Larch Street that served those of the Anglican, Methodist, Presbyterian, and Baptist faiths, as well as the Salvation Army. The Protestant presence first became visible when a frame building known as the Church of the Epiphany (Anglican) was dedicated in 1890 on a lot acquired on Larch Street from the CPR in 1886. In 1894, a bell tower was added to the Anglican Church, as was a rectory house in 1897. The original structure was torn down and replaced by a handsome red brick building in 1913.[62]

To the annoyance of the Anglican congregation, the Presbyterians then selected adjoining lots for their own church. By 1889, they had constructed a frame church building on the land they had also secured from the CPR. In 1910, they erected St. Andrew's Presbyterian Church using native stone. It was a large structure for the period, built because "the people of St. Andrew's believe[d] a Christian Church should show a nobler strength than any secular institution."[63] Another reason for its grandeur was the spirit of competition resulting from the Methodist's new church building and the cathedral-like presence of St. Anne's. In 1923, St. Andrew's constructed a new manse on the property which served until 1940.[64]

Spurred by the efforts of the legendary Silas Huntington (known as the "Apostle of the North," and after whom Huntington University is named), the Methodists erected a small frame church on Beech Street in 1889. In 1904 the congregation traded its property on Beech Street for lots on the corner of Cedar and Lisgar Streets. One lot was retained for a manse and the other for a cement block church erected in 1908 that was a major addition to Sudbury's skyline. Destroyed by fire in 1923, the Cedar Street Church was immediately rebuilt.

In 1925, the Methodist and Presbyterian Churches joined together and formed a new church body called the United Church of Canada. The united congregations continued to exist for another two years, at which time they decided to retain St. Andrew's as the official church, and rename the former Methodist building Wesley Hall. The latter continued to serve as a Sunday school and a place for midweek meetings until 1939, when it was sold to Bell Canada. Shortly thereafter, the congregation built a new Wesley Hall east of St. Andrew's, where the former manse had stood. In the meantime, a group of Presbyterians disillusioned with the union of 1925 decided in 1927 to erect their own building nearby, and called it Knox Presbyterian Church.

The same area attracted Baptists, who moved to the corner of Larch and Minto Streets in 1912, and they constructed another large church from 1918 to 1925. The final church group to appear in the area was the Salvation Army. After arriving in Sudbury in 1895, the army erected a barrack on Durham Street before shifting to Larch Street in 1922. True to its tradition, the organization evolved as the central agency for relief in Sudbury, especially during the 1930s when it provided makeshift hostels and soup kitchens.

While the Mount St. Joseph and Larch Street areas were the dominant religious zones in Sudbury, it is worth mentioning the circular Jewish Synagogue that was built in 1913 at the corner of Dufferin and Pine Streets. It remained at this site until it was sold in 1949. For a short period of time, the Pentecostals also had a downtown presence. Starting in

1937, they held services in a large tent at the corner of Larch and Minto, across the street from the Salvation Army. Services then shifted to the Finnish Pentecostal Church located in the West End, and later to another rented downtown hall. In 1941, the group took the name Glad Tidings Tabernacle and again moved, this time to the seized property of the Ukrainian Labour Temple on Spruce Street. When it was returned to the Ukrainians after the Second World War, the congregation erected its first church on Alder Street near Queen's Athletic Field.[65] While there were other Polish, Finnish, and Ukrainian congregations in the area, these were situated in Copper Cliff.

The Beginning of Social Stratification

Around the turn of the 1900s, the town began to show evidence of social stratification. A distinguishing indication of this was the rise of prestigious mansions of the Protestant elite, many of them prosperous merchants and lumbermen such as W.J. Bell. Some of these "finer residences" were erected in the vicinity of the Protestant church district downtown on Cedar, Larch, and Drinkwater Streets. A few were even constructed in the College and Baker Street area. The most stately homes were erected close to the north shore of Lake Ramsey on Elizabeth, John, and McNaughton Streets and along Ramsey Road on subdivisions developed by Andrew McNaughton and Aurora Dubois.[66] Remnants of this early pattern still remain, notably in the form of the Bell Mansion, erected in 1907 and now home to the Art Gallery of Sudbury and the adjacent Bell estate, now part of Bell Park. Another reflection of this southern "elite" association with Lake Ramsey was the creation of the Sudbury Boating Club in 1902, and the formation in 1922 of the exclusive Idylwylde Land Syndicate, the forerunner to the Idylwylde Golf and Country Club. Other expressions of settlement involved the movement of immigrant groups into the northern and southwestern reaches of the town. Beyond these zones were a number of farms that provided vegetables and milk to the local population: the Eyre, Donovan, Gatchell, Boyce, and Robinson farms, for example. While parts of the flat New Sudbury area remained as a bustling lumbering site that still utilized Junction Creek to float logs to a lumber mill in the Flour Mill neighbourhood, cleared sections were already being sold as farmland.

The First World War

The period between the start of the First World War in 1914 and the Armistice of 1918 wrought numerous changes. The war in many ways gave prosperity to the Sudbury area, as resource exploitation and commercial expansion continued. Despite the reservations the French-Canadian population had about the "imperial" nature of the conflict and conscription, this phase of Sudbury's history revealed a remarkable contribution to the Great War effort. Overall, Sudbury responded in a respectable manner when compared to the rest of the country. Double the number of local volunteers requested by the federal government joined the 97th Regiment, Algonquin Rifles. Other manpower mobilizations were also successful, and their ranks boasted a large representation of the ethnic, or what *The Sudbury Star* called the "cosmopolitan and floating" population.[67]

In 1917, Sudbury was selected as the mobilization centre for Northeastern Ontario. Though the most dramatic act of patriotism was enlistment, citizens supported the war effort in other ways. A Sudbury branch of the Canadian Patriotic Fund provided assistance for the dependents of soldiers overseas. Other organizations included the Great War Veterans Association, plus several War Relief Clubs and chapters of the Red Cross. Sudbury's response to the Great War was thus affirmative and continuous. Because of its inherently xenophobic nature, the war, however, had the regrettable short term impact of intensifying some ethnic prejudices. This was evidenced by council's decision in 1919 to reinforce the British image of Sudbury by renaming a number of streets, especially those of German or other origin, which the mayor at the time said "could only be pronounced by coughing and sneezing." For example, Berlin Avenue was changed to Windsor Avenue (now Kitchener) and later Romanet du Caillaud Street was renamed Howey Crescent (now Howey Drive).[68]

Metamorphosis into a City

During the 1920s, Sudbury began its transformation into a city. Bolstered by transportation improvements and an increased share of Inco's labour force, its population more than doubled, rising from 8 621 in 1921 to 18 518 ten years later. Within the region, its share of the population was now almost one half. In terms of its ethnic composition, it was an untypical Ontario town, with the French, British and "other" categories each accounting for roughly one-third of the population. Within the latter category, increasingly diverse ethnic groups manifested as Finns, Ukrainians, Italians, and Poles began to leave their imprints on the landscape. The town's religious character was distinctive, as 60 per cent of the population was Roman Catholic, contrasting sharply with the province's corresponding figure of about 20 per cent. The socio-cultural setting was matched by a growing tendency of the city to take on the appearance of a cluster of communities, a trend that was reinforced in the 1930s.

While relationships between the English and French-speaking sections of the population remained peaceful after the First World War, the same could not be said of other ethnic groups such as the Chinese and Finns. Fears about the "yellow perils" associated with Chinese cafés in the early 1920s were followed in 1929 by a sensational series of events associated with *Vapaus*, a Finnish left-wing newspaper, and its editor, who was charged with making derogatory statements regarding King George V. These statements attracted the attention of the province, the federal government, and the RCMP. The community feared the Finns' strong support for labour reform and the Communist Party of Canada. Radical Finns who took an overt anti-church and anti-marriage stance also angered reporters for *The Sudbury Star* and church groups. The high visibility of left-wing Finnish institutions such as the *Vapaus* newspaper, a restaurant and travel agency, boarding houses, and the Liberty Hall situated just west of the downtown, and the prominence given to this area by *National Geographic* in 1932, were other factors that allowed these feelings to fester well into the 1930s.[69] Sudburians voiced similar expressions against leftist Ukrainians, who had erected a hall on Applegrove Street adjacent to the "red" Finnish Hall. These community responses to fears about the left wing did not prove to be an isolated event: there was a similar attack on the International Union of Mine, Mill and Smelter Workers in the late 1950s and early 1960s.

As Sudbury's dominance in the region grew, it became a true mining town, completely tied to a single resource industry. The increased use of nickel in heavy machinery, mining equipment, aircraft, and, above all, in the automobile industry, expanded the consumption of nickel, which required more extraction and brought new workers to the city. The attraction of Sudbury as a residential site was enhanced by W.J. Bell's decision in 1926 to deed forty-five hectares of parkland situated west of Lakeside Park on Lake Ramsey (biography 4). Around the same time, work began on Central Park, located beside the skating and curling rink and Central Public School.[70] The name was later changed to Memorial Park, and it became the site for a cenotaph unveiled in 1928.

Another incentive for settlers to move to Sudbury was the CPR's decision in 1927 to put most of its remaining townsite properties on the market at bargain prices. As trumpeted by *The Sudbury Star.*

> One of the biggest real estate movements in the history of Sudbury was launched this week when the Canadian Pacific Railway Co. placed practically all of their remaining building lots throughout the town on sale.... At the present time there are about 400 lots remaining, and the cost of administrating this remnant of the original townsite has led the officials of the company to decide to dispose of them all and thus retire from the real estate business in Sudbury.[71]

With this decision, the fiefdom of the CPR within the community officially came to an end.

Great Depression and Its Aftermath

In 1930, Sudbury became a city.[72] During the Great Depression and throughout the early 1930s when much of Canada shuddered with misery, Sudburians managed due to the fortunes of the nickel industry. According to *Maclean's*, Sudburians weathered the situation better than those elsewhere in Canada because of their faith that the situation would soon change. As John Fenton, who was mayor at the time stated, "Things are better here than in most places, because we know it is only a matter of time … it is just a case of keep plugging until the world turns the corner. Sudbury can't miss."[73] Sudbury became a destination city for much of the 1930s as shown by its population, which grew from 18 518 to 32 203, an increase of almost 75 per cent. The city promoted its own well-being by undertaking infrastructure improvements. City council initiated a major program to install water and sewer lines, remove rocks, and extend streets. The city also purchased the old gravel pit in the West End and transformed it into an athletic field, while at the same time converting the sports field in the downtown into Memorial Park. Sudbury's attraction was enhanced

Biography 4

WILLIAM JOSEPH BELL
Lumber Baron and Public Benefactor (1858–1945)

William Joseph Bell was born on July 29, 1858 in Pembroke, Ontario.[a] He started his career at the age of eighteen as a scaler and lumberjack for a lumber company in the Ottawa Valley and remained there for five years before becoming a bookkeeper and paymaster for the E.B. Eddy Forest Products Company. In 1886 he married Katherine Skead, for whom the township in the Sudbury area is named. Bell left for Nova Scotia in 1889 and worked for three years on dock harbour installations.

After arriving in Sudbury in 1886, he gained employment with the Hale & Booth Lumber Company. He became a partner in the Arnold & Bell Lumber Company, one of the thirty-six firms cutting timber along the North Shore at the time, and then took charge of the Sable and Spanish River Boom and Slide Company. After the Spanish River Lumber Company acquired Hale & Booth, Bell rose through the ranks of the new company, eventually acquiring a controlling interest and serving as its president from 1924 until it was bought by the Poupore Lumber Company.

According to *The Sudbury Star*, "to the thousands of men who knew him intimately in the rude camps, along the river drive, on the booming out grounds at the mouth of the Spanish or in the large sawmills along the North Shore and on Lake Wanapitei, he was a 'boss' who always tried to be fair, was ever ready to lend of his munificence in time of trouble or to defend a worker whom he thought was being imposed upon."[b] Bell served as president of National Grocers and Cochrane-Dunlop Hardware Ltd. until the two companies merged. He passed away on January 12, 1945 in his eighty-seventh year, leaving behind his wife.[c]

It is for Bell's interest in Sudbury itself that his name will long be remembered. Among his legacies is the Bell Mansion, built in 1906–7 on his estate of 63 hectares adjacent to Ramsey Lake, separated from the town proper by the CPR tracks. Purchased from the estate of Robert S. Henderson, the site had previously served as the location of Deacon's Castle, a small hand-built rock-and-log outlook inhabited by a local recluse. This estate was erected using local stone with white limestone trim, and the interior was covered with rich oak panelling. When completed, it included the mansion, a coach house, a stable, and an animal shelter. Fifty teams of horses took six weeks to haul enough soil for the landscaping of the rocky site. The mansion remained as Sudbury's most elegant structure until it was damaged by fire in December 1955. Bequeathed by Mrs. Bell to the Sudbury Memorial Hospital after her death in 1954 and later acquired by the Nickel Lodge of the Masonic Order, the house fell into a sad state of disrepair and acquired a reputation for being "haunted." In 1966 the property was purchased by the Centennial Committee of the Sudbury Chamber of Commerce for use as a Museum and Arts Centre. Home of the Art Gallery since 1967, the centre was briefly owned by the Sudbury Centennial Museum Society and then operated by Laurentian University from 1968 until it became an independent entity in 1997.

In 1909, Bell became a shareholder in the Grand Opera House and later became president of the Sudbury Board of Trade. He was also the owner of the first new car in

Sudbury in 1911; his driver at the time, Percy Gardner, later became the owner of Gardner Motors in Sudbury. In line with his love of the outdoors and recreation, he became chair of the Sudbury Parks Board in 1917. During his tenure, he approved the development of Sudbury's first public park on 1.6 hectares of land from the CPR that became Memorial Park. Considered to be only a "spud patch" at that time, the decision was a far-sighted one that enhanced the quality of the downtown. In the same year, he spearheaded the acquisition of 2.8 hectares of lakefront on Ramsey Lake from La Corporation du Collège de St. Marie à Montréal owned by the Jesuits. In 1926 the park was augmented by another 1.8 hectares donated by Bell in return for the right to close several streets surrounding his mansion.[d]

Satisfied that the Parks Board was in a financial position to proceed with improvements, Bell made another generous donation, handing over 44.5 hectares of land beside the newly christened Lakeside Park, later renamed Bell Park.[e] Although his dream of a grandstand and athletic field complete with baseball diamond, football field, grass tennis courts, bowling greens, a new pavilion, and a community hall in the enlarged park proved impractical, Bell was nevertheless instrumental in preserving a large piece of waterfront property for the future enjoyment of the general public. Another contribution Bell made for the betterment of the city was his bequest of $125 000 to the Salvation Army to the build a Men's Social Centre on Larch Street.

In recognition of Bell's philanthropic work, the City of Sudbury commissioned a portrait painted by C.A. McGregor, O.S.A., which hangs in city council chambers.

---

[a]For a history of W.J. Bell and his Bell Rock mansion, refer to: Laurentian University Museum and Art Centre, *Bell Rock—Laurentian University Museum and Art Centre* (Sudbury, ON: Laurentian University Museum and Art Centre, 1990).
[b]"A City Suffers Great Loss," *Sudbury Daily Star*, January 13, 1945, 4.
[c]"W.J. Bell Dies at Home after Long Illness," *Sudbury Star*, January 12, 1945, 1.
[d]"Deeds 110 Acres Bordering Lake to Parks Board," *Sudbury Star*, February 17, 1926, 1.
[e]"William Joseph Bell, a Lumber Baron with an Obsession for Creating Parks," *Sudbury Star*, April 1, 1976, 3.

---

by the repeal of the Ontario Temperance Act (1916–27) and the establishment of a liquor store, thus offering new competition to the myriad of bootleggers in Sudbury and nearby McKim Township. The introduction of liquor and beer sales at Sudbury hotels in 1934 likewise changed the city's social landscape and made the Depression years more bearable, especially for young single men.

Due to the arrival of transients and the fact that the mining companies paid no local taxes in Sudbury, the city temporarily found itself in financial trouble. The situation apparently became so dire that the province forced the city into receivership between 1935 and 1941. Wallace has questioned this decision, noting the recovery of the mining industry by this time, the increase shown in the number of stores and services for the enlarged population, and the accompanying building boom starting in 1935 that saw the number of apartments built double in the decade.[74] Boarding houses also proliferated in many of the city's inner neighbourhoods. Because of the physical configuration of the railway lines and narrow strips of land interspersed by rocky hills, most of this population increase occurred in the areas of lower elevation (fig. 5.4).

As the new city encompassed only around ten square kilometres, its density rose to almost 3 000 people per square kilometre by the Second World War, with many residents living in substandard housing with poor municipal services. The abysmal quality of the housing stock at the time was affirmed in the report of the Housing and Community Planning Subcommittee of the federal Advisory Committee on Reconstruction.[75] In this report, Sudbury's housing ranked at or near the top of the list of Canadian cities with dwellings that needed external repairs, lacked flush toilets, bathtubs or showers, and with a high degree of overcrowding. As one observer of the local community noted, Sudbury was similar to other smelter towns: "there were the same forlorn buildings against a background of dun hills, the same baked earth and slag heaps; all vegetation killed by sulphur fumes from the smelter."[76] According to another provincial study, Sudbury's living conditions improved but still remained well below the provincial average, a situation that continued at least until 1971.[77] It is no wonder, therefore, that many residents sought refuge from these conditions by seeking properties on the fringe of the town, or further away in McKim Township.

At the same time, Sudbury's density made it more efficient for the city to supply services. Due to the lack of regional and inner city road alternatives, the concentration of residents around the downtown, and the location of grocery stores there, a vibrant central business district emerged around the intersection of Elm and Durham Streets. While favourable to downtown merchants, this node and its close association with the CPR Main and Stobie Branch lines resulted in traffic bottlenecks, especially during Inco's shift changes. According to *The Sudbury Star*, this situation was responsible for Sudbury being the worst city in North America for automobile accidents.[78] The downtown was enlivened by the emergence of Borgia Street as one of the most notorious inner city neighbourhoods in the nation. Reputed at one time to be the toughest street in Canada, it was the scene of several murders and the place where partying miners, lumberjacks, and gamblers congregated for floating games of chance. According to one vice syndicate, this area was listed as one of its prime bookings because of the high numbers of single men with cash. Rough-clad Good Time Charlies were lured to places of assignation by squads of street walkers, only to be clubbed upon arrival by the ladies' muscular consorts. For a time, a woman known as Tiger Lil ruled the area with an iron fist and sheer physical power. The mettle of rookie constables was often tested by sending them alone to patrol the street.[79]

The 1930s featured other interesting developments. Politically, the area became allied with the Liberals rather than the Conservatives. At the federal level, Edmund A. Lapierre's string of victories from 1921 to 1930 was succeeded by Dr. J.R. Hurtubise, whose tenure lasted from 1930 until 1945, at which time he became a senator. Provincially, Edmund A. Lapierre defeated the perennial Conservative favourite, Charles McCrea in 1934; in the 1937 provincial election, Liberal James Cooper won. For the most part, these politicians had undistinguished political careers. It was the newspaper mogul W.E. Mason, however, who emerged as the most dominant and controversial personality in the area (biography 5).

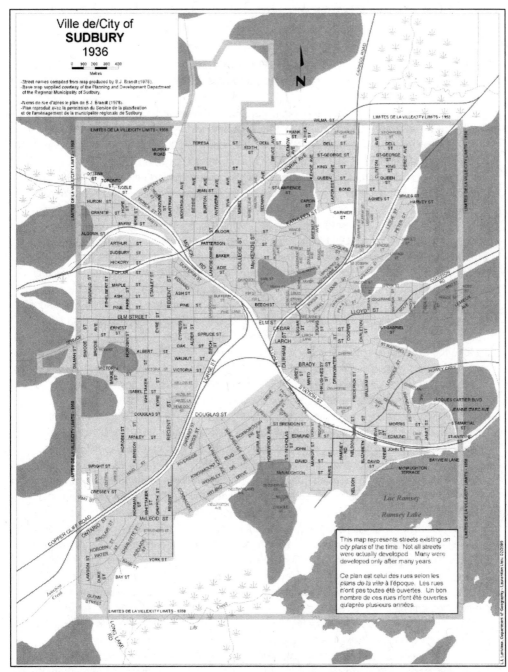

FIGURE 5.4   Streetscape of the City of Sudbury (1936). From Lionel Bonin and Gwenda Hallsworth, *Street Names of Downtown Sudbury: A Historical Directory*, 53; and N. De Santi, "The Spatial Organization of the Sudbury Transit System in the Region of Sudbury" (Honours B.A. thesis, Laurentian University, 1997).

Biography 5

WILLIAM EDGE MASON
Newspaper Mogul (1882–1948)

William Edge Mason was born on March 4, 1882 in Walkerton, Ontario.[a] After finishing school with the third highest marks in Bruce County, he began his apprenticeship as a dollar-a-week printer's devil with the *Walkerton Telescope*; in 1900 he became a journeyman printer. He got his first job in Toronto working as a proofreader, handyman, and assistant foreman for *Toronto Saturday Night*. He arrived in Sudbury with his wife, Alice Maud Tinlin, in November of 1907 and was hired as a printer for the *Sudbury Mining News* that was formed two years earlier. The newspaper was the successor to the *Sudbury News* that had operated briefly from 1894 to 1899. At the time it was in competition with the *Sudbury Journal,* which had started publishing in 1891 as a one-man operation by James Orr. Mason then joined the *Daily Northern Star* as printing foreman. Started on January 11, 1909, the *Star* claimed to be the first daily newspaper between Toronto and the North Pole. When the newspaper encountered financial difficulties, Mason persuaded ten other Sudburians to invest in it, and on April 23, 1910, the publication was reorganized as a semi-weekly called *The Sudbury Star*. While *The Sudbury Star* experienced some lean years, its future was assured by the demise of the *Sudbury Journal* in 1918 and the *Sudbury Mining News* in 1922. Starting in 1921, the *Sudbury Star* office acquired a prominent location at the corner of Frood and Elm Streets, where it remained until 1961.

Encouraged by his monopoly position in Sudbury, Mason took over the *North Bay Nugget* in 1927. In response to population growth in the Sudbury area, the publication of *The Sudbury Star* was increased to three times a week in 1935, and in 1939 it acquired daily status. In 1941 Mason also made the *North Bay Nugget* a daily. He liked his Nugget employees so much that he devised a plan for the newspaper to become Canada's only employee-owned newspaper. It retained this status until it was sold in 1955.

In contrast, Mason was not so kind to the mining employees in the Sudbury area who sought to unionize. He was for capitalism and against ethnic foreigners, and attacked union organizers relentlessly, calling them "loud mouth agitators ... inspired by communists." His support of the mining giants was unquestionable; few references were made to "sulphur fumes," and nickel was never shipped to Germany. This pro-company stance was continued by *The Sudbury Star* even long after Mason's death. Nor was he always kind to his fellow newspapermen. Mason is reported to have fired more newspapermen than any other publisher in Canada. So many reporters were shown the door that it became a mark of distinction among them to have been fired by him. To add to his reputation, Mason became part of Sudbury's folklore in 1932 when he was acquitted on a charge of arson involving the torching of the Sudbury transit building.

Mason became owner of CKSO-AM, which began broadcasting from the same building that held his newspaper offices on August 19, 1935. It remained the only radio station in Sudbury until CHNO opened in 1947. Mason also donated land upon which the Sudbury

Legion erected its building in 1947. He died on June 22, 1948. Many claimed that he literally killed himself by managing the campaign of Welland S. Gemmell, despite a failing heart. He had no children. Thousands attended his funeral, the largest ever held in the city.

Throughout his life, Mason was heavily involved with the Sudbury community. He was instrumental in founding the Sudbury Parks Board. He controlled the Grand Theatre, in which he showed the newest movies with the most advanced technology. He was president of the Sudbury Cub Wolves as well as the Sudbury Board of Trade, and served as chairman of the Public School Board. In line with his belief that Sudbury had been good to him and that he should therefore be good to Sudbury, his will established the W.E. Mason Foundation, dedicating its fund of more than two million dollars toward charitable purposes.[b] The foundation made many donations, giving land on Mackenzie Street for a new library, aiding the construction of the Memorial Park Wading Pool, paying for seats in the Sudbury Arena, assisting in the construction of the Sudbury General Hospital, securing the 24-hectare Cook Farm for land to build Sudbury Memorial Hospital, and assisting in the creation of the Sudbury-Algoma Sanatorium. Had he lived, he undoubtedly would have been unhappy with the decision by the Legion in 1965 to sell its building to Local 6500 of the United Steelworkers of America, a landmark that was unfortunately destroyed by fire in 2008.

---

[a]A biography of Mason's Life is found in "William Edge Mason: 1882–1948," *South Side Story*, January 2005, 10.
[b]For a listing of the some of the donations made by the Mason Foundation, see "Pay Tribute to Late Publisher for his Good Works," *Sudbury Star*, November 18, 1952, 3.

In the area of sports, there were great celebrations in 1932 when the Sudbury Cub Wolves won the Memorial Cup, and in 1938 when the Sudbury Wolves won a world championship in Prague. Taavi (Dave) Komonen, a Finnish-Canadian athlete, also brought attention to Sudbury in 1933 and 1934 when he placed second and first respectively in the Boston Marathon. For his marathon efforts in Canada and the United States he was selected as the Lionel Conacher Canadian male athlete of the year for 1933.

In 1935, Sudbury's first radio station, CKSO, took to the airwaves, offering music, radio plays, and local news as an alternative form of entertainment. Movie theatres such as the Grand, Capital, Regent, and the Rio showed Hollywood films and American newsreels. A YMCA opened in 1937 that had a major influence on the lives of thousands of Sudburians, not only at its Lloyd Street and Durham Street locations in Sudbury, but also at its John Island Camp and Camp Falcona facilities.[80] A year later Inco opened its Employees' Club in Sudbury, featuring a large auditorium and dance floor, badminton courts, library, gymnasium, billiard room, and bowling alleys.[81] During the summer months, Lake Ramsey continued to provide a popular venue for swimming. In the working-class areas on the fringes, the city's cultural mediocrity was offset by the construction of churches and ethnic halls. These facilities emerged as centres for immigrant cultural activities like dances, performances, and plays, and hosted celebrations such as marriages and anniversaries.

Emergence of Neighbourhoods

After the First World War, subdivisions that had started to sprout earlier in the century expanded in the Gatchell, Donovan, West End, Kingsmount, Minnow Lake, and Lockerby areas. While some residential districts such as the West End and Donovan were basically extensions of the downtown, others such as Gatchell, Minnow Lake, and Lockerby spread beyond the town limits into the surrounding Township of McKim. The latter two areas were distinctive in that they developed around Minnow (previously Black) and Nepahwin (previously Trout) Lakes.

Gatchell

The rise of the Gatchell neighourhood in the Township of McKim was a result of several factors related to its closeness to Copper Cliff, the decision by International Nickel in 1914 to limit the size of Copper Cliff, its location adjacent to the Sudbury-Copper Cliff Suburban Electric Railway Company, and the important role played by the land holdings of the Holditch and Gatchell families. The opening of the streetcar system on November 11, 1905, heralded a new era for the area straddling Copper Cliff Road (Lorne Street), as it afforded cheap access to the construction and smelting jobs available in Copper Cliff. Within a year, the shiny red streetcars were making thirty-three round trips each day, seven days a week. The system likewise offered easy accessibility to the downtown.[82] Settlement in the area remained sporadic in the early 1900s, in part because children who lived there had to walk past the Copper Cliff roast beds nearby and their associated lung-choking, thick sulphur fumes to get to school.

Starting in 1907, the history of Gatchell became intimately associated with the Holditch family and the Gatchell dairy farm (fig. 5.5). William Holditch started the development process when he acquired 6.5 hectares of land in the Big Nickel area in 1891. His son Ernest encouraged more settlement in 1909 and 1910 by constructing Kelly Lake Road and opening a subdivision west of the Eyre Cemetery. At one time he owned all of Lot 9 between Concession 1 and 2 located southwest of the downtown. He sold some of this land to the CPR in 1910 which, in turn, leased it to the Canada Creosote Company that operated from 1920 to 1963. Green railway ties and telegraph poles were brought there for preservation treatment. Many of the first homes built in Gatchell, some along Gutcher Avenue, were erected by employees of this firm.[83] Another legacy left by this company was the pollution of Junction Creek that straddled its southern boundary. In the 1920s, Inco acquired part of the Holditch family holdings for use as a slag dump. The city seized some of the properties to pay for taxes in the 1930s, but these were returned after William Holditch waived his claims to lots in the city-owned Alexandra Park subdivision in 1951.[84] In 1936, he opened the Holditch subdivision, consisting of 286 lots that were gradually sold over the next twenty years. After the Second World War, part of the Canada Creosote Company's holdings was used to start the development of the Robinson subdivision situated north of Robinson Lake.

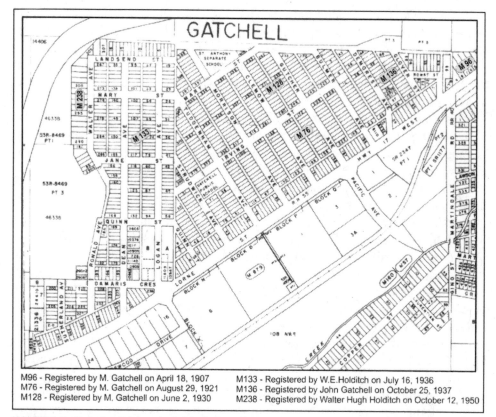

M96 - Registered by M. Gatchell on April 18, 1907
M76 - Registered by M. Gatchell on August 29, 1921
M128 - Registered by M. Gatchell on June 2, 1930

M133 - Registered by W.E.Holditch on July 16, 1936
M136 - Registered by John Gatchell on October 25, 1937
M238 - Registered by Walter Hugh Holditch on October 12, 1950

FIGURE 5.5 Early Subdivisions in Gatchell. From Gatchell History Committee, *Memories of Gatchell*
(Sudbury, ON: Gatchell Committee, 1997), 8.

In 1903, Thomas and Moses Gatchell had acquired a 78.5-hectare dairy farm situated
north of Copper Cliff Road and in the Copper Street area. As the largest dairy farm on the
fringe of Sudbury, it supported the Gatchell Dairy, whose major customer was St. Joseph's
Hospital. The property was later split into two dairy farms serving both Copper Cliff and
Sudbury Dairies.[85] While Moses did some work on subdividing his land in 1907 and 1921,
it remained until the 1930s before he started what *The Sudbury Star* claimed was Sudbury's
first real suburban development.[86] It was Moses' belief that his subdivisions along Copper
Cliff Road would one day emerge as the "Yonge Street" of Sudbury.[87] The name Gatchell
was attached to the area partly due to the fact that both Thomas and Moses served as
political representatives in McKim Township.[88]

Other activities of note included two subdivisions built near the creosote plant by
J.F. Mackey along Copper Street near the juncture of Martindale Road. By 1934, growth
in the area was sufficient to warrant the establishment of Gatchell Public School; in the
following year, a Suburban Services Board was set up to deal with sanitation concerns and

bring water, sewer, and fire services to the community. In 1941, a separate school known as Our Lady of Perpetual Help (St. Anthony) was constructed. While these schools provided the main play areas, it was the swimming hole at the sand "pit" situated on Inco property that provided Gatchell with its most unique recreational area. In addition, two churches—Gospel "Brethren Chapel" (1935) and Gatchell United (1936) were built to serve local religious needs.

The arrival of the Tosato family in 1929 heralded the transition of Gatchell into an Italian neighbourhood. Mirroring "Little Italy" in Copper Cliff, vestiges of Italian ethnicity soon appeared. By 1938, Gatchell boasted 300 homes and a population of 1 200, the great majority Italian. Surprisingly, despite this Catholic population base, no Roman Catholic Church could be found in Gatchell. It remained until 1953–64 before St. Anthony's Parish was formed; it was 1964 before the church, hall and rectory were finally completed. Prior to this time, the closest Roman Catholic Church was Our Lady of Perpetual Help near the West End. Unlike the Donovan, practically every home was owner-occupied and self-constructed.[89] A smattering of small operations catering to Italians developed, including Arturo's Restaurant, Tosato's Store, Angelo's Shoe Repair, Dagostini's Store, Ideal Garage, Gatchell Beauty Parlor, Romano Cassio's Milk Bar, and the Corona Dance Hall. Later, the Eastern Canadian Greyhound Lines divisional headquarters and garage was established in the area. Cassio's was a great favourite, and its outdoor barbeque drew large crowds of late diners. While the majority of the dwellings were small, often located on lots only ten metres in width, the existence of backyards and lanes provided ample opportunity for gardens and neighbourhood connections. During the wine-making season, many homes would be replete with the remains of pressed grapes and grape boxes. Thus Gatchell had a geography that you could not only see but also smell and taste.[90] *Bocce* was often played on the streets. For younger people, the large field where Delki Dozzi Park is now located served as a major recreational venue.

By the Second World War, the area had grown to a community of some 500 homes and 3 300 persons. The war brought with it some trying times. Following Mussolini's alliance with Hitler, Italians became the target of RCMP raids, wartime suspicion, and hysteria.[91] When the last restrictions against enemy aliens were lifted by the Canadian government in 1952, a new wave of Italian immigration occurred, at which time the Gatchell area came to acquire a stronger ethnic character.

Donovan

Another growing neighbourhood was Donovan, situated north of the downtown between Little Britain in the west and the Flour Mill in the east. Part of this area was initially purchased from the Crown by Timothy Donovan as a farm on the north half of Lot 7, Concession 4 of McKim Township. His farmstead stood near the present-day junction of College and Kathleen Streets.[92] In 1908, the greater part of his farm was acquired by George and Montague Mahaffey and subdivided into lots 10 metres wide and 36.5 metres deep.

In a full-page advertisement placed in the *Sudbury Journal*, the Mahaffeys proclaimed that these lots were solid long-term investments because of Sudbury's status as the biggest railroad centre in the north surrounded by some of the greatest mines in the world.[93] The construction of Martin's Brick Yard in the same year near Montague Street spurred more development. Utilizing clay from the Little Britain area, the operation made bricks for many homes and businesses in Sudbury. By 1912, the plant employed some twenty-five men. The opening of Sudbury High School in 1908 and College Street Public School in 1916 attracted more families to College Street. Donovan's first general store also opened in 1916 on the north side of Kathleen Street.[94] Landsdowne Public School opened later in 1929.

Following the expansion program at Frood Mine by Inco in 1926 and the merger of the Inco and Mond Nickel Company operations in late 1928, the Donovan area experienced growing pains. As the demand for miners at the Frood Mine increased, hundreds of men flocked to the area, many with Finnish or central European backgrounds. The latter included Poles, Serbians, Croatians, and Ukrainians. Thus, the Donovan from the outset had a different character from either the Flour Mill or Gatchell districts. The polyglot setting encouraged the United Church in 1930 to set up an "All People's Mission" at the corner of Antwerp and Jean Streets, initially to provide a spiritual home for the Finns, and later embracing all language groups. Due to the formation of ethnic congregations in later years, it was only nominally successful. In 1937, the Holy Trinity Church was erected on Burton Avenue to serve the Roman Catholic population. First organized as a congregation in 1932, the Finns also acquired a small building on Tedman Avenue in 1938 for use by St. Matthew's Lutheran churchgoers. The congregation shifted to its present site on Mackenzie Street in 1948.

The newcomers constructed a variety of commercial establishments and ethnic halls, adding vitality to the area during the evenings and weekends. One well-known attraction was a watering hole known as the International Hotel, established in 1934 on Kathleen Street by the Serbian Borovich family. It remained a landmark until it closed in 2000.[95] The Frood Hotel, located nearby, served as another popular drinking establishment. Across the street from these two buildings could be found the Ukrainian Grocery Co-operative Buduchnist founded in 1938 (later renamed the Sudbury Ukrainian Co-operative in 1941).[96] Other prominent businesses included City Dairy, Tolmunen's North End Bakery, Penna's Steam Baths, Frank Kangas Men's Wear, and Kari's Pharmacy. Sampo Hall was erected in 1935, and its strategic hilltop location on Antwerp Street served as a centre of cultural activity for the area's 600 Finnish residents. Religious and conservative in nature, its members' views contrasted sharply with those who were members of the two Finnish-owned Liberty and Workers' Halls found on Spruce and Alder Streets, respectively.

In 1933, the Ukrainian National Federation built a hall on Frood Road that provided an alternative meeting place to the leftist-oriented Ukrainian Labour Temple Association's hall on Spruce Street constructed earlier in 1925. The Ukrainian presence was strengthened in 1940 by the building of a Greek Orthodox Church (St. Volodymr's) on Baker

Street. It was built in response to the arrival of many Greek Orthodox Ukrainians during the interwar period. Located adjacent to the Donovan and closer to the downtown, Ukrainian Catholics, who had previously established a congregation in 1923, erected St. Mary's Church on Beech Street in 1928. The three Catholic churches on Beech Street (St. Anne's, St. Joseph, and St. Mary's) served in the 1930s as an anti-Communist bastion, countering the progressive Finnish and Ukrainian halls found on Spruce Street.

The Yugoslavs added more ethnic diversity to the district. Consisting of three groups, they had been joined together in Europe under the umbrella of the Kingdom of Serbs, Croats, and Slovenes in 1918, later renamed Yugoslavia in 1929. In the 1930s, the majority of the Yugoslavs in Sudbury consisted of Croats, most of them residing in the Donovan. Around 1930, they started to meet in Perković Hall, which was owned by the family of the same name and later became the Frood Hotel. Desirous of their own facility, they built the Croatian National Home in 1934 on Kathleen Street. Thus it was that the National Home, not a parish, became the new temple of Croatian culture. It was not until 1981 that the Croats secured their own Catholic Church; they acquired St. John's Lutheran Church on the corner of Alder and Pine Streets. Following an agreement in 1944 to separate from the Croatians, the Serbs built their own hall on Bloor Street.[97] As occurred with the Croats, the creation of a Serbian church came later. A plot of land was eventually acquired by the Serbs on Antwerp Avenue in 1955, upon which the Free Serbian Orthodox Church of St. Peter and Paul and rectory were constructed in 1964 and 1969, respectively. While the Poles erected their own hall on Frood Road in 1936, the "Polish" character of the area was never as pronounced as the other ethnic groups.[98]

A number of commonalities existed among the ethnic groups in the area. The majority consisted of recent European immigrants who knew that they had little chance of returning to their homelands, so there was no alternative other than to work hard, save money, and make the best of the situation here. As one Ukrainian immigrant succinctly stated, "we must stop kidding ourselves about making a few hundred dollars and returning to the old country. Those times have passed."[99] Many came as single men, and often frequented the halls and churches in the hopes of finding suitable spouses. Others sought refuge in the hotels, where drinking was accepted as part of the rough and respectable working-class masculine identity. This surplus male population made it possible for some families to derive alternative incomes by setting up boarding houses or bootleg establishments. For many, these businesses were started not to make extra money, but simply to survive. These businesses were important, and served as a "home away from home" for their clients. The community also tended to support local merchants and businesses. This reliance was so pervasive that *The Sudbury Star* claimed the Donovan to be a "city-within-a-city."[100]

The ethnic groupings in the Donovan existed as small communities internally divided, with different geographic spaces separating them from one another. Lacking a shared group identity and split along ideological, religious, and political lines, the concept of a collective identity that was found among the Italians in Gatchell or the French Canadians in the Flour Mills was absent here.

West End

Tucked between Gatchell, the Donovan, and the downtown district was the West End. As was the case with Gatchell, its location relative to Copper Cliff and the streetcar system along Copper Cliff Road were influences that attracted settlement to neighbourhoods such as the Nickel Park subdivision between the early 1900s and the Second World War.[101] The construction of the Sudbury Brewing and Malting factory in 1907 on Copper Cliff road provided the basis for the first spurt of settlement. After the brewery was established, newly arrived Italians who worked at the plant settled in the vicinity.

Fred Eyre, who owned a large farm farther to the west, opened his Eyre subdivision to the newcomers.[102] Part of his farm had served as Sudbury's first cemetery, where a railway worker was interred; the burial had been arranged by Florence and Dr. William Howey in 1883, so Eyre deeded the gravesite to the city in 1934.[103] In the early 1880s, only one house could be found to the west along Copper Cliff Road and another lonely dwelling along nearby Regent Street. The opening of the Sudbury Construction and Machinery Company in 1909 encouraged more westward expansion. Regent Street evolved into a small business section, bolstered by the construction of Star Bottling Works in 1918, which remained there for another fifty years.

By the First World War, settlement had grown to such an extent that Elm Street Public School was erected on the corner of Elm and Regent Streets. Population growth continued unabated into the 1930s, much of it Catholic in nature.[104] This trend brought about the construction of St. Albert's Separate School in 1928 and St. Clement's Roman Catholic Church in 1937. A Finnish Pentecostal Church was erected in 1936 on the corner of Haig and Whittaker Streets where the Caruso Club now stands. Elsewhere, the Children's Aid Society built a "Home for Little Children" in 1929 on the corner of Pine and Alder Streets that became the site of St. Charles College in 1951. Many of these West End features were framed around the large gravel pit (now known as Queen's Athletic Field) that served as a major venue for baseball and high school football.

Kingsmount

Residential development also occurred in the Kingsmount area south of the CPR railway station. While parts of this area experienced growth around the First World War period, the real impetus for building came from the expanded work opportunities in Copper Cliff around the turn of the 1930s and Inco's earlier decision to curtail housing development within the town. Another factor was the lack of upper-class residential properties within Sudbury. Due to its closeness to the downtown and its attractive setting near Lake Ramsey, the Kingsmount area by the 1930s had, according to *The Sudbury Star*, evolved as a swank residential area consisting of "better class homes."[105] Evidence of this trend began to show in 1930 when Canadian Industries Limited of Copper Cliff decided that it would erect fourteen homes in the Kingsmount area, and four would be a "more expensive site" for its workers.[106] Other indications of the area's "Rosedale" status were residents' vigorous and successful opposition to the construction of apartments in the area in

1938, and their petition in 1933 to change Olga Avenue to Roxborough Drive and Mazie Avenue to Kingsmount Boulevard.[107] The name change demands were made using the argument that new names were more appropriate for a residential area. The development of the Kingsmount area was significant, as it heralded the beginning of "neighbourhood dualism" within the city (the trend of well-off residents acquiring more attractive sites than those found in the older parts of the settlement). This southern orientation of the elite continued after the Second World War, as evidenced by the appearance of higher-income neighbourhoods in the York Street area and along the fringes of Lakes Ramsey and Nepahwin.

Minnow Lake

The desire Sudburians had to build their homes on cheaper land was also manifest east of the downtown around Minnow Lake, situated in the Township of McKim. In 1883, the lake served as the site of a sawmill, one that continued to operate for another thirty years. Sawdust and slab residues were used to form a new peninsula into the lake, remnants of which still remain. While it was swimmable up to the 1960s, the residues, combined with effects of urban and industrial pollution, eventually rendered this activity unsafe.[108]

As was the case with the Gatchell and Donovan communities, it was not until the middle and late 1930s that settlement gained momentum. The first house on Howey Drive was built in 1939.[109] Three men were instrumental in fostering this eastern thrust.[110] The first was Count Romanet du Caillaud, who initially had purchased some 405 hectares of land from the Crown surrounding Minnow Lake. Following his death in 1919, his heirs liquidated these properties. It was around this time that William Barry acquired 81 hectares from the estate that surrounded Minnow Lake on which he established a farm. While there was some interest in residential construction on his property in 1928 and 1929, it was 1935 and 1938 before he and his wife Marguerite registered subdivisions. William and Marguerite opened the first post office, operated the largest general store in the area, and donated land for religious purposes. The third developer was Alex McLeod, an automobile dealer from Sudbury who acquired 332 hectares on the northeast side of Minnow Lake. Part of his holdings, the Lakeview subdivision, was developed by Fred Eyre. Other developers included C. Lovejoy and C. Christakos. Later, the Minnow Lake area grew to include the neighbourhoods of Brodie to the west and Adamsdale further to the east.

Lockerby

The last fringe area to evolve was Lockerby. As was the case with the Gatchell and Minnow Lake neighbourhoods, it too was situated in the Township of McKim. One of the earliest property purchases here was made by George Bouchard who, beginning in 1905, acquired 259 hectares of land near the juncture of present-day Regent Street and Lily Creek. He erected a dairy on the site, and land not used for this purpose became the first small subdivision in the Lockerby area. In 1922, he created Standard Dairy and was the first dairy farmer in the region to begin pasteurizing milk. The dairy was destroyed by fire in 1936, and he built a newer one on Victoria Street in Sudbury. While operating his dairy in Lockerby, Bouchard was responsible for the creation of Walford Road in order to service the growing Finnish population that had emerged along Paris Street and around Nepahwin Lake.[111] Since the 1920s, the latter area had attracted Finns seeking a more exurban lifestyle than the town of Sudbury could provide.[112] As figure 5.6 reveals, Finns were drawn to the area during the interwar period to such an extent that it could be considered an enclave similar to those that existed in the Beaver and Long Lake districts.

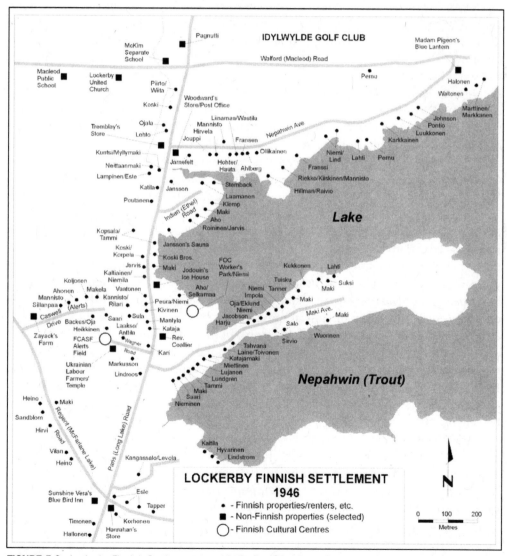

FIGURE 5.6   Lockerby Finnish Settlement (ca. 1946). From Saarinen, *Between a Rock and a Hard Place*, 67.

The tract of land south of the Idylwylde Golf and Country Club at the time consisted of virgin territory owned by William Turcotte (1904), the Leckie family (1910), and Florence McLeod (1923). In 1925, Sydney Pottle opened the first subdivision on the north shore of Nepahwin Lake. Most Finns obtained their properties from either Pottle or MacLeod. In the same year, the legendary Finnish wrestler and businessman Karl Lehto was the first Finn to register property in the Pottle subdivision now occupied by the Bel Lago Condominium building. He was followed by Dr. Heikki Koljonen, who purchased property on both sides of Paris Street.

Dr. Koljonen gave part of his property on the lake to a Finnish left-wing local of the Finnish Organization of Canada (FOC) for use as a Workers' Park (*Työn Puisto*). By 1937, the site featured a sauna, dance platform, restaurant, six-room main building, and a residential dwelling. A large diving tower and summer camp for children were added later. The site was so well known that it was visited by the famous Canadian doctor and political activist Dr. Norman Bethune. On the western side of Paris Street, Koljonen opened a subdivision and the FOC constructed a huge athletic field on behalf of Alerts (*Kisa*) Athletic Club, where thousands of Finns gathered for their annual cultural and sports festivals. Due to suburban expansion and rising taxes, the athletic field was sold in 1961 to David Caswell and Zulich Enterprises. The Workers' Park later suffered a similar fate, and it was sold to the City of Sudbury for use in perpetuity as a public lakeside park. During the Second World War, other subdivisions were developed by the MacLeod and Pernu families.

By the end of the 1930s, the texture of life had changed considerably for Sudbury and its residents. The relatively simple town of 1930 had matured into a crowded city with distinctive ethnic neighbourhoods. While economic recovery was well on its way, many indicators of Sudbury's social standards unfortunately lagged behind the rest of the province; the expanding population and the nature of the existing topography made the provision of services costly. The deep interest in sports shown by the citizenry and the existence of ethnic halls did, however, compensate for some of these deficiencies. Regionally, the sporadic outflow of population from Sudbury that had started to filter into McKim Township foreshadowed the much larger exodus that was to occur in the ensuing two decades. Inhibited by the lack of a direct highway access to Toronto, Sudbury's isolated position did not allow the community to diversify itself economically or to forge regional markets beyond the Nickel Belt.

The 1930s came to a fitting end with the visit of King George VI, Queen Elizabeth, and Prime Minister MacKenzie King on June 5, 1939. Called "one of the most glorious events in local history," Sudburians gave testimony to the ongoing endurance of their city's British heritage by allowing the royal couple a rare visit into Frood Mine, and giving Queen's Athletic Field its name in honour of the visit.[113]

CHAPTER 6

Copper Cliff (1886–1939)

From its humble beginnings in 1886, Copper Cliff started its journey as a mining camp, turning into a sulphur-barren village and then a formal town with a bustling commercial sector and distinctive settlement pattern featuring a high degree of ethnic segregation. It was, above all, an industrial centre and de facto "company town." Copper Cliff's position within the region remained dominant until 1905, at which time its population level fell below that of Sudbury. By the First World War, Copper Cliff was a compact settlement of some 3 000 residents. During the 1920s, Inco decided that no further housing would be permitted within the town's boundaries; this corporate fiat ensured the continuity of Copper Cliff's size and character for the rest of the twentieth century.

Mining Camp

After the discovery of ore in Snider Township by Thomas Frood in 1885, and its acquisition by the Canadian Copper Company (CCC), the company constructed a log boarding house. This building no longer stands: the only evidence of its location is now commemorated by a stone monument.[1] The mine was originally known as the Buttes; however, because the property featured a gossan-stained surface suggesting copper, its name was changed to Copper Cliff. The same name was later given to one of the early settlements which developed near the mine. After acquiring a number of other copper mining claims in McKim and Snider Townships, CCC's President Samuel J. Ritchie established the company's business headquarters in Sudbury and the industrial centre in Copper Cliff. His influence at the time was such that some citizens suggested Sudbury's name should be changed to Ritchie, an offer that he politely declined.[2] In the 1880s, the only connection between Sudbury and Copper Cliff was the little-used Sault Branch of the CPR.

In 1886, Copper Cliff began as a mining camp. The same year, work was started in three mines: Evans (previously Eyre), Lady Macdonald (previously McAllister or No. 5), and Copper Cliff. The Copper Cliff mine became the main focus of the CCC and the development of the mining camp. In May 1886, the company had a work force of approximately twenty-five, and by the end of the year the figure had almost tripled to sixty-five. Following the construction of a CPR spur line from the Copper Cliff mine to the Sault Branch, open-pit mining operations began. The "green" ore from the mine was roasted and further reduced at local blast furnaces before being sent as blast furnace matte to smelting operations in the United States and Europe.[3] In 1888, a rough road was opened that permitted personal transportation between Copper Cliff and Sudbury by stagecoach, on wheels in the summer and on sled runners in the winter.

The formal beginning of Copper Cliff as a company town can be dated from a survey prepared by John D. Evans, a company engineer, in August of 1886. The area covered by the planned survey became the neighbourhood of Anglo-Saxon employees and company officials ("the better class of … men") and the business district. It became known as Copper Cliff's English Town. An unplanned settlement called Shantytown also emerged, housing the immigrant labourers, many of them Finns, in rough, tarpaper shelters. Thus the tradition of ethnic segregation was firmly established at the outset. A third settlement was established in 1886 at the Evans mine, situated two kilometres southwest of Copper Cliff.

Since it was known that smelting could reduce the ore to matte by a 6:1 ratio, the decision was made to erect a smelter in the mining camp. Known as the Old or East Smelter, it was built in 1888 east of Copper Cliff, where the slag dumps are now found. An open-air roast yard used to lessen the ore's sulphur content was established in the same year between the settlement and the smelter. The smelting operation provided the focus for a fourth settlement known variously as Smelter, Smelter City, Old, or Old Smelter. Copper Cliff was the first mining operation where company officials were encouraged to bring their families to the area. To house them, twenty to thirty prestigious houses were built near the East Smelter site.[4] This area's lifespan, however, was short, and it disappeared after the plant was closed in 1902.

The planning and construction of the company town at Copper Cliff loomed large not only for its own future but for the entire Sudbury area. While the introduction of the roast yard heralded the beginning of Sudbury's reputation as a black zone of destruction, the operation of the smelters marked the transition of Copper Cliff from a mining camp to what Goltz has referred to as an "industrial village."[5]

Industrial Village

Copper Cliff's development as an industrial village was a transitory phase in its move toward becoming a full-fledged company town. Evans had a new "officers' boarding house" erected in English Town, to which he moved in 1889. The house was called the "Yellow Club" at that time. This facility, which served as a hotel for visiting dignitaries and as a dwelling for CCC's Anglo-Saxon employees, not only represented the dominance of the company within the community but also was a tangible expression of the segregation that existed within the social and cultural setting. Other "non-working class" facilities came into being as well, such as the Matte, Red, and Central Clubs.[6] The addition of a second roast yard and the expansion of the East/Old Smelter in 1889 required a larger labour force, and forced the CCC to erect additional dwellings for its own purposes, more boarding houses, and four stores. Outside of the planned area, the boarding houses had private owners, and were not run by the CCC.

In 1890, the company's main office was moved from Sudbury to Copper Cliff. While the company owned all of the land in Copper Cliff and acted both as landlord and patron, it never did establish a company-owned store. Some land was leased to individuals for housing and approved businesses. When an individual built a structure on company land and wanted to sell it, he or she first had to offer it to the CCC, which could dictate the seller's price. Thus by virtue of land ownership, the company was able to exert considerable control over the lives of all the village inhabitants. The Copper Cliff model had significant implications for the rest of Northern Ontario, as mining and lumbering company towns elsewhere adopted the same pattern.

While Copper Cliff's main business district developed on Serpentine Street, a scattering of general and grocery stores emerged in Shantytown and East Smelter. Beginning in 1890, part-time medical care for the residents was provided by a doctor from Sudbury; this service became available full time when the company hired a doctor in 1901. To avoid the need to support a Roman Catholic separate school, a "private school, on the public school principle" opened in 1890, and a Public School Section formed later in the year. The school building, built at the intersection of Balsam Street and Evans Road, was used as a school, church (by Anglicans, Methodists, and Presbyterians), and community hall. In 1893, students whose parents worked at the Evans Mine in Snider Township were permitted to attend the school, necessitating an expansion to the building. In 1892, the CCC reluctantly gave the Roman Catholic Mission in Copper Cliff permission to use a remodelled house as an interim chapel on the edge of the East Smelter site, which disappeared when that smelter was dismantled. To satisfy the growing religious needs of its residents, the CCC allowed four new church buildings to be erected in Copper Cliff village and Shantytown between 1898 and 1901 by Roman Catholic, Presbyterian, Methodist, and Anglican congregations. While a Lutheran congregation had formed in 1897, it did not have a church building.

Out of deference to the Polish community and the Jesuits who had served the Copper Cliff mission since 1886, the company granted permission to the area Catholics to build the St. Stanislaus Catholic Church in 1898. Following its designation as a parish in 1900, the church's administrators faced the daunting challenge of providing religious services for three identifiable groups: French Canadians, Poles, and those of Irish descent. Despite the difficulties of this situation, church officials managed to meet their parish obligations for the diverse community. The Honourable Michael Starr, who served as Minister of Labour in the Diefenbaker Cabinet, was born at Copper Cliff in 1904, and his baptism is recorded in St. Stanislaus' church records.[7] St. Stanislaus was also responsible for developing a mission in nearby Creighton, which had a rapidly growing Catholic community. Known as the St. Michael the Archangel mission, it succeeded in erecting its own St. Michael's Church in 1917. In 1935, this mission came to an end when St. Michael's was raised to parish status. In the meantime, St. Stanislaus established a parish cemetery in Waters Township, where the Iron Ore plant now stands.

By the turn of the century, the four Copper Cliff settlements had come to reflect a varied cosmopolitan character. Finns, Poles, and Ukrainians had a substantial presence, although their numbers lagged behind those of the Anglo-Saxons and French Canadians. Whereas the Finns, Poles, and some Ukrainians resided mainly in Shantytown, the majority of the French Canadians and another group of Ukrainians lived at the East Smelter. In addition, an Italian cluster known as the "Hill" and later called the "Crow's Nest" by 1899 had started to evolve northeast of the village. This was the fifth such ethnic enclave in the Copper Cliff area, and virtually every Italian lived there.[8] Despite the admonition of the federal Minister of the Interior Clifford Sifton, that no steps should be taken to assist or encourage Italian immigration to Canada, more Italians continued to arrive.[9] While the first Italians were from a part of southern Italy known as Calabria or The Boot, immigrants from central and northern Italy followed later. By 1906, the majority of the residents in the Crow's Nest were Italians, and by the First World War the Italian presence was so prevalent that the area was simply known as Little Italy. Since Italy had been reunited only in 1870, many of these Italians had to learn to get along with one another as well as with their new countrymen. To this end, the adaptation process was facilitated by the community's self-sufficiency, its *borgo* (village) setting and the presence of local *padronas* (hostesses) who served many valuable functions.[10] Situated "across the tracks and under the stacks," the area's character was easily visible: it was compact, with small gardens and streets that bore names such as Venice, Genoa, Milan, and Florence.[11] Numerous businesses flourished, such as A.J. "Red" Pianosi's grocery store that operated from 1915 to 1966. For his generous spirit, Pianosi received the Order of Canada. Eddie Santi ran the same store later on, until 1996.[12] The subsection known as the Crow's Nest, however, gradually began to disappear after 1912, due to Inco's need for more industrial land.

While Copper Cliff's ethnic groups formed their own religious and fraternal organizations, they often had links to organizations based in Sudbury. The symbiotic relationship between Sudbury and the Copper Cliff settlements was facilitated by the road constructed in 1888. A formal organization that differed considerably from the others was the Golden Rule Temperance Society, Branch No. 80 of the Finnish National Temperance Brotherhood of America. Erected in 1895, the society built the first Finnish hall in the Sudbury area, one that was later shared with the Finnish Lutheran congregation. This society and its building subsequently became the focal point for many disputes within the Finnish community and with company officials. Policing for Copper Cliff was provided by a Sudbury constable, and in 1900, a Justice of the Peace and Notary Public took their places in the local judicial system. As the community grew, these measures provided the community with a growing sense of law and order.

Company/Corporate Town

Starting in 1899, the population and economic setting changed substantially. The Evans mine closed, and the residents who lived in the surrounding area relocated elsewhere. CCC built a new West Smelter and roast yard. In 1900, the Orford Copper Company built the Ontario Smelting Works, which became the focal point for another small settlement known as Orford Village. This same year, a second public school was erected at Union and Rink Streets, and a new clubhouse known as the Gorringe (later Ontario) Recreational Club came into being. It was a male Anglo-Saxon preserve, replete with tennis courts and a croquet lawn. The economic status of the community was enhanced in 1901, when the Bank of Toronto opened its first North Ontario branch in the downtown area. Another important building in the town was the hall built by the Catholic Order of Foresters on Collins Drive in 1903. The Foresters' Hall stood until 1947, serving not only St. Stanislaus Parish but also the wider Copper Cliff community as a socio-cultural centre.

By the turn of the nineteenth century, the Township of McKim, now encompassing some 1 600 residents in Copper Cliff alone, was about the same size as the Town of Sudbury. The position of the community was such that the CCC viewed Sudbury as simply an extension of Copper Cliff.[13] The CCC's domination of township politics since 1885 notwithstanding, there was a growing sentiment among company officials to become separated politically as a town. While this opinion was partly fuelled by the desire for greater independence from Sudbury, other reasons were important. Company officials felt that incorporation would allow the new municipality to restrict its tax base to the local community rather than the township at large, thereby reducing the taxes CCC had to pay. Another advantage they envisioned was the ability to shift certain expenses then assumed by the CCC to the town council. Since existing legislation meant that an incorporated town would get more grants from the province, CCC management reasoned that monies freed up by this change in political status could be used to improve the town's fire, health, and police protection, which would especially address poor sanitary conditions and the proliferation of local bootleggers. Incorporation could also be used to enhance CCC's power, enabling the company to have more control over the village in every way. Finally, such independence was seen by the company as a way controlling the decisions of the courts, especially where coroners' inquests were involved. Sudbury juries after 1895 were beginning to return unfavourable verdicts to the CCC in wrongful death lawsuits, and officials thought that the trend could be reversed if more "dependable" jurors could be selected by police officers who were employed by the company.[14]

Following a petition to the provincial government, the legislature passed An Act to Incorporate the Town of Copper Cliff on April 15, 1901.[15] Created from both McKim and Snider Townships, the new town came into existence on January 1, 1902, with a population in excess of 2 000 distributed among five clusters: Shantytown, English Town, Crow's Nest/Hill/Little Italy, East/Old Smelter, and Evans Mine (fig. 6.1). The new town had a mayor and six councillors, and the new officials began to initiate police and fire protection and appointed a local officer of health; the provision of telephone, electricity, street lighting, water, sewer, cemetery and medical services, however, still remained with the company.

This incorporation simply reinforced the community's company town status, since the majority of the town council consisted of company officials. In fact, from 1906 until 1970, every Copper Cliff mayor was a senior member of Inco's management.[16] While control of housing and land leases had already given the CCC considerable power over its employees, town status provided virtually total domination and furnished the company with the legal means to control the police force, arbitration and compensation for accidents, coroners' inquests, and juries. Copper Cliff's town status also gave the CCC access to the Canadian military without requiring provincial or federal authorization, which they could use as a potential weapon against the threat of labour radicalism. Any effort on the part of merchants or others to oppose company candidates could also be quickly squashed through the elimination of their leases.[17]

The town of Copper Cliff at the time of incorporation was stark and ugly. A Toronto-based reporter in 1902 described the settlement as being without grass, trees, or flowers, with backyards "as bare and forbidding as a billiard table before the green baize is glued on."[18] Following the formation of the International Nickel Company in the same year, the visual appearance of the community underwent major changes as it became the "showcase" of the new corporation in Canada. Unfortunately for McKim Township, the separation of Copper Cliff made it a casualty of resource development, and its population remained small until the middle of the 1930s.

The new status of the town encouraged the growth of the downtown district and branch stores from Sudbury appeared. Copper Cliff thus began to manifest as an autonomous community with little need for its residents to travel to Sudbury other than to go to hotels, which were not allowed locally. The addition of a weekly newspaper in 1902, the *Copper Cliff Courier*, and the implementation of a major construction program between 1912 and 1914 to build 100 homes, along with boarding houses, residential clubhouses, a new hospital, the Rex Theatre, additional stores, a business block, and post office added to this sense of self-reliance. No longer afraid of competition from Sudbury, company officials and local merchants agreed to open a street railway system in 1915 between the two communities. The connection for some families was useful, as it allowed Copper Cliff students graduating from the public school built in 1914 on School Street to now attend Sudbury High School. The shift of the roast yards to O'Donnell in 1916 and the reopening of the public library in 1918 were other improvements for local residents.

The removal of the roast yards set the stage for land reclamation and the development of picturesque Nickel Park in the heart of the town.[19] Of great import to the town was the opening of the Copper Cliff Club located on Creighton Road in 1916. Equipped with billiard tables, a bowling alley, swimming pool, kitchen, library, maids' quarters, ladies' rest rooms, and three large multipurpose rooms, the building was an attractive addition to the town's architecture and social setting.[20] Unfortunately for the non-Anglo-Saxon population, this superb facility was declared to be off limits. This ongoing rejection

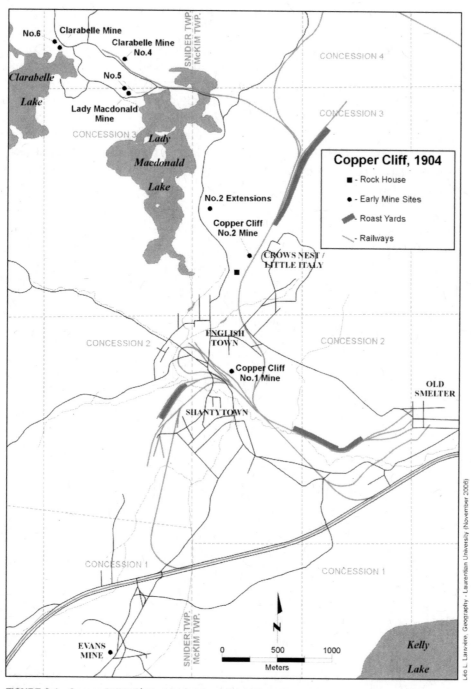

FIGURE 6.1 Copper Cliff (1904). Adapted from Goltz, *Genesis and Growth of a Company Town*, 95. Used with permission.

of the French, Finns, and particularly the Italians was yet another manifestation of the ethnic segregation and stratification that marked much of Copper Cliff's history.

Some ethnic groups, notably the Finns and Italians, adapted to this situation by developing their own social, cultural, and recreational features. One notable response came in the form of the prestigious Finland Hall, described as being "one of the finest opera halls in New Ontario."[21] Anglo-Saxon discrimination likewise inspired the Italians to build their own Italian Club. Some indication of the extent of spatial segregation that emerged out of these developments is shown in figures 6.2 and 6.3. As was the case for the Crow's Nest settlement, the ethnically diverse East Smelter fringe neighbourhood declined after the cessation of its smelting activities; after the First World War, the site was abandoned and used as a slag dump.

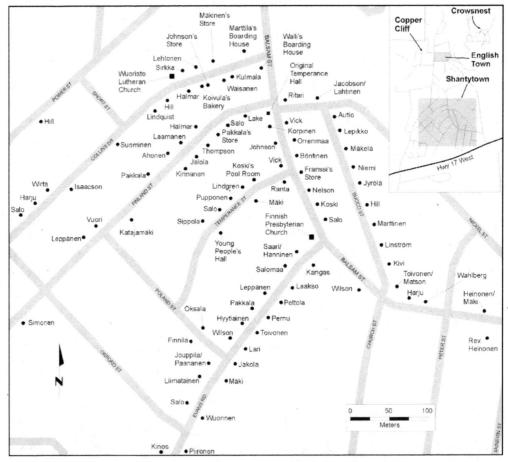

FIGURE 6.2 Finnish Settlement in Copper Cliff's Shantytown (1915). From Saarinen, *Between a Rock and a Hard Place*, 36.

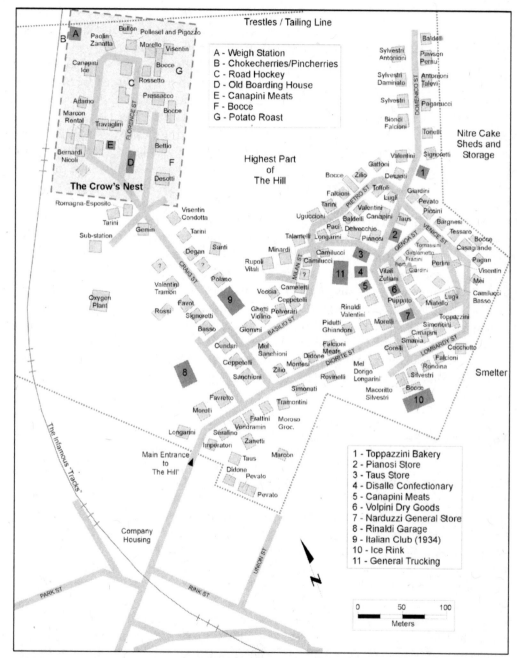

FIGURE 6.3 Little Italy and the "Crow's Nest" in Copper Cliff (ca. 1930–40s). Adapted from Copper Cliff Italian Heritage Group, *Up the Hill: The Italians of Copper Cliff*, 4–5 and 26–27, and The International Nickel Co. of Canada Limited, *Map of Copper Cliff* (Copper Cliff, ON: International Nickel Company of Canada Limited, August 1932).

During the early years of incorporation, five religious congregations existed in the town, all endorsed by the company. In 1908, the Finns built the first Lutheran Church in the Sudbury area on Collins Drive. The growth of this congregation resulted from the influx of workers following the opening of the smelter in 1904, the rebuilding of the Ontario Smelting Works in 1905, and increased activity in the roast yards. Ruthenian Ukrainians who followed the Eastern Catholic tradition erected St. Nicholas Greek Catholic Church in 1909, the first of its kind in Ontario, on the same street as the Finns' Lutheran Church. This church had a short existence and was closed in 1923, as the majority of the Ukrainians by this time had left Copper Cliff for Sudbury following the postwar economic Depression. Anglicans replaced their existing church in 1909 on Balsam Street with a more imposing structure. The growing Italian labour force was sufficient in 1914 to encourage the establishment of the St. Elizabeth Mission on Craig Street in the midst of Little Italy, which now supported a population of some 600 people. The mission, however, lasted only until 1921. Italians allied with the mission again shifted their religious allegiance to the St. Stanislaus parish.[22] In 1934, the site was transformed into the Italian Club.

From time to time, churches and missions were used by the company as agents of assimilation among the various Copper Cliff immigrant groups. The first of these was the Finnish Presbyterian Mission begun in 1913. Its church on Balsam Street served a variety of religious, social, and educational purposes. One objective of its managers was to blunt the impact of the radical Finns who were posing problems for International Nickel in the workplace. Another mission supported by the company among the Italians in 1914 was the Methodist Mission Board. While cordially received by the Italians, it did little to change religious preferences.

The First and Second World Wars

While sporting and leisure activities were integral parts of Copper Cliff's development from the outset, they became more firmly entrenched in the interwar period. These pursuits and their related facilities were strongly supported by the company not only for their public relations impact but also for the way they minimized the negative effects of the workplace and company town environment. These initiatives can be traced back to 1888, when the CCC built a tennis court and sports field. By the turn of the 1900s, a lacrosse team was organized on the basis of an "import system" that drew athletes to Copper Cliff from other parts of the community with the offer of a company job.[23] Company officials later extended this practice to all of International Nickel's other company settlements, and as time passed, drew athletes from elsewhere in Canada. After the Monell Cup was established in 1914 by the Nickel Belt Baseball Association, the Copper Cliff Redmen emerged as a powerhouse in the league. Large crowds supported the team, especially in 1925 when it captured the Ontario Senior Championship.

Copper Cliff also started its own hockey league in 1914, and teams were established for various company divisions. The opening of Stanley Stadium in 1935, with seating capacity for 1 100, ushered in a new era for ice hockey and competitive skating, serving as the nucleus of the town's athletic endeavours. In 1940, Bert McClelland appeared on the coaching scene, and in subsequent years he established one of the most enviable high school records in the country.[24] Curling was popular as well; the curling club was housed in a wooden frame building in 1915, and this structure was replaced in 1950 by a new seven-sheet curling rink beside Stanley Stadium.[25] Another amenity popular among citizens was the Copper Cliff Memorial Community Hall built in 1936, boasting a lounge, auditorium, and gymnasium. It was home to the "skirl of Scottish pipe music" by the Copper Cliff Highlanders Cadet Corps formed in 1917. Through the years, this group has been the recipient of numerous awards. The opening of Copper Cliff's own high school in 1937 added another component to both the educational and sporting milieu.

Aside from a few additions to the town's landscape, such as a new company research building in 1937, the opening of Copper Cliff dairy, and the rebuilding of the McIntosh Block shortly thereafter, the infrastructure described in this chapter remained much the same for decades to come. While the townsite was surrounded by barren land, the slag dumps, and the tailings area (aerial photograph 6.1), the town itself became more attractive with the ongoing beautification of Nickel Park and the implementation of a tree-planting program. The latter was supported by International Nickel's own greenhouse, where more than 40 000 plants were started each year. In the meantime, the relationship between Copper Cliff and Sudbury remained the same. While close in terms of geographical distance, the compactness of the company town and its strong sense of municipal and psychological independence kept the two communities far apart until the formation of regional government in 1972.

AERIAL PHOTOGRAPH 6.1   Copper Cliff (1946). *Source*: Ministry of Natural Resources Records, Archives of Ontario (RG 1-429). Photo taken for Ontario Department of Lands & Forests (1946).

CHAPTER 7

From Local to Global Monopoly

The Merging of Inco and Mond (1902–1928)

The urban setting associated with Sudbury and Copper Cliff prior to the Great Depression was intimately associated with the growth of the mining sector. The formation of the International Nickel Company in 1902 marked the beginning of a significant phase in its march toward global dominance. By 1905, Sudbury had become the chief supplier of nickel to the world, replacing the island of New Caledonia located in the South Pacific Ocean east of Australia. New Caledonia, however, remained important for International Nickel as a political foil. Whenever the Ontario government threatened its position, the company always hinted at the possibility of shifting operations to New Caledonia.[1]

Following reorganization in 1912, the company's name was changed to *The International Nickel Company*. In keeping with its intention to build a nickel refinery in Canada, the name was changed in 1916 to The International Nickel Company *of Canada*. While local competitors such as Mond Nickel and British America Nickel appeared along the way, they were eventually swallowed up by this mining colossus. British America went into liquidation in 1924, and its assets were acquired by Inco (the registered trademark of the company since 1919) in the following year. As a result of a quirk in the location of the Frood deposit, Mond came under Inco's control in 1929, thereby ensuring the company's global monopoly. While Falconbridge Nickel Mines was incorporated earlier, in 1928, it remained until the 1950s before it attained any real significance in the nickel industry.

Mond Nickel Company

One of the few companies to share Sudbury's mineral wealth with International Nickel was Mond Nickel, a British company under the control of the Mond family. Ludwig Mond, the founder, had acquired some properties in Sudbury in 1899; he founded Mond Nickel the following year. The ability of Mond to penetrate the industry was due to its friendly relationship with International Nickel. In part, this relationship was necessary to International Nickel because of Mond's wealth, its established business networks, and most importantly, its superior refining process. Since Mond showed interest mainly in selling to the British government rather than joining the North American market, and declared no intention of erecting a refinery in Canada, it was able to negotiate a cartel form of market share and price-fixing arrangement with its competitor, one that lasted up to the outbreak of the First World War. In short, Mond served as the perfect ploy for International Nickel to refute accusations that it was a foreign monopolist in Canada.[2] From the beginning, Mond was thus not a competitor, but rather a useful ally of International Nickel.[3]

International Nickel did not have this cozy relationship with other companies that attempted to penetrate the American market. In 1904, the Consolidated Lake Superior Corporation run by the energetic F.H. Clergue was forced into bankruptcy because of the strong financial control which J.P. Morgan and Company held over corporate undertakings related to its steel and nickel interests in the United States. Morgan likewise used his financial clout to bring about the demise of the Nickel-Copper Company in 1907, and the Canadian-based Dominion Nickel-Copper in 1912. The latter company, however, succeeded in selling its properties, which in 1913 were brought under the umbrella of the British America Nickel Corporation.

Following its acquisition of the Victoria Mine (formerly McConnell) in 1899, Mond proceeded in 1901 to construct a roast yard to reduce the sulphur content found in the raw ores, and a smelter to remove any of the remaining metals for further processing. This also necessitated the construction of a townsite at the Victoria Mine station on the CPR railway, and the completion of a 3.4-kilometre-long aerial tramway (a raised line of moving buckets containing raw ore) connecting the mine to the roast yard and smelter site. Remnants of this tramway can still be seen there today. The company also purchased eighteen mining properties, including Garson Mine in 1907. Ores from Garson were sent to the Victoria smelter. In line with its understanding with International Nickel, Mond started a new refinery at Clydach, Wales. The power required for Mond's operations came from the Wabageshik Dam situated on the Vermilion River built in 1909.[4]

In 1910, Mond bought the Frood extension. Aware that its existing smelter at Victoria Mines was poorly sited with respect to its Frood and Garson ore supplies, a new smelter was erected at Coniston, the junction of the CPR and the CNoR (Canadian Northern Railway). When the Coniston smelter was completed in 1913, the existing plant at Victoria Mines was abandoned. As well, Mond acquired the Levack mine and developed the Worthington mine. Mond completed its second power plant at Nairn Falls in 1916 on the Spanish River. Disaster struck when the Worthington mine collapsed in 1927; fortunately, there was no loss of life.

The Mond Company made the decision in 1920 to compete head to head with Inco in both Canada and the United States, thereby reversing its twenty-year-old policy. To achieve this end, Mond acquired control of the Nickel Alloys Company of Clearfield, Pennsylvania, and converted it into the America Nickel Corporation. Here it started to manufacture malleable nickel in its various forms including Mond 70, Mond's answer to International Nickel's Monel Metal. These new nickel alloys were prized for their resistance to highly corrosive conditions, especially for equipment used at sea. In 1923, Mond and Inco reached an arrangement acknowledging Mond's right of access to the North American market. But Mond's objective of overtaking Inco during the 1920s then foundered as the family-owned and managed firm proved unable to replicate Inco's carefully integrated planning strategies and extensive use of impartial, independent consultants. The bottom line was that Mond had failed to increase significantly its share of the American market as the decade progressed.[5]

Up to this time, both companies had been working on their respective portions of the Frood ore body. Mond owned the centre of the deposit and Inco owned both ends. When it was discovered in 1925 that the lower levels of the Frood mine contained high levels of copper and precious metals, Inco and Mond made the historic decision in 1928 to merge their operations. The official rationale for the merger was that it would permit the orderly and economic development of the Frood ore body as a single mine and offered other advantages, such as the integration of sales organizations in America and in Europe. There was some truth in this assertion. By 1928, Inco and Mond had both invested heavily in developing the Frood ore bodies. As well, the Ontario Mining Act then in force required that in such situations, the two companies could either develop the property in tandem or leave a column at least thirty feet wide along their common boundary, permanently immobilizing tons of extremely valuable ore. Bray and Gilbert, however, suggest that the above reasons were not the main causes for the fusion of the two companies. In fact, Mond still had considerable potential to remain competitive with Inco. This potential, though, required a drastic reorganization of the company both in England and in North America. This was something that Lord Melchett, formerly Sir Alfred Mond and the major shareholder in the Mond Company, was unwilling to do. Lord Melchett himself spelled out the reasons for his decision:

> Our company can only continue by itself after drastic reorganization both in London and out here [Canada]. To carry this out and to make our organization work successfully means that I personally shall have to devote a great deal of time and effort involving frequent journeys to this side for some years. I have come to the conclusion after this visit that I cannot do this without imposing a strain on my health, which at my age I do not propose to do. Nor is there anyone in our organization at present who really could undertake this task.[6]

It can thus be argued that the merger was not really based on any rational strategy, but rather reflected the managerial failure of a family-based English enterprise to compete effectively with the entrepreneurial approach adopted by its more aggressive American rival. With this consolidation Inco became truly international in scope, its operations located in Canada with subsidiaries in the United Kingdom and the United States.[7] The move was politically strategic as well, as it enhanced the company's protection from anti-trust action in the United States and from monopoly charges elsewhere.

British America Nickel Corporation

It is worthwhile recounting the brief history of the British America Nickel Corporation, as it serves as an excellent example of the aggressive means used by International Nickel to acquire its monopoly position in the world. When British America was formed in 1913, it not only acquired the assets of the Dominion Nickel-Copper Company but also the North American rights for the Hybinette refining process, which had been proven successful at the Norwegian Nikkelwerk refinery located in Kristiansands. While this gave the company a more efficient refining process, it was at a distinct disadvantage because of the lower quality of its ore grades when compared to International Nickel or Mond. Lacking markets, its only hope was to secure government contracts for nickel on the basis of its Canadian origin. Unable to secure financing in Canada, the company tried to get backing in the United States and England. J.P. Morgan & Company, however, managed to prevent the entry of British America into the United States.

With the outbreak of the First World War, the allied governments became the chief consumers of nickel. There was therefore encouragement for opening up new sources of supply and a willingness on behalf of governments to finance nickel companies; British America's continuance as a European competitor was facilitated by wartime events. As part of a 1916 British arrangement with the Nikkelwerk refinery to prevent supplies of nickel reaching Germany from Norway, the British government agreed to provide $3 million to British America Nickel, in which the group held a substantial interest. Part of the agreement included a trading contract with British America for 6 000 tons of nickel a year for ten years.

Assured of a market and financing, British America began to develop its operations in the Sudbury area. While construction began in 1917, labour difficulties and material shortages delayed completion of its new smelter at Nickelton, near the Murray mine until 1920, long after the war had ended. In the same year, the British government cancelled its contract with British America. Another blow came as a result of the Washington Arms Conference of 1921–22, which seriously curtailed military markets for nickel. Left with no markets and high stocks of metal, British America had no choice but to close down its operations in Sudbury in 1921. It reopened in the following year under Norwegian and Canadian control and immediately attempted to grab a share of the American market. The company's aggressive action of attempting to secure US contracts in 1923 startled Inco into action. The company chose to fight. The resulting price war of 1923–24 demonstrated the ability of Inco to force out an unwanted competitor. Even so, the ultimate collapse of British America was not only a result of the price war but also its shaky financial position and the lack of effective leadership.[8]

Unable to pay its debts, British America was forced into liquidation in 1924, and its plants were closed down. Most of the company's assets were acquired by Inco through its subsidiary, the Anglo-Canadian Mining and Refining Company, and sold off. The company wisely retained the rights to the Hybinette refining process, since it was more economical than the one originally devised by Orford Copper. For decades thereafter, the concrete ruins of this failed enterprise situated west of the historical Murray Mine plaque on the Sudbury-Azilda highway were a visual reminder for local residents of this nickel battle gone by.[9] The structural outlines resemble an ancient Roman ruin, aptly depicting the ruthless power of the world's greatest mining enterprise in Canada. Some remnants still remain.

From International Nickel Company to Inco, and Merger with Mond (1902–1928)

Between 1902 and 1928, International Nickel prospered from the pre-war European demands for nickel in armour plate, the military needs of the First World War, increased peacetime uses for nickel in the United States, and the impact of the roaring twenties. By 1903, nickel production from Sudbury exceeded that of its main rival, New Caledonia. This dominance became continuous after 1905. The control of Sudbury's wealth was paralleled by the dominance of International Nickel within the nickel industry. Through the use of long-term contracts with its consumers, the company was able to thwart competitors from entering the market, especially in the United States. Its ability to meet the growing global demand for nickel was facilitated by the opening of Creighton mine in 1901 and the growth of this operation by the First World War into the world's largest operating mine.[10] Its output far surpassed that of the company's other major source, Crean Hill. Also significant was the opening of a new smelter by the CCC in Copper Cliff in 1904 which heralded the appearance of the first of three great smokestacks which dominated the Sudbury skyline for years to come. These smokestacks served to disperse the sulphur fumes released during the smelting process into the atmosphere. Around the same time, the former East and West smelters were taken out of production. Officials reorganized to increase the capital structure of the company in 1912, at which time its name was formally changed to *The* International Nickel Company. The opening of the Frood ore body in 1913 further solidified the company's economic position.

International Nickel's growth into a powerful monopoly soon gave rise to serious concerns in towns near their operations in Canada and the United States. In Canada, the issues were the absence of export duties on nickel and matte, and the company's lack of a refinery in the country. Politically, this became known as the "Nickel Question." Although local politicians made several appeals to the Ontario and Dominion governments, their concerns fell on deaf ears. The Ontario Metal Bounty Act (1907–17) was a timid response to the calls for Canadian refineries that had no effect whatsoever on securing nickel refining in Canada. As well, the federal investigation of mining practices undertaken by the

Committee on Mines and Minerals in 1910 led to naught. While these feeble political attempts affirmed the ability of International Nickel to manipulate the two higher levels of government, they nonetheless laid the foundation for the renewal of attacks against the company in 1914. Additional criticism resulted when International Nickel opened a new refinery at Bayonne, New Jersey in 1913. In the United States, the issue was related to the threat of antitrust action.

Following the outbreak of the First World War, International Nickel closed down its mines and smelters, whereas Mond continued its operations after a brief shutdown. This led to a public outcry that International Nickel was more interested in serving Germany, the largest importer by far of nickel products from the United States, than meeting the needs of the Allied governments.[11] Reassurances that this was not the case provided by the Dominion government and the company late in 1914 did little to satisfy the press or political opposition. The call for a refinery in Canada continued, and even the *Canadian Mining Journal*, normally a staunch apologist on behalf of International Nickel, was in favour of Canadian refining. The ensuing political uproar was such that the Ontario government announced the formation of an official commission in 1915 to deal with the matter. In the same year, the Dominion government took action by establishing two Munitions Resource Commissions to study the question of base metal refining in Canada. When the second commission came out firmly in favour of government aid to British America Nickel to assist the company in erecting a refinery in Canada, International Nickel saw for the first time the potential for genuine competition to replace its token and largely prearranged rivalry with Mond.[12]

The criticism that International Nickel was a "foreign monopoly under the control of German interests" came to a head when it was reported that the German submarine *Deutschland* had made two trips to the United States in 1916 and taken tons of nickel, presumably of Canadian origin, back to Germany. When knowledge of these sensational events became public, International Nickel knew that the die was cast, and the company immediately decided to build a new refinery at Port Colborne in Ontario. It likewise announced the formation of The International Nickel Company *of Canada* as a subsidiary of International Nickel to carry out this task. Thus, when Ontario's voluminous Royal Nickel Commission report came out later in 1917, the commission saw no need to make any recommendation regarding the need for the company to establish a Canadian refinery. Significantly, the report condemned the idea of nationalizing International Nickel or giving government aid to British America Nickel. The conservative position taken by the commission destroyed any prospect of International Nickel facing competition from either a private or public corporation. Indeed, it only took another seven years before its only real competitor, British America Nickel, disappeared from the scene.

The refinery at Port Colborne was completed in 1918, at which time all of International Nickel's assets in Canada were transferred to The International Nickel Company of Canada. As a result of this reorganization, the CCC lost its corporate identity, thereby ending its thirty-two year association with the Sudbury area that extended back to 1886. A shift in the company's new public relations strategy then became evident in 1919, when its first advertising campaign appeared in the *Saturday Evening Post*; another landmark that year was the registration of its trademark name, Inco.[13]

After the First World War, the demand for nickel fell and the cartel temporarily collapsed. This decline was exacerbated by the limitations placed on naval and other armaments by the Washington Arms Conference in 1921–22. The new circumstances forced both Inco and Mond to change from pure mining companies to ones where research and the creation of new markets were critically important. These developments allowed the economic situation in Sudbury to improve during the 1920s (fig. 7.1).

During this period of reorientation, Robert Stanley led the charge for Inco (biography 6). The Bayonne refinery in the United States was closed in 1921, and a new rolling mill for converting Monel (a nickel-copper alloy) into ingots for rods and sheets was erected at Huntington, West Virginia in 1922. By the middle of the 1920s, the growing demand for nickel plating by the automobile industry, and the adoption of new chrome-nickel-iron alloys for use in kitchens, cables, and telephone and radio communication devices, all combined to encourage production from the mines in Sudbury. To meet the power demands for its present and future needs, Inco constructed a new hydroelectric plant at Big Eddy on the Spanish River in 1927.

Indispensable in war, nickel had now become even more indispensable in peace. In the Sudbury area, the growing demand for nickel focused attention on the problem of Frood mine. Since different parts of the Frood ore body were jointly owned by Inco and Mond, the latter company's officials suggested that "what was needed, if the mine were to be properly developed and all the ore recovered, was single ownership and not a duplication of effort."[14] Joint ownership, Lord Melchett argued, would have many advantages: it would conserve ore, reduce costs by approximately 19 per cent, and save huge sums on capital expenditures and equipment. From Inco's point of view, the strategic importance of the Frood deposit was not only related to the fact that it was a "deposit of dimensions" whose ores had a high content of precious metals, including platinum but also because the future outlook of its own Creighton mine was uncertain. Robert Stanley readily accepted Melchett's offer. It was simply too good for Inco to refuse at the price agreed upon. All the usual benefits of corporate concentration, near-market monopoly, economies of scale, and large profits were inherent in the merger, without any obvious disadvantages. Given the buoyancy of the market, moreover, the timing was ideal.[15]

Company officials reached an agreement to merge the two operations effective January 1, 1929. The merger involved a series of stock exchanges whereby Inco became the parent organization and Mond was relegated to a subsidiary position to control operations in Great Britain. A new company known as International Nickel Company was formed to take control of all operations in the United States. With this action, Inco acquired the following properties: the Frood Extension; the Garson and Levack mines; the smelter at Coniston; a nickel refinery in Clydach, Wales; a precious metals refinery at Acton, London;

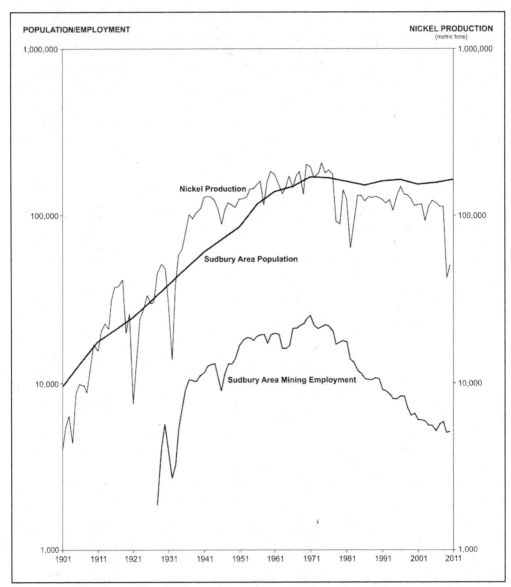

FIGURE 7.1 Population, Mining Employment, and Nickel Production Trends in the Sudbury Area (1901–2010/11). From the Ontario Ministry of Northern Development and Mines, *Ontario Mineral Score* (Toronto: Ontario Ministry of Northern Development and Mines, 1981–1994); Statistics Canada, *Censuses of Canada 1901–2006*; and Natural Resources Canada, *Canadian Minerals Yearbook* (Ottawa, ON: Natural Resources Canada, 1996–2011). *Note*: This diagram illustrates relative rather than absolute trends.

Biography 6

ROBERT CROOKS STANLEY
Saviour of Inco (1876–1951)

Robert Stanley was a driving force that helped to build Inco into one of the world's most successful mining and metallurgical enterprises.[a] He was born in Little Falls, New Jersey, in 1876, and received a mechanical engineering degree from the Stevens Institute of Technology in 1899. From the Columbia School of Mines he received a degree in mining engineering in 1901. He then went to work for the S.S. White Dental Company, where he began to conduct experiments on various alloys. When an officer of the company transferred to the Orford Copper Company, he persuaded Stanley to go with him. In 1903, he was made superintendent of the America Nickel Works, but was transferred in the following year to the Bayonne Refinery as assistant general superintendent. Stanley again experimented with alloys, and in 1905 he discovered Monel, an alloy made of nickel, copper, and other metals that was stronger than steel, malleable, and resistant to corrosion.

The discovery attracted the interest of the United States Navy much the same way that nickel steel had some twenty years earlier, only this time for the purposes of propellers, not armour. The prospect of increased Monel uses prompted the company to erect its own conversion plant and rolling mill at Huntington, West Virginia. Under Stanley's direction, this Bayonne plant was successful in stepping up its production to meet wartime needs.

In 1912, Stanley became general superintendent of Orford Works. Following the First World War, the demand for nickel fell back to what it had been at the turn of the century. After becoming a director of International Nickel in 1917 and vice-president in 1918, Stanley came to the realization that new peacetime markets for nickel had to be developed to replace its traditional wartime uses. He set up a Development and Research Department to find new uses for nickel and nickel alloys; a sales and advertising program for products was likewise initiated in 1919 and another launched in 1927 for nickel products that were underutilized. These launches illustrate the rapidly changing nature of Inco's consumer markets. A research laboratory was set up at Bayonne. After the First World War, Stanley supervised the moving of refining operations from Bayonne to the new nickel refinery at Port Colborne.

When Stanley became Inco's president in 1922, the company was at its lowest ebb. To offset this situation, he spearheaded a comprehensive mine and plant expansion totalling more than $50 million in 1926 that was completed by 1931, including development of the Frood deposit, the expansion of nickel sales in Europe, and the acquisition of Mond. He continued to emphasize research, development, and sales, especially as they related to the automobile and durable consumer goods industries. As he stated in 1931, "the Company's main activity has ceased to be that of mining. Its works and business are day by day becoming more commercial ... and the Company's success now depends less on ore reserves than on the ability to find, or make, markets which will take the manufactured product of the Company."[b] Inco thus became one of the first companies in North America to pursue research vigorously. It was also at Stanley's instigation that the company built improved recreational and educational facilities for all of its company-owned townsites.

Stanley became Chairman of the Board of Directors in 1937. Throughout the Second World War, he was responsible for increasing nickel production for wartime uses. The thorny issues he had to deal with included negotiating the payment the Soviet Union (USSR) had to make for the Petsamo nickel deposits in Finland, which Inco had acquired in 1934. These deposits were taken over by the USSR during the Second World War and never returned. A protocol was eventually signed involving compensatory payments to Inco. In the aftermath of the war, he laid out the program for the company's conversion to all-underground mining. In 1949, Stanley relinquished the presidency but remained as chairman of the board, a post he held until his death on Staten Island, New York in 1951, at the age of 74.

Stanley's legacy was substantial. J.F. Thompson, a former Inco executive committee chairman, has gone on record stating that "he was the man who saved the Nickel Company in 1922."[c] When Copper Cliff's new hockey arena was built in 1934, it was named Stanley Stadium in his honour. Another tribute was made as recently as 2005 when a journalist made the following statement while commenting on the contemporary performance of Inco Chairman Scott Hand: "Not since Robert Stanley, Inco president and CEO from 1922 to 1949, has a corporate leader had such a profound impact on the company's future."[d]

---

[a]A summary of Stanley's role with International Nickel can be found in: Executive and Operating Staffs, "The Operations and Plants of International Nickel Company of Canada Limited," *Canadian Mining Journal* 58 (November 1937): 581–89, and 67 (May 1946): 309–18; "Robert C. Stanley," *Sudbury Daily Star*, February 16, 1951, 1; and "Robert C. Stanley," *Inco Triangle*, March 1951, 2.
[b]Thompson and Beasley, *For the Years to Come*, 226.
[c]Ibid., 357.
[d]Stan Sudol, "Merger Puts Focus Back in Mining," *Sudbury Star*, November 11, 2005, B7.

---

and Henry Wiggins & Company, which operated rolling mills, fabricating, and chemical plants in Birmingham. Prior to the merger, Inco took the step of making the parent company a Canadian corporation. One reason for this move was to avoid antitrust action in the United States because of the elimination of Mond as a "competitor" in that country. While Inco became nominally Canadian, executive control remained firmly in the hands of its American directors. The merger gave the new company worldwide control over 90 per cent of the nickel market, thereby solidifying Inco as a global monopoly and Stanley's position as head of Inco.

The US control of Inco was paralleled by the country's concern about controlling nickel as a strategic wartime mineral. For example, during the 1920s and early 1930s, paranoia existed in certain US circles concerning the possibility of a theoretical war between the United States and the British Empire that would use Canada as a theatre of warfare. Added to this paranoia was the belief held by many Americans in the expansionist ideology of "Manifest Destiny:" that the United States had the inherent and inevitable right to control all of North America. To prepare for this unlikely scenario, the United States government even prepared a Joint Army and Navy Basic War Plan in 1924 that included the option of invading Canada and preparing the provinces and territories for statehood. This plan was approved in 1930, and updated in 1934 and 1935. One of the critical aspects of this plan was its repeated emphasis on the need to control Sudbury's vital nickel deposits via an invasion route through Sault Ste. Marie. As noted in the report, "Unless nickel can be obtained from the Sudbury mines, serious shortage in that important warmaking material will develop ... within a few months after the war begins." The document is instructive, as it reveals the enormous strategic global importance of the Sudbury nickel deposits at the time.[16] Another part of the strategy involving northern Ontario was a proposal to generate hydroelectric power at James Bay. While nothing official came out of this plan, it is pertinent to note that the idea of hydroelectric generation at James Bay has since been revisited on several occasions, notably in the form of the Grand Canal Project proposed in 1959 involving the diversion of water to dry areas in the United States.

Another outcome of the 1929 merger was the consolidation of the hydroelectric plants previously owned by the two companies. While they were among the biggest customers for power generated by Ontario's Hydro-Electric Power Commission, they also relied on local sources of hydroelectric power. As early as 1902, International Nickel established a subsidiary known as the Huronian Power Company for the purpose of developing power sources along the Spanish River. A plant, known as High Falls, was built on one side of an island in 1905; it was the first to supply the mining industry with power in Northern Ontario.[17] In order to regulate the flow of the Spanish River, the company acquired the rights to eleven crib dams reaching as far north as the river's source at Biscotasing Lake. A second High Falls plant was built on the other side of the island in 1917. The two plants supported a small community of eighteen homes and a one-room elementary school in a setting that looked "more like a slumbering New England village than a nerve-centre of Northern Ontario."[18] Completed in 1920, with a length of 358 metres and a maximum height of 45 metres, the Big Eddy dam was the largest in Canada at that time. It also transformed the physical setting by creating a new lake (Agnew), some forty kilometres in length.[19] After the merger in 1929, Mond's subsidiary, the Lorne Power Company, which had built power plants at Wabageshik on the Vermilion River in 1909 and Nairn Falls on the Spanish River in 1915, was absorbed into the new entity.[20]

CHAPTER 8

Beyond Sudbury and Copper Cliff

Railway Stations, Mining Camps, Smelter Sites,
and Company Towns

While the early history of the area was closely associated with Sudbury and Copper Cliff, developments took place on the periphery that set the stage for the rise of a wider regional economic setting and settlement pattern. In contrast to other parts of Canada, however, this region did not develop in the form of circular suburban population belts spreading from the Sudbury/Copper Cliff core; rather, the expansion of settlement took place in leapfrog fashion, centring on smaller nodes at considerable distance from the two centres. This nodal pattern can be linked to the unique location elements associated with the area's railway stations, mining camps, smelter sites, and company towns. Thus, the seeds for the geographical evolution of the region into a "constellation" urban setting were planted early.

Railway stations stretching along the various lines of steel constituted the region's first outliers of settlement. While several acquired other functions that permitted them to evolve into small villages or towns, only Capreol situated on the CNoR mainline evolved into a major transportation centre. Mining camps (including roast yards) later sprouted along the inner rim of the Sudbury Igneous Complex (SIC), with some acquiring smelter status. The majority of these communities—Blezard, Chicago Mine, Crean Hill Mine, Gertrude Mine, Happy Valley, Mond (near Frood Mine), Mond (near Victoria Mine) Mount Nickel, Nickel City, O'Donnell, Murray Mine, Stobie, Victoria Mines, and Worthington—disappeared by the Second World War and became the first of the region's many "ghost towns." Others, such as Creighton, Garson, the Towns of Frood Mine, Coniston, Levack, and the Falconbridge townsite managed to evolve into "company towns" of varying size. One of these towns, Sellwood, was unique in that it was associated with iron and not the nickel/copper ores of the inner rim of the SIC. However, like their mining camp

forerunners, the company towns of Creighton, Frood Mine, and Sellwood eventually suc-
cumbed to the abandoned town phenomenon in the fifties and sixties. A company town
with an altogether different character was the Canadian Forces Station that existed near
Falconbridge through the Cold War years.

Railway Stations

The first traces of white settlement in Sudbury's hinterland can be linked to the CPR's
construction camps and railway stations.[1] Maps dating from 1883–85 note the existence
of small stations such as Stinson, Wahnapitae, Romford, Azilda, Chelmsford, Larchwood,
Onaping, and Windy Lake. Because of their restricted locations, few of these stations
developed further. In 1885, the CPR acquired a square mile of land at "Cartier Division
Station." Its role as a divisional point—later boosted by timber operations—made Cartier
one of the important early communities in the region. While it had more than 300 hundred residents and is commonly associated with the Sudbury area, it was never part of
any incorporated municipality. The Algoma line of the CPR was started in 1882 and
extended to Sault Ste. Marie in 1888, giving rise to hamlets at Naughton, Whitefish, and
Worthington. Following its incorporation in 1888 and expansion to Little Current in 1913,
the Manitoulin & North Shore Railway (M&NSR) constructed stations along its 147-kilo-
metre stretch of line at Elsie Junction, Nickelton (Murray Mine), North Star, Creighton,
Crean Hill, Mond, Stone, Worthington Road, Drury, Turbine, and Nairn.[2] Meanwhile, the
M&NSR became part of the Algoma Eastern Railway (AER) in 1911, an arrangement that
lasted until 1930, when the AER was leased to the CPR for 999 years and become part of
the Sault Ste. Marie Branch (fig. 8.1). Passenger service to Little Current ceased in 1963,
and the line beyond Espanola became a "ghost railway" due to the end of inbound coal
shipments and outbound iron ore by boats.

The CPR mainline stations sited in the heart of good timber and agricultural lands
developed a more diversified economic base. Among these can be included Azilda, which by
the turn of the nineteenth century had a population of around 100. By this time, other stations, including Cartier, Chelmsford, Nairn, Wahnapitae, and Whitefish had acquired even
larger populations, ranging up to 200. Wahnapitae was an interesting settlement in that it
had a connection northward via the Emery Lumber Co. Railway. In several of these communities, street plans prepared by the railway companies provided some semblance of order.

Capreol

Out of all the townsites, only Capreol evolved as a full-fledged transportation centre.
At one time, Capreol served merely as a mile-post on the Canadian Northern Railway
line completed in 1908 that ran north from Toronto through Sudbury to Sellwood, a
nearby mining community. During the planning stage for the transcontinental Canadian
Northern Railway, a divisional site was deemed necessary in the Sudbury area to link

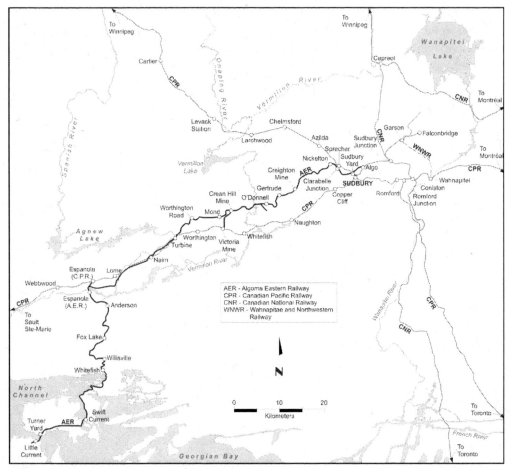

FIGURE 8.1 Algoma Eastern Railway (1930). From Wilson, *Algoma Eastern Railway*, 4. The WNWR refers to the Wahnapitae & Northwestern Railway, constructed in 1888.

its Montreal–North Bay and Toronto–Winnipeg lines. When Frank Dennie, the Scottish owner of a hotel in Hanmer got wind of this plan in 1912, he found the surveying post for the proposed location in the Township of Capreol, and bought 121 hectares of surrounding land. This situation forced Sir Donald Mann of the Canadian Northern Railway to meet with Dennie to resolve the ownership problem. Rather than selling his land outright, Dennie opted for a legal agreement whereby he gave the railway company 40 hectares of land in exchange for a promise that the company would make the site a permanent divisional point, with shops, a roundhouse, and other railroad facilities. The importance of this agreement became evident in 1958, when it was used to stop an attempt on the part of the company to move its repair shops to North Bay.[3] In the meantime, Dennie subdivided his remaining 81 hectares.[4]

The divisional point became operational in 1915. While a townsite plan was prepared in 1913, it remained until 1915–20 before the first three subdivisions came into existence.[5] By 1918, the community had a population of 500 people and some sixty houses; this growth spurred its incorporation as the Town of Capreol. Since the CNoR provided the main source of employment and owned much of the land, Capreol evolved as a CNoR company town; only the Marshay Lumber Company mill nearby provided alternative employment in logging and cutting. Unlike its rural counterparts in the Valley, however, Capreol's population consisted largely of anglophones rather than francophones. The community also had a distinctive Italian presence. Many Italians were employed by the railway company; others, including the Mazzuca, Colasinome, and D'Appolonia families, established successful businesses.[6]

Prior to the Second World War, Capreol had acquired many of the features of a modern town, including water and sewer facilities, one of the "finest" YMCA buildings in all of Canada (1921), an enclosed arena, an elementary and high school, numerous churches, cultural amenities, and a vibrant downtown.[7] By 1961 the town had a population of more than 3 000 and a service base bolstered by residents who lived in the nearby townships of Capreol and Norman. Between 1959 and 1978, Capreol was also affected by the rise and fall of the National Steel Company. The town remained an isolated political unit until 1973 when it was absorbed as part of the Regional Municipality of Sudbury. At this time, the former village of Milnet and the Sellwood townsite became part of Capreol. Up to 2012, both Milnet and Sellwood remained ghost towns. In this year, however, the Sellwood property became the proposed site for Cliffs Natural Resources ferrochrome smelter. The completion of this project, scheduled for 2015 and its accompanying workforce of 450 will have a significant impact on the economic diversification of the Capreol area and the eastern part of the Valley.[8] Much of Capreol's history as a railway town is now preserved in its Northern Ontario Railroad Museum & Heritage Centre, incorporated in 1993.

Mining Camps

Numerous mining camps developed in the area prior to the First World War. As early as 1902 there were some forty in existence. Many ceased production at one time or another, only to be resurrected at a later date. Since mining was labour-intensive, residential activity developed around the camps. While some sites were little more than a ramshackle collection of small cabins, shanties, and boarding houses surrounding mining headframes, such as those associated with Errington Mine prior to its closure in 1931, others prospered as isolated villages in attractive wilderness settings. The oldest camp was Worthington. Started in 1885, its ore body was one of the richest in the district, averaging 8 per cent nickel and 18 per cent copper. Situated along the Sudbury–Sault Ste. Marie highway, by the 1890s it had acquired the reputation of being "quite a village," boasting private housing, a fine hotel, a post office, and a school. Despite several setbacks, the village of some

800 residents continued to exist until 1927, when a dramatic cave-in permanently collapsed the mine. A report of the Bureau of Mines later estimated the extent of the cave-in to be 91 metres in diameter and 14 metres deep. Fortunately there was no loss of life; the forty-six miners underground were all safely evacuated.[9] The townsite was abandoned by the company, and whatever remained of the community dwindled to a small service outlet on the old "Soo" highway. Following the realignment of the highway to the south in the 1950s, its service function disappeared, and today only a small lake remains as evidence of the cave-in that ended the town's existence.

The Mond townsite was situated southwest of Victoria Mine and north of the Mond Roast yard (fig. 8.2).[10] Its lifespan coincided roughly with that of the mine, from 1901 to 1923. Containing at least three hundred residents, Mond was said to be an active "League of Nations," with Finns, Ukrainians, Poles, Italians, French-Canadians, and British families living in relative harmony.

Virtually every resident had his or her own small barn and farm animals, and potato gardens provided a degree of self-sufficiency. The close proximity of the townsite to both the AER and the Algoma branch of the CPR provided good access to Sudbury.

Another mining camp was Crean Hill. Discovered in 1885 and developed as the first "engineered" mine by the Canadian Copper Company (CCC) in 1905, it fell along with the mine's fortunes, closing in 1919. The townsite existed until 1937, at which time 137 dwellings and other buildings were torn down and taken away by the purchasers.[11] Work on the mine again commenced in 1951 and after twelve years of development, production was renewed in 1964. However the mine had only a temporary revival and it was closed in 1978.

Smelter Sites

Among the peripheral mining camps, there were six that, at one time or another, grew to such an extent that they could be called "smelter towns."[12] These included Chicago Mine, Victoria Mines, Murray Mine/Nickelton, Blezard, Mount Nickel, and Gertrude Mine. The Drury Nickel Company established a "model" camp in 1891 at its Chicago mine (also called the Inez or Travers mine). This was the first attempt by mining officials in the Sudbury area to provide better housing, water, and sanitary arrangements for its workers.[13] While a smelter was erected in 1892, it had a sporadic existence and was abandoned in 1894. The site was refurbished in 1896; however, it was again deserted in the following year.

The most important of the smelter towns was Victoria Mines.[14] Following the discovery by Ludwig Mond of the Victoria Mine in 1899, a roast yard was built, as well as a smelter that was used to process ores from the company's Garson and Worthington Mines located several kilometres to the south, adjacent to the CPR Sault line. The site was favourable because of its access to water from Ethel Lake and the availability of wood supplies for roasting purposes. After the ores were processed here, the remaining matte was sent to Clydach, Wales for refining and selling. A dominant image associated with the townsite was the

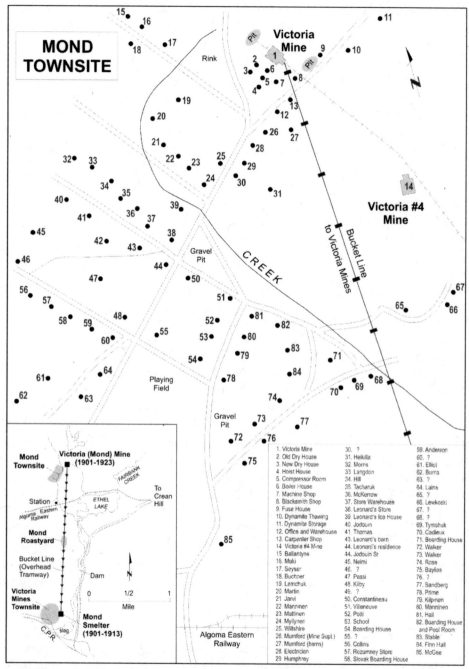

FIGURE 8.2 The Mond-Victoria Mines Setting (ca. 1917–1918). From W.H. Makinen, "The Mond Nickel Company and the Communities of Victoria Mines and Mond," in *Industrial Communities of the Sudbury Basin*, 26–31.

existence of an aerial tramway linking Victoria Mine, the Mond roast yard, and the smelter at Victoria Mines. The tramway, one of the first in Canada, transported ore in large buckets from Victoria Mine to the roast yard. Once there, the ore was unloaded and roasted for three to four months before being reloaded and shipped to the smelter for further processing.[15]

Starting in 1901, Victoria Mines was a bustling settlement based on a 129-hectare "state of the art" gridiron townsite plan that grew around the smelter, which at its height employed around 200 men. Until 1913, the number of residents varied between 300 and 1 000. Among this number was Hector "Toe" Blake, who later became the famed Montreal Canadiens coach. In contrast to the more ethnically diverse Mond townsite to the north, the majority of the Victoria Mines residents were anglophones. The settlement featured boarding houses, one apartment, more than fifty single dwellings, grocery stores, service facilities, a post office, Mond offices, two elementary schools, a jail, and three churches. The CPR station, completed in 1904, reduced the isolation of the community by providing daily passenger service to Sault Ste. Marie and Sudbury.

The changing regional dynamics of the mining industry brought the smelter and its surrounding townsite to an end. The growing inadequacy of the Victoria Mines townsite was made clear in the report of the Royal Ontario Nickel Commission:

> From 1901 until 1913 the company conducted its smelting at Victoria Mines, but with the final perfection of its refining plant, and good prospects of increasing business, the limitations on the Victoria Mines site, which was not adapted to further expansion, were felt to hamper the company's operations. There was railroad connection with but one line, and after the purchase of additional mines, the smelter was found to be too far from the centre of the ore supply.[16]

Following the decision to abandon the Victoria Mines smelter and townsite, there remained the problem of selecting another site for Mond's new smelter. After careful investigation, company officials selected a site at Coniston. After Victoria Mines' demise, many workers moved to the new site. The remaining churches and company houses were sawed into sections and moved either to Worthington or Coniston. By the advent of the Great War, the townsite only had a few vestiges of its past; even its location on the "Soo" highway, which was passable by the late 1920s and paved in 1931, was not sufficient to maintain its existence. As was the case with Worthington, the shifting of the highway farther south in the 1950s increased the isolation of the site. By this time, only two houses remained to reflect the former presence of the Mond Company.

Murray mine, the site of the first mining discovery in the Sudbury area in 1883, at one time offered considerable promise as the setting for a major townsite. H.H. Vivian & Company briefly operated a smelter at the site from 1890 to 1894. The British America Nickel Corporation acquired the mine in 1913 in the hope of becoming a major competitor to International Nickel; it drew up plans in 1915 to erect a village at the site known as Nickelton. This endeavour was later downsized, and the company only built some forty dwellings and a boarding house when the new smelter opened in 1920. Nickelton, now home to 300 persons, thus became the region's newest industrial community after the First World War. This prosperous beginning proved illusory, as the village was soon deserted following the company's bankruptcy in 1924.[17]

While there were smelter sites at Blezard (1890–95), Mount Nickel (1899–1901), and Gertrude (1902–3), they were smaller and had only a limited existence.

Company Towns

Around the turn of the century, a number of company towns similar to Copper Cliff emerged in the hinterland. The first to appear was Creighton (and O'Donnell). Others, including Garson, Frood (and Stobie), Levack, Coniston, Falconbridge (and Happy Valley) followed in its wake. These towns were similar in that the mining companies had a dominant influence in shaping their economic, cultural, and political scene. The Sellwood townsite situated in Hutton Township was different in that it was based on the extraction of iron ore deposits unrelated to the SIC.

Creighton (and O'Donnell)

The significance of A.P. Salter's compass deviation in his survey of 1856 did not become apparent until 1885, when the first mining claims were staked on what would eventually emerge as the "Pit" or Creighton mine. Although the site had been acquired by the CCC in 1887, it was not until 1900 that development took place. In this year, the company started construction on what it called a "first-rate mining camp" at Creighton. In order to house its more than 900 miners, the company built a small settlement featuring cottages, three general stores, a school, and a post office, laid out in a regular arrangement along streets.[18] Throughout its history Creighton was never incorporated, and it existed as a townsite owned and run by company officials within the unincorporated Township of Snider.

The completion of the Manitoulin and North Shore Railway (M&NSR) established a vital link between Creighton and the roasting and smelting complex at Copper Cliff. Few realized at the time that the first shipment of Creighton ore in 1901 marked the start of production from what would emerge as the world's greatest nickel mine.[19] With the assistance of hydro power from the CCC's High Falls plant in 1906, Creighton Mine easily met the growing demands brought about by the First World War. The mine's increased production caused the labour force to grow to more than 1 200 and the overall population to 2 350, resulting in the construction of additional homes in a section dubbed by local residents as New Town.[20] One highlight in 1917 was the visit of the Duke of Devonshire to the community.

During the interwar period, the community prospered and new families arrived. The growing population reflected a diversity of ethnic groups and, as was common locally, a voluntary internal separation of the population. Various nicknames were given to parts of the town, such as the Dardanelles, Spanish Town, and Dogpatch. The ethnic groups were made up of immigrants from Italy, Ukraine, Poland, Romania, Holland, Ireland, and Finland. This diversity was reflected in the community's businesses and institutions, including boarding houses that catered to specific ethnic groups, such as Lietala's Finnish Bath House, Frank Anderson's Dairy Plant, Italian barber, butcher, and general store establishments, Jaworski's General Store, J.C. Kelly's Store, the Ukrainian, Cabrini, and Finnish Halls, and the Romanian Orthodox Church. In keeping with International Nickel's usual practice, the company provided adequate health, police, and educational facilities. Near the end of the Second World War, the need for more accommodation led the company to build new homes in a subdivision known as Victory Hill, later named "Snob Hill" because it was occupied mainly by higher management officials.

Creighton was arguably the most sports-minded company town in the Sudbury area. Spurred by the support of company officials, sport was advanced as a means of promoting community spirit and cultural cohesion. Many aspiring baseball players were hired by Inco to represent their company towns in the Nickel Belt Baseball League. Tom Davies, the Chairman of the Regional Municipality of Sudbury who was born in Creighton, bragged that "the quality of competition which came to Creighton Mine during thirties, forties and fifties was unparalleled in North America."[21] Other sports included soccer, hockey, bowling, tennis, and badminton. The Finns were also active in promoting gymnastics, wrestling, skiing, and track and field events. This range of activities could be traced back to the formation of the *Yritys* (Endeavour) Athletic Club in 1916. Given the setting, it is no wonder that some like George Leck, a company worker, could boast in 1934 that "Creighton was the best town owned by International Nickel and the best he had ever lived in."[22]

Mining in Creighton expanded in 1950 with the erection of a new low-grade ore concentrator and pipelines to transport the concentrate to Copper Cliff, thereby eliminating the need for shipments by rail.[23] The new concentrator, intended to utilize Creighton Mine's lower-grade ores, was deemed necessary because of the fear that the open-pit operations at Frood Mine would soon be exhausted.[24] Later in the decade, the mine expanded to become the world's deepest nickel mine. As a result of the tremendous postwar growth of the nickel industry, however, it became clear that the Creighton townsite could no longer support additional residential expansion. When the decision was reached by Inco to construct the new Town of Lively nearby in 1949, the die was cast for the future of Creighton. Residents began a slow exodus from the community. The decline in population, along with the need for costly upgrades to its water and sewer systems, compelled the company to close the town in 1988. All resident buildings were torn down or removed. Aside from a historical plaque and a number of abandoned streets, there is little reminder that this was once home to more than 2000 residents. No one at the time, though, could foresee that a new chapter in the history of Creighton would begin when part of the minesite became the home of the world-famous Sudbury Neutrino Observatory.

While in existence for only thirty years, Creighton's sister village of O'Donnell, situated some seven kilometres to the west along the M&NSR line, deserves mention. Here, a village of thirty families developed alongside the O'Donnell Roastyards, which in 1915 replaced the one that had been in Copper Cliff. Initially providing employment for 200 employees, this number was gradually reduced to forty. In the village, besides the homes (some of which were fashioned from railway cars) there was a one-room school, the town hall, a general store, office, ice house, rink, train station, and a clubhouse for single men. Other required services came from the Creighton Mine townsite. The community was abandoned in 1930 when the new Copper Cliff smelter replaced the need for outdoor roast beds.

Garson

While the discovery of mineral deposits at Garson was made by John Cryderman in 1891, it remained until 1899 before the property was transferred to Mond Nickel Company. The first shipment of ore to Victoria Mines began in 1908. Another nearby mine known as the Kirkwood had been discovered as well, but it was subject to little development. Unlike other company towns, the Garson area had previously been the scene of trapping and lumbering activity due to the abundance of post-glacial deposits and huge stands of red pine "that were three feet at the butt" and "as thick as the hair on a dog's back."[25] The Emery Lumber Company, which possessed numerous holdings in the area, had its operational headquarters at Headquarters Lake along a railway line known as the Wahnapitae & Northwestern Railway that ran from the village of Wahnapitae to a site near the present-day Sudbury Airport.[26] This was the first site of permanent settlement in the area. Some of the land cleared by forestry operations was used for agricultural purposes while the Garson minesite was in its formative stages. Many farmers raised grain crops to be milled into flour at the Manitoba and Ontario Flour Mill plant in Sudbury. Agricultural activity peaked around 1921, after which time it underwent long-term decline.

The Garson townsite had a solid economic base; indeed, the Garson Mine eventually became the top producer for the Mond Nickel Company. Forty company houses for employees were constructed during the mine development phase. It was the core of an area that was known as Company Town and later Inco Town. Since the land north of the railway was leased to Finns, this part of the townsite was referred to as Finn Town. The remainder of the village lived in what was called the Business Section, Polack Town, or Italian Town (fig. 8.3). While Mond and later Inco played a major economic role in the development of the townsite, it never was a total company town, as only the original townsite was company built. All the downtown businesses and remaining residential areas were privately owned. Politically, the townsite existed as part of the Township of Neelon-Garson, which elected its first council in 1910. The council also represented the Coniston area in Neelon Township until it was incorporated as a town in 1934. Unlike the other company towns, the council was largely responsible for the local police force and other services such as fire, water, and sewers.

Life in the community was enhanced by the arrival of electricity in 1937, running water in 1948, and a sewage system in 1962. In the early years, transportation facilities were notably absent, with only a Saturday train or horse-drawn carriages providing access to Sudbury until the new Falconbridge highway was constructed in 1937. Aside from Saturday visits to Sudbury, the Finnish Hall and the basement of the public school provided the main venues for entertainment and social gatherings. The social setting changed in 1950 with the opening of the Garson Employee Club. Similar to other established Inco employee clubs in Sudbury, Copper Cliff, Creighton, and Levack, it was a recreational venue featuring an auditorium, badminton courts, billiard room, bowling alleys, and a gymnasium.[27] There was a lively sports scene.

Soccer was especially popular until the end of the Second World War; thereafter, sports such as hockey, broomball, baseball, softball, badminton, tennis, and bowling prevailed.

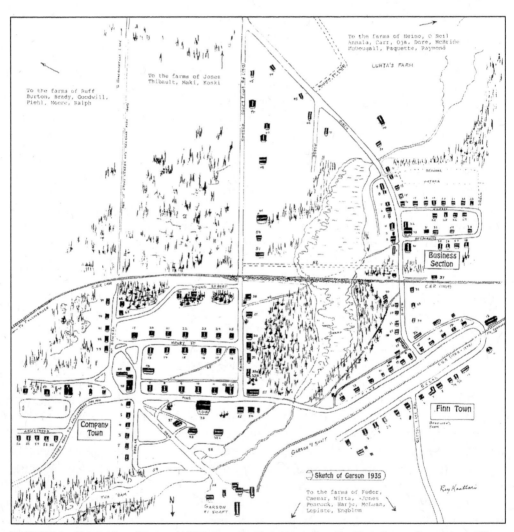

FIGURE 8.3 Sketch Map of Garson (1935). From Kaattari, *Voices from the Past: Garson Remembers*, 37.

The community suffered during the Great Depression years, but recovered quickly. A sand pit was opened to provide flux for the Inco converters in Copper Cliff and a second shaft was opened. By the Second World War, Garson's population was approaching 2 000, a figure that forced Inco to build another thirty-five homes. After the war, the opening of the New Sudbury Shopping Centre brought about the demise of Garson's existing downtown. Businesses that remained in the area had disappeared or shifted their location to the suburban plazas that had sprouted along the Falconbridge highway leading to Sudbury by the 1970s.

The opening of the Garson-Falconbridge Secondary School (later St. Charles College) in 1963 marked the beginning of a new era for the townsite. Another boon occurred in 1973 when Garson was selected as the location for the Town of Nickel Centre in the Regional Municipality of Sudbury. Due to its favourable regional location and abundance of developable land, Garson then experienced considerable growth as a suburban outlier for Sudbury. The underlying strength of the community was such that the closure of the Garson Mine by Inco in the 1970s due to safety concerns did not have any permanent long-term impact. The Garson Mine, however, reopened in 1994 and in 2007 production was started on the Garson Ramp Mine.

Town of Frood Mine (and Stobie)

While Frood mine was discovered by Thomas Frood in 1884, it remained until 1889 before mining operations were started by its new owner, the CCC. After a brief suspension of operations from 1903 to 1911, a company townsite was laid out in 1915 which included a water system, forty dwellings, a company store, and a public school. In 1916 the site was incorporated as the Town of Frood Mine. In the meantime, an extension of the same Frood ore body had been acquired by Mond Nickel in 1911, and by 1924 the company had started development work on what would become the Frood Extension Mine. During this interval, corporate ownership of the property had shifted from the CCC to International Nickel in 1918. Thus it came to be that both of the major mining companies in the Sudbury area were developing the same ore body.

While a small townsite was also started by Mond at the Frood Extension Mine, its lifespan was short. In 1925, both Inco and Mond stated publicly that future employees of their respective Frood and Frood Extension operations would henceforth have to live in Sudbury. Inco even provided a $500 000 loan package to help its workers acquire housing in Sudbury. After the amalgamation of Mond and Inco in 1928, the Mond townsite was removed and it became the site of an open pit. For a brief time during the late 1920s, the Frood site supported some 800 persons. After the Second World War, the population declined as many workers did not want to live at the site, preferring to live in Sudbury. Following the town's annexation to the City of Sudbury in 1960, the exodus continued. Frood mine closed for good in 1998, and its buildings were removed in 2005.

Northeast of the two Frood mines was Stobie Mine, CCC's most productive mine prior to 1900. The fledgling village featured more than ten houses in 1888, and by 1902 there was telegraph and mail service, a school, a sawmill and a general and grocery store, all of which enabled it to persist even though the mine itself had been closed the year previous in favour of the larger Creighton deposit.[28] Vestiges of settlement remained there until the 1920s.

Coniston

As was the case with the Garson townsite, Coniston's early history was rooted in agriculture and lumbering. The area first came to the attention of settlers following the passage of the CPR mainline in 1883. Beginning in 1902, the flat terrain adjacent to the railway attracted farmers, and by 1905 some twenty families had established themselves in the area. The regional setting changed dramatically in 1908, when both the CNoR and CPR built railway lines through the area. The CPR branch line from Toronto joined the CPR mainline at Romford Station, situated just west of the existing farms. Another CPR station was then built along the mainline in the townsite proper. The CNoR line was constructed farther to the east and crossed the CPR mainline, where a small station was erected. To meet the district's demand for forest products, a lumber mill known as Coniston Woodworks was built.[29] By this time, a small residential area known as "Old Coniston" had evolved.

It was largely as a result of Coniston's strategic railway setting that Mond Nickel began the process in 1911 of transferring its smelter operations from Victoria Mine to Coniston. The company believed this transfer was beneficial for several reasons: the inability to expand its Victoria Mine site, the availability of land in the Coniston area, a plentiful water supply from the nearby Wanapitei River, the availability of electrical power from the Wahnapitae Power Company's dam at Stinson, better export opportunities offered at Coniston for the shipment of matte to Montreal and Clydach, Wales, and finally, the nearness of the site to the Garson and Frood Extension Mines acquired by Mond Nickel in 1905 and 1911, respectively. In 1911, Sir Alfred Mond acquired some 1 500 hectares of farmland for the company, bringing agricultural activity to an end.

Edgar T. Austin was hired to survey the town and to blueprint a "model" community based on a gridiron layout. Some of the first residential buildings were shipped by rail from the old Victoria Mine townsite. Many of those who formerly worked at Victoria Mine transferred to Coniston. Included among the facilities in place by 1913 were a separate and public elementary school, eleven executive homes, forty-six cottage-type houses, a fire hall, a combined municipal building and jail, and a company boarding house which could accommodate forty to fifty men. Over the next few years, numerous churches and commercial establishments appeared in the planned part of the town. In 1925, the Coniston Continuation School was constructed and provided secondary education until 1962.

A new smelter and a roast yard were completed in 1913, and many of the construction workers opted to remain in the settlement. The townsite remained stable until the boom of the 1920s brought in more workers and created pressure for residential expansion. The company allowed workers to construct privately owned homes in neighbourhoods that soon acquired Italian, Polish, and Ukrainian characters; others settled along the fringes of the town on farmland that its owners considered useless for agricultural purposes because of the pollution from the smelter's stacks and roast yard that operated until 1918. Through this process the town became divided into ethnic neighbourhoods known as Polack Town, Italian Town, Old Coniston or French Town, and English Town (fig. 8.4). Buildings in these ethnic enclaves such as the Ukrainian Labour Farmer Temple, Prosvit Hall, and Club Allegri became important landmarks.

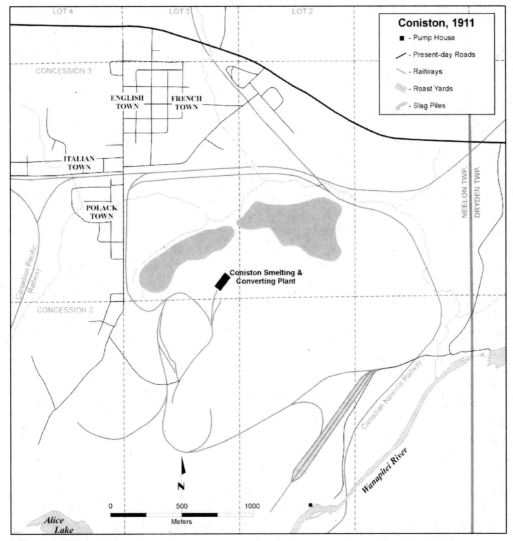

FIGURE 8.4 Sketch Map of Coniston (1930s). From Léo Larivière, Cartographer and Map Librarian, J.N. Desmarais Library, Laurentian University.

The merger between Mond Nickel and Inco in 1928 had major repercussions for the townsite. While Mond Nickel had planned to use Coniston as its Canadian headquarters and showplace, this plan was rejected by the new Inco officials, who favoured the Town of Copper Cliff. Sensing an end to company paternalism on the part of Inco and aware of its policy to only manage and maintain rather than grow the community, the residents of Coniston opted for incorporation as a town in 1934, severing its ties with the Township of Neelon and Garson. The newly incorporated municipality had a population in excess of 2 000. While the Great Depression hit the community hard, many laid-off workers survived by earning income from picking blueberries. It was a common sight to see hundreds of baskets of blueberries being shipped from the CPR station in Coniston to Toronto.

Until 1944, only company-sanctioned people ran for local political office. Following union certification at the smelter in 1944, a dramatic shift in the political scene took place when three hourly rated candidates won seats on council; the shift was complete in 1952 when a candidate who was not sponsored by the company won the mayoralty race. A new smelter stack installed in 1954 further away from the town replaced two smaller stacks, and resulted in cleaner air for Coniston residents by sending the sulphur gases higher into the atmosphere. A sanitary system installed in 1958 provided further improvement to the community. The local scene was enlivened by numerous sports activities that included baseball, soccer, women's softball, hockey, bowling, curling, and weightlifting.

Aside from the arrival of some Ukrainians in the 1950s, Coniston, unlike Garson, never enjoyed a major influx of population in the postwar years. With the exception of the construction of a new community centre and library in 1971, few changes occurred prior to its incorporation as part of the Town of Nickel Centre within the Regional Municipality of Sudbury in 1972. The loss of community identity was compounded that same year with the announcement that the Coniston smelter would be shut down. Reasons given for the closing included the projected costs of meeting new provincial environmental restrictions and the outmoded nature of the plant. The company sold the houses it owned to its employees, demolished most of the smelter in 1976, and retained part of the site for use as an industrial park.

Levack

The passage of the CPR mainline through the north rim of the Sudbury Structure also attracted prospectors. In 1899, James Stobie and Rinaldo McConnell discovered an ore body in the Township of Levack that was later sold to Mond Nickel in 1913. Framed by the Onaping River, Moose Creek, and rugged hills on the north and south, the ore site had an alpine setting that was visually different from other mining locations in the region. In 1913 the first families arrived and, according to one account, "dozens of carpenters, many of them Finnish, arrived to build the mine and town."[30] After the mine became operational in 1914, a small hamlet evolved consisting of tent dwellings, bunkhouses, and wooden frame houses. On the cleared land between the village and the mine, Finnish, Polish, Ukrainian, and Russian families settled in what became known as Warsaw. This section later became a popular bootlegging destination. In Levack proper, Mond Nickel built most of the homes and rented them to their workers; however, a few homes were built by private owners on land leased from the mining company. By the middle of the First World War, nearly 300 people resided in the community.

The building of the Finnish Hall in 1925 finally gave Levack a community centre that could be used for cultural purposes. Life in the settlement was subdued in the early years; *The Sudbury Star* made only two brief references to events here: one regarding a local wedding and another about roving canine packs attacking deer.[31]

During the Great Depression, Levack mine remained closed until 1937. Some local workers were fortunate because they were given the opportunity to work at other properties such as Frood and Stobie Mines or the Coniston Smelter. Within the community, a sense of alienation emerged following Mond Nickel's absorption by Inco; residents felt that decisions were now being made by "outsiders."[32] A new era began with the incorporation of the Town of Levack in 1938, and the arrival of electricity the following year. Between 1948 and 1952, the community underwent major expansion as a result of Inco's transition to all-underground mining and its decision to allocate $10 million to improve employee housing in its townsites. The housing program at Levack involved the construction of new houses incorporating six different dwelling styles and a new curling rink.[33] A welcome addition to the town was the opening of Levack District High School in 1957.[34] It remained in existence until 2001, at which time it became home to Levack Public School.

Falconbridge (and Happy Valley)

The history of Falconbridge mine is rich in historical lore, involving stories about an unsuccessful search on the part of Thomas A. Edison, its subsequent purchase by the E.J. Longyear Company, and finally, its acquisition by Falconbridge Nickel Mines Limited led by the dynamic entrepreneur Thayer Lindsley. The townsite began as a few bunkhouses around the first shaft dug by the company. Lindsley, however, had loftier ambitions for the settlement's future. In order to ensure an orderly development of the townsite, he hired a civil engineer by the name of Ernie Neelon to develop a master plan where everything was laid out, such as storm and sewage mains, a future sewage treatment plant, water mains, and the placement of hydro and telephone lines. In this plan, the use of back lanes was encouraged to eliminate the need for garbage collection on the streets. The street plan provided for broad roads lined with wide sidewalks. As well, allowance was made for some 400 single-family and apartment lots with dimensions that were spacious enough to allow for front and back lawns.[35]

By 1931, the community had a population of more than 400. The major features of the plan, which included a medical centre and elementary school, were completed in 1939. Since there was no secondary school, older students had to attend either Sudbury High School or Garson-Falconbridge Secondary School. A unique recreational facility for the youth of the community, known as Falcona Camp, was developed on the shore of Nelson Lake situated in Bowell Township along the northern rim of the Sudbury Structure.

As the townsite grew after the Second World War, it acquired additional facilities such as a bowling alley, community hall and church centre, an indoor arena and curling rink, a community centre with a swimming pool, tennis courts, and a baseball diamond. A commercial sector came into being, dominated by A. Earle Hodge's store, warehouse, and garage. It remained until the 1940s before a road connection was completed between the Falconbridge and Garson townsites. The political setting changed in 1958, when the Township of Falconbridge was incorporated. By this time the population had grown to almost 1 400. Falconbridge retained this status until 1973, at which time the township became part of the Town of Nickel Centre within the Regional Municipality of Sudbury. Throughout this period (1958–73), the Township of Falconbridge held the unusual distinction of only having one Reeve, J. Franklin.

The evolution of the Falconbridge townsite was accompanied by a fringe development of twenty-three houses to the south known as "Happy Valley." Situated on land owned by Inco, the location was occupied by squatters who preferred to live in an unregulated area rather than the planned townsite. Originally, Happy Valley was a habitable environment with plenty of land and water. However, the configuration of the valley unfortunately proved to be an excellent settling place for the noxious gases spewed by the Falconbridge smelter. The natural vegetation disappeared, and Falconbridge Nickel Mine and the Province of Ontario were both flooded with complaints about air pollution in the Valley. Other problems included declining water levels in local wells, a lack of fire protection, and the poor state of roads and drainage facilities.[36] The residents of Happy Valley eventually filed a lawsuit against Falconbridge Nickel, but lost the case. The Ministry of the Environment responded by setting up an air pollution monitor in 1971. When the atmosphere became so polluted that it was hazardous to local residents, the government required the Falconbridge smelter to limit its operations. Falconbridge therefore opted to purchase the residential dwellings in the valley in 1973 and 1974 and pay for associated moving costs. According to one reporter, Happy Valley was "the first community to be wiped from the map of Canada to make way for continued air pollution."[37] The site was gradually abandoned, with the last resident leaving in the late 1980s. It is now off limits to the public and protected by a large steel fence.

Sellwood

Unlike the other company towns, the company townsite of Sellwood had no relationship with either Inco or Falconbridge Nickel Mines. Located about twenty kilometres from Capreol, it was known before 1900 that there was a major magnetic anomaly at Moose Mountain, situated northwest of Capreol in Hutton Township. By 1901, John W. Gates and Associates of New York began to develop the property.[38] When the presence of iron was detected, Sir William McKenzie and Sir Donald Mann, owners of the CNoR, created the Moose Mountain Mining Company and acted as major shareholders. McKenzie and Mann were instrumental in having the northern terminus of the railway line's Toronto branch extended beyond Sudbury to the Moose Mountain site in 1908. This permitted the shipment of iron that had been stockpiled since 1906 southward to Key Harbour via a CNoR spur line. A crushing plant to reduce the large rock to smaller sizes and a concentrator to separate the metal from the ore were later constructed at Sellwood. McKenzie and Mann had such high hopes for the iron deposit that they planned to build a smelter and refinery at Key Harbour. This did not happen, as mining operations fluctuated over the next few years.

After the Warren Lumber Company opened in 1909, the townsite population grew to 1 500. While the settlement was administered by the mining company, many of the 100 homes and lots were privately owned. The townsite boasted numerous facilities, such as a luxurious clubhouse, the three-story Warren Hotel with 100 rooms, eight stores, two bakeries, a post office and schoolhouse, four poolrooms, a bowling alley, two restaurants, a Chinese laundry, an Orange and Finnish Hall, a Protestant church and Catholic Mission, and a theatre. There were even eight bootleggers who offered home delivery service twenty-four hours a day.[39]

As a result of the mine's erratic activity, residents began to leave the townsite in 1916 in favour of a new life at Capreol. The downturn in the demand for iron after the First World War forced the mine to close in 1920. While the property was officially abandoned in 1923, the lumber mill continued to operate for several more years. By 1930, not a soul was left in Sellwood. Lowphos Ore then undertook further exploration and production between 1947 and 1956. When the National Steel Company opened a new mine and pelletizing plant in 1959, whatever remained of Sellwood was finally demolished to make way for the new mine, mill and pelletizing operations.[40] This new technology formed fine-grained iron into small balls known as pellets that were suitable for further processing. In 1978, all operations ceased.[41] For a time, the site was used for making gravel.

In 2011–12, Sellwood was given a new lease on life, when Cliffs Natural Resources acquired the site and selected it as the location for the proposed ferrochrome smelter. This smelter will be used for processing the company's chromite ore from its Black Thor minesite situated in the Ring of Fire in Northwestern Ontario. This deposit of chromite, the largest ever discovered in North America, is to be shipped from the minesite through an integrated transportation system, including an all-weather road to Nakina, and from there to Sellwood via the existing CNoR line.[42]

Canadian Forces Station Falconbridge

The Canadian Forces Station (CFS) at Falconbridge was a military radar station that was active from 1952 to 1986.[43] Situated on a rugged hill northwest of the townsite of Falconbridge, the site was established in 1952 as part of NORAD's (North American Aerospace Defence) Pinetree Line of thirty-three radar stations that stretched from coast to coast. Their main purpose was to detect and identify unknown aircraft and then direct interceptor aircraft to these targets. Known earlier by other names, it was given its CFS designation in 1967, following the unification of Canada's military into the Canadian Forces. Locally, it was simply known as the Radar Base. CFS, like its counterparts, was a self-contained community. In addition to shared quarters, the station included 101 homes, water and sewage facilities, a school, library, church, infirmary, and a variety of recreation facilities: bowling alley, recreational hall, gymnasium, sauna, and children's wading pool. In the mid-1960s, it was re-designated as a radar scanning station. Later, it was used as a training facility.

On November 11, 1975, the station acquired fame of another kind when a visual and radar UFO incident occurred that was witnessed by numerous credible sources, including Ontario Provincial Police (OPP) officers. Despite sending two NORAD F-106 aircraft, Air Guard helicopters, and Strategic Arm Command helicopters to the scene, no positive identification of the object was ever made. Military reports eventually attributed the UFO sightings to either "the lights of Jupiter" or "atmospheric phenomena."

As a result of technological changes, Falconbridge was closed down in November 1986. The site was offered for sale and in following years it was acquired by several companies. Remnants of the base still exist. Although some of the homes have been renovated and rented, the remainder of the facility has either been boarded up or left in a deteriorated state.[44]

CHAPTER 9

Beyond Sudbury and Copper Cliff

*Forestry, Agriculture, Indian Reserves,
and the Burwash Industrial Farm*

The economic activities mentioned in the previous chapter were not the only ones foster-ing a pattern of dispersed population settlement in the Sudbury area. Other influences came into play that reinforced this trend, such as the exploitation of raw materials associ-ated with forestry, the rise of agriculture, the creation of two Indian reserves, and finally, the establishment of the Burwash Industrial Farm. Up to the First World War, lumbering remained significant. This was especially true for places such as Milnet, Wahnapitae, and many CPR railway stations and Valley settlements. In the Skead area, its impact remained significant even after the Second World War. The postglacial deposits found in the Valley and in pockets of the Precambrian Shield east, south, and west of Sudbury and Copper Cliff likewise gave rise to agricultural landscapes and rural enclaves that were mainly occupied by francophones or Finns. Outside of Sudbury proper, there existed two Indian reserves and a provincially run prison facility known as the Burwash Industrial Farm. Their periph-eral locations notwithstanding, these outliers reinforced Sudbury's position as the hub for the area, and contributed to the constellation trend associated with its hinterland.

Forestry

Forestry served to transition the Sudbury area from its reliance on the fur trade into a min-ing economy. According to Historian A.R.M. Lower, by the end of the 1800s the Sudbury area was well positioned to become a major centre of forest activity. Situated on the North Shore of Lake Huron with several large rivers draining into Georgian Bay, in 1870 this vast landscape contained more virgin pine than any other part of Canada.[1] Although timber from the North Shore was shipped to the United States as early as 1864, the local forests remained undeveloped until the limitations resulting from isolation disappeared with the arrival of the CPR mainline, and the forests in the Ottawa Valley and northern Michigan had been "thinned out." Important as well was the growing demand for lumber from Chicago and the American Midwest. It was for this reason that the commissioner of Crown Lands, acting on authority from the province, opened land north of Georgian Bay and the North Channel between Parry Sound and Sault Ste. Marie to the lumber industry, administering timber berths and timber licences. Despite these measures, the early export of lumber from the timber limits remained small; in 1880, for example, the commissioner of Crown Lands wrote that "no timber has been brought from the upper waters of the Wahnapitae [sic] and the only venture of taking timber down the French River was last winter."[2]

Within a space of less than twenty years, however, the Georgian Bay watershed was swept over by American and Canadian lumbermen who made inroads into the area by setting up logging camps to reach the headwaters of distant rivers. The shift of lumbering into the Sudbury area was prompted by three considerations: first, the province considered it expedient to sell pine reserves before mining began; second, it was deemed advisable to dispose of the forests ravaged by a large fire in 1891; and third, the mining industry itself was proving to be a growing market for timber of all kinds. The effect on the local area was remarkable: "The rapidly expanding reconstruction era within the United States created a lumber market with which the American Lumber Industry was incapable of coping. Turning their hungry eyes northward to Canada's virgin pine growth, the Americans moved in lock, stock and barrel. In a very short time, the Sudbury District was the scene of large-scale lumber operations never equalled in its history."[3]

Production was initially geared to supply the construction materials the railways needed: red and white pine square timbers, waney timbers (planks with defective edges), and sawlogs. To satisfy these needs, small lumber mills were erected locally on Minnow Lake, the north shore of Lake Ramsey, Junction Creek, the Whitson River, and at Larchwood, Coniston, and Capreol.[4] By the turn of the nineteenth century, square timber production had ended and there was an increase in the production of other products such as pulpwood and mining timbers. This resulted in a broadening of the species cut to include spruce, balsam, and jack pine.

Sudbury's position in the lumber industry was significant: it served as a major destination for those seeking employment in the adjoining forests, a function that continued well into the 1930s.[5] As the editor of the *Copper Cliff Courier* observed in 1903, "lumbering operations in this district are, or should be, considered our first line of defense" with respect to the regional economic base.[6] Prior to the Great Depression, the regional impact of forest employment in the Sudbury area still exceeded that of mining.[7] According to one account, some 235 companies and more than 11 000 men were employed in the mills and bush around Sudbury.[8]

Many of these operations were located on the fringes of Greater Sudbury in places such as Espanola, Nairn, Webbwood, Massey, Cartier, Benny, Burwash, Noelville, and

Markstay; the logging companies frequently employed francophone or Finnish-Canadian farmers or jobbers during the winter months. Timber licence holders, including Green and Co., Haight and Dickson, W.C. Cochrane, Larchwood Lumber Co., Labarge Lumber Co., Pineland (formerly Acme) Timber Co., W.B. Plaunt and Son, White and Plaunt, Marcotte-Murphy-Seguin, Leon Portelance, Noel Maurice, J.A. Lapalme, T. Pajala, Arthur Johnson, C.A. Fielding, Kormak Lumber Co., and Arnold and Bell situated their administrative headquarters in Sudbury.[9]

For this reason, this period could be considered the era of the lumberjack. Tim Sheehan was one of this hardy breed. A boss for the Morgan Lumber Company, he was mentioned in one account as being "a typical Uncle Sam with goatee, long boots, wide brimmed hat ... and was quite a conspicuous figure around the Hotels when he would drive in from his camps in Morgan Township, back of Chelmsford."[10] Sudbury was more strictly a logging town in the nineteenth century, as mining had not progressed to the point where certain metals were in demand and the industry's technology was advanced enough to excavate efficiently.

The arrival of the railways affected both the demand and supply of forest products. Construction and maintenance needs for the CPR mainline, its Sault Ste. Marie Branch line, the CPR and Canadian Northern Railway (CNoR) lines from Toronto to Sudbury and Capreol, the Manitoulin and North Shore Railway (M&NSR) from Sudbury to Little Current, and the extension of the CNoR northeast of Capreol created huge lumber demands for the erection of bridges and the laying down of axe-hewn ties. The CPR at that time did not hesitate to use its prerogative to gather timber from up to thirty-two kilometres on either side of the railway route; indeed, the Province of Ontario claimed that the railway company abused this right (granted to it under Section 19 of An Act Respecting the Canadian Pacific Railway passed in 1881) to such an extent that it stripped many regions of "enormous quantities" of timber.[11] While the railway companies generally acquired their wood supplies from independent sources, some, like the CNoR, established their own divisions for cutting purposes. In terms of the timber supply, railway construction itself opened up virgin forests and caused mills to appear along the lines of steel.

Within the present boundaries of Greater Sudbury, a variety of forestry-related industries emerged. First, there were operations linked to the growing needs of the mining industry. Second, the major rivers were used to drive logs to mills elsewhere or to export. Third, a number of mills and lumbering centres were created. Finally, there were the cutting operations associated with the clearing of agricultural land in the Valley and other rural pockets west and south of Sudbury.

The timber needs of the mining industry were substantial. The CCC had cutting rights in several townships, which it contracted to others to supply wood for its roast yard operations. Even after the roast yards were eliminated, the company's demand for wood remained high. Later, Inco became the single biggest purchaser of lumber in the area because of its need for ties, poles, posts, and lagging. As the *Inco Triangle* noted in 1944, "There are 100 lumber mills operating in the Sudbury District, and most of them sell a proportion of their output to Inco.... In other words, Inco takes 45% of their output."[12] According to one author's estimate, in the late 1940s Inco provided year-round work for 1 000 men in the bush.[13] Early in the 1950s, Inco acquired the largest logging company in the Sudbury District, George Gordon and Company, to ensure itself of an ongoing supply of timber. Another major producer, M.J. Poupore, supplied Falconbridge Nickel with much of its lumber requirements until the company and its timber rights were acquired by the latter in 1952.

For a period of time, the Vermilion, Wanapitei, and Whitefish Rivers were used to transport logs through the region to sorting areas on the North Shore of Lake Huron, and from there to large mills in Michigan and southern Georgian Bay. These log drives necessitated the construction of sluices to move lumber around some of the hydroelectric facilities erected on these rivers. The passage of legislation by the Province of Ontario in 1898 banning the export of pine timber was significant in that it encouraged the construction of new mills along Georgian Bay at places such as Blind River, Cutler, Byng Inlet, and Parry Sound. Mills were likewise erected in the Sudbury area. Green & Company, for instance, built a sawmill along the Vermilion River near the Town of Capreol. Others appeared at Larchwood, on Wanapitei Lake, Milnet on the Vermilion River, and on Whitefish Lake near Indian Reserve #6. The legacy of a sawmill built on the shores of Minnow Lake remains to this day because it dumped all of its sawdust from 1883 onward into the shallow lake (only three to four metres at its deepest), which slowly turned it into a marsh.[14]

Skead and Boland's Bay

In some locations, mills spurred the creation of small settlements. Situated on the east and west shores of Lake Wanapitei, the villages of Skead and Boland's Bay came into being as a result of logging.[15] The lake was favourably situated, and floating logs displaying the stamp hammer work of companies such as Victoria Harbour, Graves and Bigwood, Emery, and Emery and Graves Lumber were often visible. W.J. Bell, owner of the Spanish River Lumber Company, began milling operations at Skead in 1924. To attract workers, he brought in crews to construct boarding houses, general stores, a post office, and a dance hall. Two small pockets of settlement emerged at Skead and at Boland's Bay. When the Skead mill was closed in 1930, the village disappeared. After a fire destroyed the M.J. Poupore Company's mill in Gogama, the firm relocated to Skead, where it took over the site, opened a new sawmill in 1943, and rebuilt the village. While it only had a permanent winter population of no more than fifteen families, the village's regional setting nonetheless included a separate and public school and a Roman Catholic Church. The Skead mill ceased operations in 1956. On May 7, 1957, a devastating fire destroyed the nearby village of Boland's Bay. This fire, which was started by a line-cutting survey crew working for a mining company near Garson, "covered nearly 12 000 acres in a single afternoon to become one of the fastest-travelling and most dangerous conflagrations in the province's history." It attracted the attention of the international press and later became the subject of a book.[16] The fire, which threatened the local Canadian Forces radar base and the Sudbury Airport, drew

1 200 firefighters from places as far away as Kenora, Thunder Bay, and Sault Ste. Marie. Twenty homes were destroyed along with a lumber yard, and eighty-eight people were left homeless.[17] With the assistance of the Poupore Lumber and Falconbridge Nickel companies and the province, the community was rebuilt and it continued to flourish because of the growing importance of tourism on Lake Wanapitei, the rise of commercial fishing in 1961, and the availability of mining employment. It now has a population of some 600 and supports its own community centre.

Wahnapitae

The origins of the village of Wahnapitae can be traced to its strategic location at the juncture of the CPR mainline and the Wanapitei River. It quickly developed into a bustling village that, for a brief time, showed signs of competing with Sudbury for supremacy in the area.[18] The village's position was enhanced in 1888, when the Emery Lumber Company (later Holland and Emery, and then Holland and Graves) started to haul logs by means of a thirty-two-kilometre log train known as the Wahnapitae and Northwestern Railway that started near the intersection of the CPR mainline and the Wanapitei River and then wound northward to the company's headquarters in Garson and beyond to the site near the present-day Sudbury Airport. This railway continued until 1902, at which time it was sold to J.J. Gartshore. Since the lumber company was one of the largest in the district, with numerous bush operations in the Lake Wanapitei area, the village served as accommodation for woodsmen, sportsmen, and prospectors excited by the potential of a local gold boom.[19] Places such as the Queen's Hotel, Gauthier Hotel, and Commercial House became well known, and a number of retail stores sprouted.[20] Its geographical isolation and connection to Sudbury was improved by the construction of a winter road in 1891, and the completion of a highway between Sudbury and North Bay in 1912. While the decline of the lumbering industry in the early years of the twentieth century had a negative impact on the village, other developments, such as the construction of the Wahnapitae Power Company's hydroelectric plant nearby and the growth of Coniston after the First World War, helped to sustain the settlement. After the Second World War, the village became synonymous with Wahnapitae Lumber and Building Supplies and its well-known, community-minded owner, Ernie Checkeris. Its suburban location continued to attract so many new residents that a community centre was built in 1965. Today it exists as a quiet village situated on Highway 17 between Coniston and Hagar.

Forestry and Agriculture

Timber-cutting served as an integral part of early agricultural activity both in the Valley and the rural enclaves surrounding the Town of Sudbury. During the construction phase of the CPR mainline, French-Canadian labourers found the flat lands of the Valley more attractive than railway construction. Many opted to remain and pursue agriculture as an occupation. To this end, they were supported by the Roman Catholic clergy, who saw farming as the "noblest of professions."[21] In order to meet their obligations under Ontario's Free Grant and Homestead Act passed in 1868, striving farmers had to clear at least six hectares of property in order to receive a patent for their land. Thus their first cash crop consisted of trees, with sawlogs being the most important commodity. Some settlers became jobbers, or intermediate producers, who took contracts from lumber companies to cut and transport sawlogs or pulpwood. When agricultural lands were cleared, many farmers sought winter employment in cutting operations further away in Morgan, Norman, Hanmer, and Blezard Townships.[22] Logs from Morgan Township were especially valuable, as they were considered to be "some of the largest and of the best quality logs in the entire northland."[23]

In his study of French-Canadian farming in the former townships of Valley East, Gaudreau identified three phases of forest activity. The first involved the cutting of the best pines for sawlogs to be shipped to destinations outside of the region by lumber company employees; this phase happened prior to agriculture taking hold. The second centred on both settlers and local subcontractors cutting sawlogs and pulpwood for export by means of railway lines. The final phase took place after the pulpwood had disappeared, involved the production of various wood products from trees of lesser quality, and more co-operative approaches to harvesting by groups of farmers.[24]

Lumbering also provided a source of winter income for Finnish farmers situated in the Lorne Township, Long Lake, and Wanup areas.[25] This activity made it possible for Finns to acquire and keep agricultural properties that otherwise would have been unprofitable to operate. While many took on lumberjack roles at places such as Lake Panache, others served as drivers along the Vermilion and Spanish Rivers. Some farmers worked as part-time jobbers for companies such as the CPR, the Kalamazoo Vegetable and Parchment Company, the Spanish River Pulp and Paper Company, and worked in local camps run by the White and Plaunt Lumber Company. Others worked as private subcontractors responsible for setting up bush camps, hiring men, and delivering wood to the companies. It was not uncommon for local farmers to work in lumbering operations as far away as Biscotasing, Nemegos, and Kormak.[26] Finnish women also worked in the lumber camps; they were sought after as dishwashers, cooks, bakers, and kitchen staff.[27]

The cutting of saw logs remained significant for many rural townships up to the First World War. The Royal Ontario Nickel Commission in 1917, however, conceded that local forestry was "a disappearing business ... with only one crop of trees."[28] By the 1920s, only the cutting of pulpwood, the delivery of fuel wood for roast yards, and the harvesting of second growth timber were worthwhile. The remaining stands of timber did manage to provide some ongoing economic sustenance for certain urban centres. James McCreary operated a sawmill in Larchwood that had the capacity to produce lumber in excess of one million feet per annum. Local cutters likewise prospered from the opening in 1920 of the Canada Creosote Company located west of Sudbury. At this site, green railway ties and telegraph poles were filled with creosote oil until the plant's demise in 1963.[29] Despite the local decline in lumbering, Sudbury remained as a seasonal home and distribution point for lumber crews for the next few decades.

Agriculture and Rural Settlement

The origin of agricultural activity in the Sudbury area resulted largely from the three types of land acquisitions that were open to settlers. First there was the Free Grant and Homestead Act of 1868 that gave land grants to potential farmers. It was intended to counteract the American Homestead Act, which had also offered free land so long as the deed holder met certain conditions. Second, farmers had the option of purchasing Crown land directly. Finally, there were squatter's rights. While Free Grant Districts were created in Algoma and Nipissing, homestead townships were intended for those areas outside the settled south not particularly valuable for minerals or pine timber.[30] While free grants were given in certain areas, lands within the Temiscaming, Sturgeon Falls, Warren, Sudbury, Massey Station, and Dryden agencies were initially designated only as "for sale" lands, listed at the price of fifty cents per acre.[31] McKim was the first local township to be opened for such sales in 1884; this was followed, two years later, by the establishment of a Land Agent and a Land Agency known as Algoma District No. 5. By 1931, another twenty-one townships had been opened for sales. In addition, land was put up for sale in Lorne and Nairn Townships from 1889 to 1895, under the terms of the Railway Aid Act of 1889.

While free grants were given in the Sudbury area as early as 1898, they were not listed in the annual reports of the Commissioner of Crown Lands of the Province of Ontario until 1906. In this year, the first listing of eleven townships was made available under the free grants initiative. These grants gave sixty-five-hectare lots to settlers who fulfilled a five-year residency requirement, cleared six hectares, and constructed a house at least thirty square metres in size.[32] Free areas by 1912 included the townships of Balfour, Blezard, Broder, Capreol, Garson, Hanmer, Neelon, Rayside, Dill, and parts of Morgan and Lumsden Townships. The Sudbury lands were late in being opened up for free grants due to a belated effort to prevent undue speculation and monopolization of large landholdings.[33] Some Crown lands were even withdrawn from public sale from 1915 to 1921 in townships west and east of the mining operations to minimize sulphur damage from the O'Donnell and Coniston roast yards, and to reduce payments by mining companies for agricultural losses.[34] Later, other lots were granted under the Returned Soldier Settlement Act of 1917, set up to assist servicemen coming back from the Great War in acquiring agricultural land, and later, in the 1930s, under various relief programs.

A recurring problem in the Sudbury area was the existence of "squatters" on lands not sold by the Crown or released as free grants. According to the Ontario Director of Colonization in 1900, it was "a difficult matter to arrive at the actual number of settlers who located on land ... for the reason that in some sections, considerable settlement has taken place where the land has not been formally opened for settlement."[35] While this was common in places such as the Valley and Lorne Township, it was most apparent in the Wanup area. Squatting increased considerably during the Great Depression; according to one account in *The Sudbury Star* in 1932, "40 Finnish families have squatted in Dryden and Cleland Townships, and displaying the spirit of the early pioneers, are hewing out homes for themselves rather than to accept relief."[36] The squatters took comfort in the fact that the Ontario government had never evicted a squatter from a farm for illegally occupying the land. As the Supervisor of Settlement stated, "all credit to these squatters. They have no legal right to the land they are on, but it is doubtful if the government will put them off."[37] So they remained.

The key factor in the location of agricultural activity was the spread of railway operations. The first and most attractive area was the flat and fertile soil of the Valley that drew the French Canadians from the Ottawa Valley who had arrived laying track for the CPR in 1883. Their success drew others into the area. To the west, railway construction along the Algoma branch of the CPR attracted Finns in Snider, Waters, Denison, and Drury Townships. Agriculture south of the Town of Sudbury developed later as a result of the construction of the CPR and CNoR railway lines in 1908. The cutting of timber in adjacent areas spread the range of agricultural activity into these pioneering zones. Another factor shaping farming was the influence of promotional literature by the CPR, the Province of Ontario, and the Algoma Land and Colonization Company. Pamphlets extolled the North Shore and land around Sudbury as "wonderfully productive" areas.[38] In these areas, production started on a subsistence basis and then expanded into commercial agriculture.

By the late 1880s the Valley's agricultural potential had attracted fifty-seven families, totalling 400 persons. According to the 1891 Census of Canada, agriculture in the Valley was devoted mainly to oats (largely for horse feed), hay, peas, potatoes, and turnips. With the aid of colonization roads, agriculture spread from Rayside and Balfour into Blezard and Hanmer Townships. The same census indicated that agriculture was also important in the fringe areas surrounding Sudbury and Copper Cliff.[39] For a short period of time, the core townships of McKim, Neelon, and Waters served as the largest farming community in the region, with more than 1 618 hectares of improved land and some 100 farms. Agriculture in these areas, however, suffered from the sulphur smoke that emanated from the roast yards, which damaged crops as early as 1888.[40]

The forces that drove agricultural settlement in the Valley were active west of Sudbury in the 1890s between Whitefish and Worthington, and after the turn of the century in places such as Lorne and Louise Townships. By the First World War, land was being opened south of Sudbury in places such as Broder, Dill, and Cleland Townships. The 1911 Census is revealing in that it contrasted the ongoing strength of the rural sector in the Shield districts compared to the Valley. While approximately 60 per cent of the total number of farms and total acreage could be found in the Valley, the remaining 40 per cent were situated in the more rugged townships found west, south, and east of Sudbury. There were specialization differences between the Valley and the outlying farms, as the former relied more on field crops, whereas the latter focused mainly on pasture and livestock holdings.[41] The strength of the agricultural sector at the time was such that it prompted the government of Ontario to appoint W.H. Ross as its first District Agricultural Representative in 1912. This office played a major role in encouraging better agricultural practices and the use of superior seed and livestock. Around the First World War, there were two visible trends in agriculture: farmers were planting more potatoes because of their resistance to sulphur damage, and they were producing more poultry and dairy products due to the growing local market.

The establishment of the Northern Development Branch (NDB) within the Department of Lands, Forests and Mines was an important step in advancing the spread of agriculture. Created to administer the Northern Ontario Development Act of 1912, it provided funds to construct roads and bridges, advance settlement, and assist settlers. The NDB in 1926 became the Department of Northern Development (DND).[42] In the Sudbury area, new roads were extended to the fringes of settlement; busier routes like the Burwash Road were upgraded to Trunk Road status. The branch also provided direct agricultural aid to farmers, including low-interest loans, quality seed and pesticides, and emergency supplies in years of crop failure. This assistance was complemented by the Ontario Department of Agriculture's promotion of better livestock, seed, and farming techniques. At the federal level, the Agricultural Aid Act and later legislation helped to fund agricultural representatives who advanced modern agricultural techniques. With these types of provincial and federal assistance, agriculture grew to become a multi-million-dollar economy by the early 1920s.

A broad overview of the development of agriculture in the area from 1911 to 1971 is shown in table 9.1, which illustrates the long-term trends regarding the number of farms in the area. For comparative purposes, similar data for the entire Sudbury District is provided as well. In the Sudbury area, the number of farms grew until 1921. Even if missing data for Broder, Dill, Cleland, and Capreol is factored in, it appears that farming experienced decline in the 1920s. With the exception of places such as Dryden and Cleland Townships that were impacted by the Relief Land Settlement Scheme, there is little evidence of any major local back-to-the-land movement during the 1930s. Nonetheless, many urban residents did go to the farms of families, friends, or acquaintances for temporary employment and/or accommodation. Beginning in 1941, most rural areas, especially those in McKim Township, began to experience permanent decline. Only Broder, Dill, and Cleland Townships resisted the overall trend. The farming areas most greatly affected by decline during the 1950s and 1960s were in the immediate vicinity of the City of Sudbury, the Town of Copper Cliff, or in Valley townships undergoing subdivision because of the expansion of the mining industry. By 1971, only two townships showed future promise for major agricultural activity: Rayside and Balfour. When compared to the Sudbury District as a whole, it is clear that the relative importance of local agriculture in the area declined substantially. Whereas in 1921 the area supported more than 40 per cent of the district's farms, by 1971 this figure had dropped to around 25 per cent.

French-Canadian Settlement in the Valley

The rural economy fostered the rise of many French-Canadian settlements in the Valley. These featured large, contiguous "hay and oats" fields, and buildings erected alongside straight and intersecting township roads. Clusters of these buildings often coalesced into hamlets, such as Azilda, Boninville, Brunetville, and Blezard Valley, or into larger crossroad villages such as Chelmsford and Hanmer. The dairy farms in these areas produced milk not only for companies such as Crown, Levack, Modern, Standard, and Palm but also for up to a dozen cheese and butter factories.[43] The establishment of a farmer's market on Borgia Street in Sudbury in 1914 was significant as it provided the Valley farmers with a weekly venue for the sale of meat and root crops. This important source of revenue continued until the 1950s. While production methods left much to be desired during the 1920s, the introduction of new grading techniques for potatoes, the establishment of model farms in Blezard and Rayside Townships, and the growth of a four-year rotational cycle of land use for cereal production, hay, and natural pasturage improved efficiency considerably. The quality of seed potatoes, especially from Morgan Township, gained national fame. After the Second World War, the introduction of electricity, growing use of the tractor, and good sandy soil conditions for the production of sulphur-resistant potatoes and strawberries heralded some promise for those farms that survived the transition years of the 1950s.[44]

Town of Chelmsford

Of the rural settlements that arose in the Valley, two rose to prominence: the Town of Chelmsford and the Township of Hanmer. A third, Azilda, had some early impact, but its influence was gradually overshadowed by that of Chelmsford, situated ten kilometres to the west.[45] Strategically situated in the heart of the Valley and alongside the CPR main-line, Chelmsford quickly became a focal point for lumbermen, farmers, merchants, and Catholic missionaries. Its origin can be traced back to 1883, when a sawmill was erected alongside the Whitson River to produce wood products for the CPR. At that time, some twenty families resided in the area under the provisions of the Free Grants and Homestead Act. Forestry operations and the new sawmills in the vicinity attracted more people. By 1889, a two-storey school and Jesuit chapel had become part of the local scene. In the following year, the CPR erected a railway station and the village became part of the incorporated Township of Balfour. A Jesuit presbytery was also constructed. In 1906, the school/chapel building was expanded to serve as a convent that would be home to three Grey Sisters of the Cross from Ottawa who provided French-language instruction. The building was structurally unique, designed in the Queen Anne Revival Style featuring a hip roof, ornate wood cornices, and symmetrical window spaces.[46]

Until the 1950s, the history of Chelmsford and that of the Roman Catholic Church was inseparable. Residents faithfully followed the dictate of Father Lionel Séguin: "On ne pourrait que bien difficilement s'imaginer comment un groupe de Canadiens-français songerait à s'établir, avec la pensée de survivre, sans son église, son presbytère, son curé

TABLE 9.1 Number of farms in the Sudbury Area, 1921–1971 (Selected Townships Only)

Townships (S)	Year								
	1911	1921	1931	1941	1951	1956	1961	1966	1971
Balfour	106[a]	135	108	101	98	44	34	40	28
Blezard		64	70	78	54	48	14	10	n.a.
Broder, Dill & Cleland	87	130	n.a.	80	171[g]	74	72	n.a.	n.a.
Capreol	82[b]	73	n.a.	62	20	10	14	14	n.a.
Dowling	38	48	32	28	33	20	20	16	11
Drury, Denison & Graham	27[c]	53	53	58	34	31	17	10	n.a.
Fairbank	n.a.	13	n.a.	n.a.	n.a.	n.a.	n.a.	11	16
Hanmer	173[d]	111	87	99	85	83	40	28	n.a.
McKim	41	24	22	17	3	2	n.a.	n.a.	n.a
Neelon & Garson	102	94	98	49	52	4	8	6	n.a.
Rayside	150[e]	151	130	153	122	148	77	59	57
Waters	45[f]	71	80	73	39	25	15	12	n.a.
Total (Sudbury Area)	851	967	681	796	711	489	311	206	112
Total (Sudbury District)		2 267	2 148	2 045	1 634	1 402	841	658	416
% of Total Sudbury Area vs. Sudbury District		42.6	31.7	38.9	43.5	34.8	36.9	31.3	26.9

Source: Statistics Canada, *Censuses of Canada (1911–1971)*.
Notes: [a] Includes Morgan Township
[b] Includes Norman and Rathbun Townships
[c] Includes Trill Township
[d] Includes Blezard Township
[e] Includes Lumsden Township
[f] Includes Creighton and Snider Townships
[g] Includes Hawley Township

et ses écoles" [one cannot imagine without difficulty how a group of French Canadians would think of settling in one area with the thought of survival without their church, their own presbytery, their own priest, and their own schools].[47] Following creation of St. Joseph's parish in 1896, Father Stéphane Côté arrived on the scene and was the dominant figure in the community until 1941. His legacies included attracting the Grey Sisters from Ottawa to serve as local teachers, and spearheading the completion in 1913 of the most outstanding monument of worship in the Valley, the Church of Saint-Joseph. A sense of the deep religious character of Chelmsford is conveyed by the fact that it provided the Roman Catholic Church with six priests, two brothers, one seminarian, and seventeen sisters between 1906 and 1945.[48]

Father Côté's arrival in 1906 likewise strengthened Chelmsford's regional position. He was influential in promoting agriculture, which he considered to be the noblest profession in the world and a natural symbiosis to the Roman Catholic faith. He was also a tireless promoter of the need for improved transportation links between the community and Sudbury. By 1910, Chelmsford had a population in excess of five hundred, a figure sufficient to allow it to become an incorporated town separate from the Township of Balfour. The majority of the residents were from the Ottawa Valley, with strong kinship ties to others in the town. As one citizen commented, the community "consisted of one huge family."[49] By the First World War a lively downtown had emerged, offering a variety of services for surrounding farmers. The town also supported a cheese factory and cheese co-operative. As aerial photograph 9.1 shows, agriculture in the area remained important until the end of the Second World War, and the town continued to be clearly demarcated from its rural surroundings.

In the 1920s, the Chelmsford area became a centre of mining, and a rash of staking activity took place. Nine mining companies came into being, providing new employment

AERIAL PHOTOGRAPH 9.1   The Chelmsford Area, 1946. Source: Ministry of Natural Resources Records, Archives of Ontario (RG 1-429). Collected by the Ontario Department of Lands & Forests (1946).

opportunities. Ore was originally discovered by James Stobie on the banks of the Whitson River in 1897; the deposit was sold to Orford Copper Mines, but never fully developed. It remained until 1922 before interest in mining was again spurred by the discovery of coal in nearby Larchwood; this enthusiasm disappeared when it was discovered that the coal would not burn well. In 1925, the Treadwell-Yukon Mining Company began to mine the copper, lead, and zinc deposits associated with Errington Mine situated in Balfour Township. The company constructed a small village at the minesite that lasted until 1932, at which time operations were suspended due to extraction problems. The site then became associated with other companies, such as Ontario Pyrites (later Giant Yellowknife Mines), Sudbury Basin Mines, and Consolidated Sudbury Basin Mines. Interest in mining was so great at the time that *The Sudbury Star* was spurred to comment, "Seldom has such an array of important mining interests been found in any camp in Northern Ontario."[50] This optimism proved to be short-lived. While the Errington Mine did resume production in 1952, it lasted only five years. Further north, Sudbury Offsets (later Nickel Offsets) worked sporadically from 1926 until 1957, mining nickel and ore deposits situated in Foy and Bowell Townships.[51]

By 1951, Chelmsford had developed into a thriving town of more than 1 000 residents with its own water and electrical system. The majority of its residents were now employed by Inco. In response to this growth, modern amenities were incorporated into the settlement, such as new sewer lines, a secondary school, a new town hall, and a Mine Mill Local 598 centre. The introduction of husky racing in 1951 brought widespread attention to the community. To accommodate its growth, Chelmsford annexed part of the Township of Balfour in 1953; later the Township of Balfour annexed the Town of Chelmsford in 1968, along with the Townships of Creighton and Morgan. In 1972, the Township of Balfour amalgamated with the Township of Rayside in turn to form the Town of Rayside-Balfour within the Regional Municipality of Sudbury. Another consequence of amalgamation was that Chelmsford and Azilda, its largely French-speaking neighbour with a population of almost 4 000, were brought under one political body. Traditionally in the shadow of Chelmsford, Azilda developed rapidly after 1948 following the approval of its first farm subdivision. This event opened the floodgates to settlement, and can be considered as the real start of the village of Azilda.[52] The selection of the community as the site of the Rayside Secondary School in 1971 was another factor that contributed to its growth.

Starting in the 1970s, Chelmsford became a residential exurb of the City of Sudbury with its own malls, numerous retail outlets, and amenities close by, such as the Chelmsford, Colonial, and Forest Ridge golf courses. In 1974, Sudbury Downs located nearby hosted its first live harness racing program. In 1999, the facility became part of the Slots at Racetracks program sponsored by the Ontario Lottery and Gaming Corporation. Throughout the years, Sudbury Downs has provided direct or indirect employment for 250–300 people, spurred the creation of many commercial stable and training facilities, and provided significant tax revenues for Greater Sudbury.[53] The proposed cancellation of the slots program by the province in 2013, however, will undoubtedly have a negative effect not only on Chelmsford but also on other rural parts in the Valley.[54]

Township/Village of Hanmer

While the early history of the Township of Hanmer was intimately intertwined with the construction of the CPR railway line, lumbering, and agriculture, it differed from Chelmsford in that major growth did not occur until the late 1950s. Following the passage of the CPR through Azilda and Chelmsford, French-Canadian settlers spread eastward in the Valley, reaching Hanmer in 1898. The population soon grew to almost 200, and in 1904 the Township of Hanmer was incorporated.[55] Prior to the First World War, the area had acquired an almost total French-speaking population (fig. 9.1). Within the township, a small village also known as Hanmer began to emerge. Featuring a hotel, post office, blacksmith shop, and elementary school, it served as the nucleus for what was to become the dominant French-speaking rural settlement in the eastern part of the Valley. As was the case with Chelmsford, the religious element was always present. In 1905, the Hanmer area was constituted as a separate parish from Blezard, and a small church erected. While plans were made in 1916 for the construction of a major church to rival that situated in Chelmsford, this dream did not materialize until 1960. The CNoR line was completed in 1908, running northward from Sudbury through Hanmer Township to ship iron ore from Moose Mountain Iron Mine to Key Harbour. Shortly thereafter, the same line was used for regular passenger traffic from North Bay to Capreol. As a result of the new rail line, the community experienced a growth spurt. In 1942, the Grey Nuns of the Cross arrived on the scene to serve as teachers. A convent was then erected which complemented the parsonage which had previously been built.

In the years preceding urban settlement, both mining and forestry existed. While there was mining to the southwest associated with the Blezard and Cameron Mines owned by the Dominion Mineral Company and those of the Sheppard, Mount Nickel, and Little Stobie Mines, this activity had little long-term impact. Lumbering was more important. Companies such as Loveland and Stone from Saginaw and Bay City, Michigan, and lumber barons including R.B. Booth and A.B. Gordon from Pembroke, Ontario, became active in harvesting red and white pine, tamarack, and spruce trees throughout the township.[56] The clearing of the forests led to the pursuit of agriculture, as an estimated 80 per cent of the cleared township consisted of "very good" agricultural land.[57] Over time, a prosperous dairy industry flourished, and the opening of the Farmer's Market in Sudbury encouraged vegetable production as well. Some 60 per cent of the vegetables sold at this market were said to have come from the eastern part of the Valley.[58] While there was hope that the Manitoba and Ontario Flour Mill Company's flour mill and grain silos constructed in Sudbury in 1911 would spur grain output, this did not prove to be the case. Soil conditions favoured the cultivation of potatoes; indeed, Théodore Despatie, a farmer from Hanmer, was named World Champion Potato Grower in 1949. Not to be outdone, his two sons earned first and second places in this championship in 1950. The 1950s, unfortunately, were not favourable to local agriculture due to competition from the Maritime Provinces and the shift of regional milk production eastward toward Verner and Warren. More recently, the decline in the number of farms in the Hanmer area has been partially

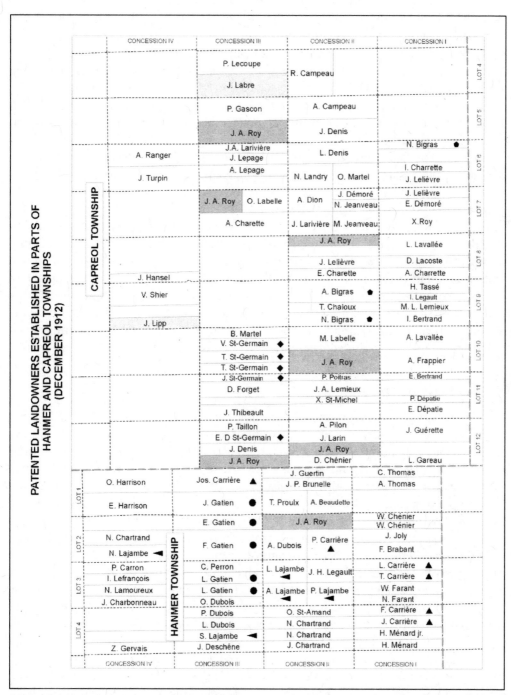

FIGURE 9.1 Land ownership in parts of Hanmer and Capreol Townships (1912). From Léo Larivière, Cartographer and Map Librarian, Laurentian University.

offset by the appearance of larger commercial enterprises in nearby Blezard Valley, such as Martin's Potatoes and Garden Centre/Valley Growers Inc. As Lebelle makes clear, the transformation of Hanmer from a sleepy rural village to a suburban bedroom community was a phenomenally quick process: "Assessment rolls for Hanmer reveal an interesting story in shift of occupations. In 1945, 3.6% of the heads of households worked in mining while 15 years later, it had gone up to 52.5%. Farming also underwent a dramatic shift, from 82.1% of household heads being involved in agriculture in 1945, to only 4.8% in 1960, dropping further still to 2.2% of the population in 1964."[59]

Unlike the sprawl of Val Thérèse, Val Caron, and McCrea Heights, the village of Hanmer was fortunate in that it managed to retain a more compact character. This advantage notwithstanding, Hanmer's growth was hampered by the fact that part of the settlement had spilled over into Capreol Township. This situation was not resolved until 1967, at which time the Townships of Hanmer and Capreol were amalgamated into one political entity; two years later, the municipality was enlarged to encompass Blezard Township and part of Capreol Township to form the Township of Valley East. In 1973, the township became a town within the Regional Municipality of Sudbury. With the completion of L'École Secondaire Hanmer in 1970, the inauguration of the Centennial Arena in 1971, the construction of the Howard Armstrong Recreational Centre in 1983, and a medical centre, shopping plaza, and the Bishop Alexander Carter Catholic Secondary School, Hanmer gradually evolved into a self-assured community that has continued to attract new residents.

Finnish Rural Enclaves

The Finnish enclaves found in the Precambrian Shield differed substantially from Chelmsford and Hanmer, as their setting featured a topography dominated by rocky hills, dispersed lowlands, numerous lakes, rivers and swamps, and forested areas. Geographically, these enclaves could be found in a broad west-to-east swath of territory that included Lorne Township (Beaver Lake), Louise Township (Whitefish), Waters Township, Broder Township (Long Lake) and Dill, Cleland, Secord, and Burwash Townships (Wanup). The farms in these areas featured houses built with dove-tailed construction techniques, saunas, numerous outbuildings, and mixed farming methods that included the use of rye as a food grain, offering a strong visual contrast to those farms found in the Valley. Dairy farming served as the dominant form of production. This milk production, in turn, supported the creation of urban-based dairies such as Copper Cliff Dairy, City Dairy, the Sudbury Producers and Consumers Co-operative, and one linked with the Anderson farm. The need for costly equipment and a shift to mining employment, however, brought dairying to an end by the turn of the 1950s. Today, only recreational forms of farming and tree farms remain as vestiges of this agricultural past. Another contrast with the Valley settlements was the absence of the church and its replacement by left-wing ideologies and "body and soul" features such as the "hall."[60] The wide array of cultural activities involving theatre and sporting activities was also different. Whereas sporting organizations were in existence in virtually every Finnish enclave from the outset, it remained until 1949 before references were made regarding any form of athletic association in the Township of Hanmer.[61] Finns constituted the earliest ethnic group in most of these settlements, but such was not the case in the Long Lake area. The first homesteaders here were mainly French-speaking, and some of Ukrainian descent. The French presence was acknowledged by the existence of the Rheault Post Office.[62]

Beaver Lake

A typical Finnish enclave was the community of Beaver Lake associated with Lorne Township (fig. 9.2). Permanent settlement began here in 1912, and by 1921 the area supported 279 Finns, a greater population than that found in either Sudbury or Copper Cliff.[63] The community left no doubt as to its cultural identity: there were Finnish road names, local structures covered with home-made red iron paint, and meetings of the local school board and statute labour board were conducted in Finnish. A dispersed agricultural economy developed that lasted until the 1940s, based on beef and dairy farming and various forms of co-operative endeavour.

Farming, though, never developed on a full-time basis, due in part to the nature of the terrain, soil limitations, and small farm sizes (aerial photograph 9.2). After the Second World War, a shift to mining employment and costly provincial regulations brought an end to dairy farming. Fortunately, the regional location of Beaver Lake proved to be favourable, as it was in the heart of an economic zone that offered diverse employment prospects related to the construction of hydroelectric generation stations, mining, lumbering, the trapping of beaver and mink, general construction, carpentry, local road construction, and the provision of local goods and services. A variety of enterprises surfaced that made the enclave virtually self-sufficient. Due to its isolation, the close proximity of the farmsteads to one another, and the nature of the road patterns, the community exhibited a high degree of unity and cultural cohesion. This cohesion was enhanced by a presence of a strong core consisting of a hall, school, sports club and track and field facility, and a co-operative store. The Finnish character of Beaver Lake remained intact until a new highway was constructed through the area in 1951–53. This transportation link drew Beaver Lake into Sudbury's exurban web, attracting residents of differing ethnic backgrounds, and diminished the existing organizational structure. A vestige of the ethnic history of Beaver Lake still remains in the form of the Beaver Lake Recreational Hall and a local historical plaque erected in 1989.

Whitefish Lake and Wahnapitae Indian Reserves

The schedule accompanying the Robinson-Huron Treaty No. 61 signed in 1850 provided for Chief Shawenakishick and his band a tract of land "now occupied by them and contained between rivers called White Fish River and Wanabitasebe, seven miles inland," to be known as Whitefish Lake Indian Reserve #6. Another smaller reserve, the Wahnapitae Indian Reserve #11, was established at Post Creek where it entered Lake Wanapitei.[64]

While an Order-in-Council dated July 14, 1851, called for the survey of all the North Shore reserves, this task was not completed until much later. The Whitefish Lake Indian Reserve, according to the survey in 1884, was 17 704.5 hectares in size (fig. 9.3).[65] The reserve reflects a typical Shield topography of low, wooded hills and a multitude of lakes and streams well suited to the traditional hunting and gathering practices of the Ojibway.

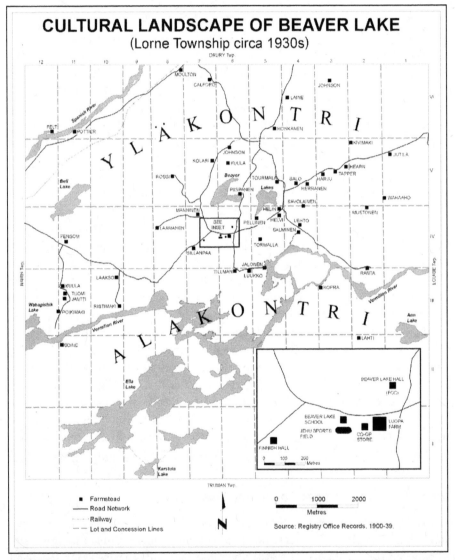

FIGURE 9.2 Beaver Lake (Lorne Township) cultural landscape, ca. 1930s. From Saarinen and Tapper, "The Beaver Lake Finnish-Canadian Community," 180. *Yläkontri* and *Alakontri* are Finnish terms referring to the northern and southern parts of Lorne Townships respectively.

Little of the terrain was suitable for agriculture. The principal village for the band was once on Lake Panache, but it moved in the late 1700s to Whitefish Lake at a site now known as Second Beach. It was located on the portage from Whitefish Lake to Black Lake. Here, a settlement of about thirty-five log cabins and wigwams was established. Extensions of the village also developed on Black and Vermilion Lakes. The community later supported a Roman Catholic church and school in 1916. When the Brunne Lumber Company started its operations on Lake Panache in the 1930s, the residents gradually moved there from the Whitefish Lake village. By 1942, the Whitefish Lake village had been abandoned and a new school built on the shore of Lake Panache. When the lumber company ceased operations a few years later, the villagers moved to the present site near Simon Lake that formerly was part of the Papequish farm. This move was complete by 1952.

Life in the village had an annual rhythm dictated by the seasons. Agricultural and economic pursuits revolved around gardening, hunting, trapping, fishing, guiding, and the sale of cranberries, maple syrup, woven baskets, paddles, and blueberries. Purchases

AERIAL PHOTOGRAPH 9.2   The Beaver Lake Area, 1946. *Source*: Ministry of Natural Resources Records, Archives of Ontario (RG 1-429). Collected by the Ontario Department of Lands & Forests, 1946.

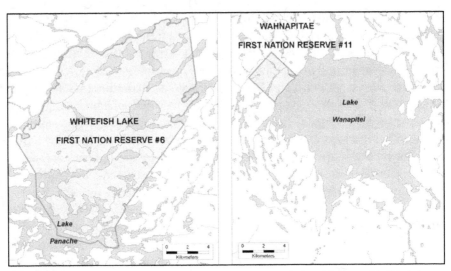

FIGURE 9.3 First Nation Reserves in the Sudbury Area. Adapted from Higgins, *Whitefish Lake Ojibway Memories*, xx, and Léo Larivière, Cartographer and Map Librarian, Laurentian University.

were made from the Hudson's Bay post until 1887, and thereafter from Sudbury or the Gemmell stores located at Whitefish and on Lake Panache. Lumbering began following the construction of the Algoma Eastern Railway from Sudbury to Algoma Mills in 1883. Disagreements ensued between the band and the railway company regarding the cutting of trees on reserve property. The reserve at the time functioned as a conduit for the movement of supplies and lumber by the Victoria Harbour Lumber Company via a 6.5-kilometre tote road. This operation continued into the 1920s. It was succeeded by the Brunne Lumber Company, which provided work for a number of the band members until the operation went bankrupt after the Second World War. Beginning in 1952, the reserve entered into an agreement with the Fielding Lumber Company for the cutting of ties. A number of band members also cut logs, pulpwood, and firewood on a smaller scale for sale to mills, contractors, and individuals. In 2000, the registered population of the reserve was 767, of whom 453 resided offsite.[66] The 314 residents on the reserve were housed in 99 units.

The Wahnapitae Reserve #11 on Lake Wanapitei comprises 1 036 hectares. Largely abandoned after 1850, the site started to revitalize in 1968, when former chief Norman Recollet moved back to the reserve. It now has a registered population of about 450, of whom 370 live offsite. Including non-members, the reserve has a population of some 100. Future plans for the site include the construction of a $4.5 million multipurpose facility called the Centre of Excellence that will include a meeting hall, medical clinic, a sustainable development department, and business incubator services.[67]

While these federally based reserves were never part of any provincial/municipal structure, their closeness to Sudbury fostered ongoing Aboriginal interaction with local centres. This interaction was originally through the fur trade, and then later as Natives became consumers of local products and services. More recently, migration from other reserves in Northern Ontario has expanded Aboriginal demand for housing, education, social, and health services in Greater Sudbury.

Burwash Industrial Farm

While situated south of Greater Sudbury, the Burwash Industrial Farm nevertheless had considerable impact on the region, especially after the road link to Sudbury was completed in 1933. Its creation was a unique and daring experiment in the history of prisons in Ontario because it was considered a jail "without bars or cells."[68] While this statement is not entirely true, the experiment was noteworthy and reflected the provincial trend of penal reform in the 1930s. It was thought at the time that a change from the negative attractions of cities to the natural atmosphere of the forest and land would serve as a reformative influence.

Burwash evolved to become one of the largest reform institutions in twentieth-century Ontario. The reasons for locating the industrial farm in the Sudbury area included the closing of the Ontario Reformatory at Guelph, the fact that one-third of prisoners at Guelph were from Northern Ontario, and the importance of Sudbury's strategic railway setting, which gave it access that other potential sites in the North lacked. The fact that there was no major highway in the area minimized the likelihood of easy escape from the institution. The selection of a site in Burwash, Laura, and Servos Townships was based on the presence of a broad valley some eleven kilometres long and five kilometres wide situated between the CPR and CNR lines that ran from Toronto to Sudbury (fig. 9.4). Following the site's selection, an Order-in-Council was passed in 1914 establishing the Burwash Industrial Farm. By the end of 1915, the provincial government had acquired a total area of 14 164 hectares for this purpose. Over time, the farm leased another 40 469 hectares of land to accommodate inmates and prison staff families. Construction of the first building was completed in 1916, and inmates arrived shortly thereafter.

Between 1916 and the demise of the farm in 1975, the facility grew to house three permanent campsites, several temporary ones, and a town for prison staff and their families with a population of 600 to 1 000 people.[69] The farm accommodated between 180 and 820 offenders with sentences ranging from three months to two years less a day. The self-sustaining minimum and medium security facility allowed inmates to grow their own food, tend crops, develop agricultural skills, and provide food to other provincial institutions. The institution also ran a lumbering operation, including a mill, and even had a tailor shop that provided clothing for prisoners and shirts for prison officers. It boasted its own twenty-bed hospital, assembly hall, newspaper, and entertainment and sports facilities.

The economic influence of the farm was substantial, since for a long period of time it was the Sudbury area's fourth largest employer; it likewise influenced the growth of nearby settlements. As one resident stated, "Estaire is made out of Burwash."[70] Due to changes in correctional practices in the early 1970s and the fact that 90 per cent of the prison inmates now came from Southern Ontario, making it difficult for their families to visit, the facility (then known as the Burwash Correctional Centre) was closed, and the townsite all but disappeared. Its future became a hot political issue. Plans for the federal government to use part of the property as a maximum security prison came to naught in 1978, as did the province's plans to use it as a provincial park or rehabilitation centre for the Workmen's Compensation Board. Other unsuccessful proposals included the transfer of the property to the Regional Municipality of Sudbury, the creation of a retraining school for Aboriginals or welfare recipients, and the leasing of part of the land for use by the ill-fated Sudbury Angora Co-op. The failure of these initiatives resulted in the buildings being torn down in the mid-1980s, and the property was transferred to the Department of National Defence. Little at the site remains today to reflect the farm's history other than a provincial plaque erected by Ontario Heritage Trust.

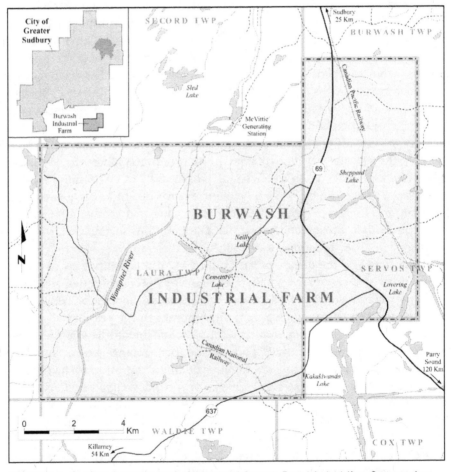

FIGURE 9.4 Burwash Industrial Farm (Correctional Centre). From Léo Larivière, Cartographer and Map Librarian, Laurentian University.

CHAPTER 10

From Falconbridge Nickel and Inco to Xstrata Nickel and Vale Canada (1928–2012)

The years between the Great Depression and the early 2000s brought dramatic changes to the Sudbury mining industry. While there were a few economic bumps along the way, the long-term trends for both mineral production and mining employment were positive until the early 1970s. These trends, in turn, encouraged an increase in Sudbury's regional population, one that saw the city acquiring a skewed metropolitan status. Underlying these developments was the territorial control that the two mining giants had over the mining lands in the Sudbury area. "King" Inco dominated production well into the 1950s; by the end of that decade, however, Falconbridge Nickel Company also emerged, with strong financial backing from the United States government, as a multinational corporation capable of being a competitor. Throughout this period, the industry underwent major technological changes and, from time to time, experienced labour conflict with local unions. While the picture at the turn of the 1970s appeared to be rosy, hidden factors emerged that brought the boom years to a sudden and dramatic halt. By the end of the twentieth century, both Falconbridge Limited and Inco were forced by global events to shift into survival mode; despite major internal transformations and several attempts at mergers, the two companies eventually found themselves at the mercy of forces they were unable to control. The end result was the takeover of Falconbridge Limited by the Swiss mining group Xstrata Nickel and that of Inco by the Brazilian mining company CVRD. The events that shaped mining in the Sudbury area around the turn of the twenty-first century were unparalleled in the annals of Canada's mining history.

In the ensuing period, another interesting trend emerged involving the rise of smaller mining companies, some with the intention of acquiring mining properties and others seeking to become partners with the two existing giants. In 2012, the mining-based economy of the area acquired a new dimension following Cliffs Natural Resources decision to locate its ferrochrome smelter north of Capreol.

Mining Land Ownership

The ownership of land along with its mineral and surface rights provided the basis for the creation and growth of the mining industry. The principal objective of mining companies in the Sudbury area was therefore to acquire as many mining properties as they could to ensure the long-term viability of their operations. In 1913, Ontario amended its Public Lands Act to ensure that any previous title granted by the Crown would automatically include the right to both the property's surface and its underground mineral deposits. Mineral rights granted by the Crown thereafter were dependent on how the title was worded. After 1913, mining companies in the Sudbury area made every attempt to assure that their properties retained maximum rights. While there was considerable shifting of mining land ownership from one company to another over time, the overall pattern conformed to the oval shape of the Sudbury Igneous Complex. The pattern of ownership during the 1960s (illustrated in fig. 10.1) is similar to that which existed for much of the twentieth century. This important map reveals that the mining companies

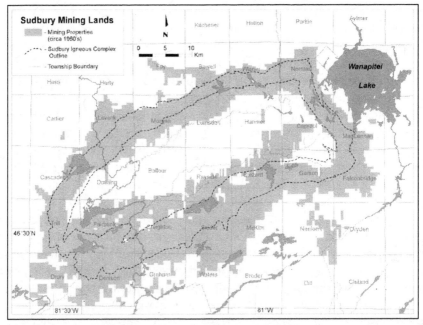

FIGURE 10.1 Mining Lands in the Sudbury Area, ca. 1960s. From Sawchuk & Peach Associates, *Nickel Basin Planning Study* (Toronto, ON: Department of Municipal Affairs, 1967), map following page 46.

at the time owned some 132 272 hectares of land around Sudbury. In geographical terms, it is perhaps more meaningful to describe it as being equivalent to the size of fourteen townships.[1] Since this land has been retained mainly for mining purposes, the result is that almost 40 per cent of the territory of Greater Sudbury has been excluded from other development purposes. To complicate matters further, the spatial configuration of mining ownership resulted in a huge "divide" between the settled parts in the south along Highway 17 from those situated in the Valley, giving rise to a population and municipal setting unique in Canada.

Falconbridge Nickel Mines Limited (1928–1970s)

The origin of Falconbridge Nickel Mines extends back to 1901, at which time the world-famous inventor Thomas A. Edison opened an office in Sudbury while exploring for nickel. He needed it to produce an alkaline storage battery cell that would power a car for Henry Ford. Following a survey along the Eastern Range, he secured promising properties in Falconbridge Township and began sinking a shaft in 1902–3. Upon encountering quicksand, he discontinued his work, not realizing that he had stopped just short of reaching a deposit of nickel-copper ore. The property reverted back to the Crown in 1915. In the following year, the Minneapolis and Michigan Development Company acquired the property and then turned it over to the E.J. Longyear Company for development. No buyer could be found for the deposit, and these claims remained idle from 1918 to 1926, when they were sold to Falconbridge Mines Limited. The property was subsequently resold to Falconbridge Nickel Mines, a new enterprise created on August 28, 1928, jointly controlled by Ventures Limited and Sudbury Basin Mines Limited.[2] The transaction included the mineral rights to 3 221 hectares.[3] At the time, the $2.5 million price tag was the highest ever paid for a mine in the Sudbury area.[4]

Led by Thayer Lindsley, a geologist, mine finder, and promoter, a mine and smelter complex was completed by 1930. By this time, refining technologies had advanced to the stage where the process of open air roasting for ore was no longer required. In 1933, a concentrator and sintering plant (for making nickel powder) had been added and the smelter expanded. Meanwhile, in 1929 the company purchased the Nikkelverk refinery in Kristiansand, Norway, along with the rights to its Hybinette nickel refining process. The transaction allowed the matte from Falconbridge to be shipped to Norway in whisky (later metal) barrels for processing. While it was an expensive proposition to ship matte across the Atlantic Ocean for refining, the Mond Nickel Company had previously shown that it was economically viable because of the importance of the European market. The cost of operating in Norway was also offset by the low cost of electricity there and the fact that nickel in Europe commanded higher prices than it did in the United States.

Lindsley, in effect, put together pieces of the old British America Nickel Corporation, including some of its experienced staff. These actions were tolerated by Inco because, unlike the British America Corporation, Falconbridge Nickel Mines was small and mainly interested in serving the European market. Even when the Great Depression led Falconbridge to dump nickel on the United States market in 1932, Inco refused to retaliate against the upstart, largely because its officials feared antitrust action.[5]

In order to ensure future production, Lindsley initiated a program of land assembly under the umbrella of the Sudbury Nickel and Copper Company (until 1934) and later through the guidance of the company's own Geology Department. Assisted by a program of systematic mapping of the Sudbury Structure and the use of a magnetometer, he adopted the policy of acquiring every available piece of ground, especially along the North Rim. This move was to pay off handsomely during Falconbridge's later expansion. By 1935, the program yielded results with the discovery of ore southwest of Levack on what was to become the Hardy Mine, a property acquired from Sudbury Basin Mines along with McKim Mine.[6] The Mount Nickel and Blezard properties were purchased in1936 from the Ontario Nickel Corporation. To accommodate these developments, Falconbridge's refinery in Kristiansand was expanded.

After weathering the effects of the Great Depression, Falconbridge, under the "umbrella" price maintained by Inco, prospered. Sales and net profits increased during the 1930s. Following the loss of its Norwegian refining operations to the Germans on April 9, 1940, Falconbridge operated at a reduced rate and stockpiled its matte production. Under government mandate, arrangements were made for Inco to refine the company's mining output, thereby allowing for another round of mill, sintering, and smelter expansion to take place at Falconbridge in 1941 and l942.[7] As a result of these events, the company's local employment surpassed 1 000 in 1943 (table 10.1), and in the following year, its output reached record levels. In 1944, the company acquired the Strathcona property from Strathcona Nickel Mines Limited. With the cessation of the Second World War in 1945, Falconbridge's matte output was again diverted to Norway. In 1948, Falconbridge purchased the Fecunis Lake property from the McIntosh estate and added a new roaster building at its site. The McKim Mine opened in 1950 and was notable for producing the company's first ore in the area that did not come from Falconbridge Township. Due to the limited capacity of its Nikkelverk refinery, the company had to sell its surplus matte. Falconbridge began development work on the Hardy, Falconbridge East, and Fecunis Lake ore bodies in 1950 and 1951, and also started a modernization program for its Norwegian refinery. These postwar developments were all accomplished following the hiring of Horace John Fraser as Mine Manager of Falconbridge Nickel in 1945 (see biography 7).

With these developments in place, Falconbridge was strategically situated to take advantage of new contracts with the Defense Materials Procurement Agency (DMPA) of the US government following the outbreak of the Korean War. Between 1951 and 1957, the US government took control of the nickel market and began to acquire the metal for its national strategic stockpile. A 1951 contract called on Falconbridge Nickel Mines to supply 50 million pounds of refined nickel and 1.5 million pounds of cobalt over a period of ten years, with an option to supply an additional 25 million pounds of nickel.

TABLE 10.1 Employment trends for Vale Canada and Xstrata Nickel (Sudbury operations: 1928–2011)

Year	Inco/Vale Canada	Falconbridge/Xstrata Nickel	Total
1928	1 807	48	1 855
1929	3 824	252	4 076
1930	5 396	236	5 630
1931	3 812	251	4 063
1932	2 246	446	2 692
1933	2 966	253	3 219
1934	4 863	411	5 274
1935	6 320	648	6 968
1936	8 320	671	8 991
1937	9 889	610	10 499
1938	9 647	657	10 304
1939	9 585	647	10 232
1940	10 442	588	11 030
1941	10 706	819	11 525
1942	11 617	934	12 551
1943	11 942	1 005	12 947
1944	11 795	1 236	13 031
1945	10 014	1 044	11 058
1946	8 008	906	8 914
1947	10 239	1 111	11 350
1948	11 596	1 295	12 891
1949	11 614	1 232	12 846
1950	12 482	1 414	13 896
1951	14 981	1 840	16 821
1952	15 935	1 992	17 927
1953	16 373	2 235	18 608
1954	15 842	2 573	18 415
1955	15 301	2 815	18 116
1956	16 111	2 834	18 945
1957	16 587	2 853	19 440
1958	16 684	2 761	19 445
1959	14 314	2 787	17 101
1960	16 464	2 847	19 311
1961	16 914	2 896	19 810
1962	16 684	2 840	19 524
1963	13 923	2 158	16 081
1964	13 470	2 733	16 203
1965	16 675	3 044	16 719
1966	17 907	3 274	21 181
1967	17 468	3 639	21 107
1968	17 976	4 056	22 032
1969	18 542	4 049	22 591
1970	18 908	4 895	23 803
1971	20 134	4 985	25 119
1972	17 339	4 682	21 981
1973	16 799	4 500	21 279

continued

TABLE 10.1   (continued)

Year	Inco/Vale Canada	Falconbridge/Xstrata Nickel	Total
1974	17 166	4 583	21 749
1975	18 034	4 096	22 130
1976	17 945	4 001	21 946
1977	16 854	3 623	20 477
1978	14 069	2 884	16 953
1979	13 882	3 642	17 524
1980	13 838	3 886	17 724
1981	13 598	4 034	17 632
1982	11 106	2 717	13 823
1983	10 641	2 637	13 278
1984	9 213	2 684	11 897
1985	8 600	2 602	11 202
1986	8 360	2 272	10 632
1987	8 200	2 214	10 414
1988	8 000	2 397	10 397
1989	8 100	2 576	10 676
1990	8 100	2 436	10 536
1991	6 900	2 246	9 146
1992	6 800	2 171	8 971
1993	6 500	2 036	8 536
1994	6 037	2 036	8 073
1995	5 997	2 005	8 002
1996	6 205	2 111	8 316
1997	6 158	2 065	8 223
1998	5 249	1 926	7 175
1999	4 555	1 811	6 366
2000	4 734	1 799	6 533
2001	4 614	1 358	5 972
2002	4 500	1 530	6 030
2003	4 425	1 488	5 913
2004	4 129	1 420	5 549
2005	4 400	1 181	5 581
2006	4 000	1 168	5 168
2008	4 000	1 638	5 638
2009	4 000	1 839	5 839
2010	4 000	1 024	5 024

Source: Community & Strategic Planning Section, City of Greater Sudbury; figures supplied by Vale Canada and Xstrata Nickel.
Note: Recent figures for total employment by the companies are understated as they do not include non-permanent employees such as contractors.

As well, Falconbridge was authorized to treat the output from both the East Rim Nickel and Milnet Mines. In yet another provision, the US granted Falconbridge a credit of up to Cdn $6 million for the early delivery of nickel and cobalt. This, however, was only a prelude to the big breakthrough in 1953. Under contract DMP-60, the US government agreed to buy 100 million pounds of nickel over nine years at market prices, with options

Biography 7

DR. HORACE JOHN FRASER
Dynamic Falconbridge Leader (1905–1969)

Born in 1905 in the Prairie hamlet of Irvin, Saskatchewan, Horace John Fraser went on in life to excel in four successful careers as an outstanding earth scientist, a respected university professor, an efficient government administrator, and finally, a top mining executive.[a] In 1918, he completed his grade 8 education in the one-room Irvin School. In order to provide better educational opportunities for their children, the Fraser family moved to Swan Lake, Manitoba where Fraser sped through high school. Prior to his fifteenth birthday he was accepted in a pre-medical course at the University of Manitoba. He later switched courses, and graduated in 1924 with a Bachelor of Science degree in chemistry. During his summer months at the university, he was employed as a field assistant with the Canadian Geodetic and Geological Surveys. He acquired a teaching position at Harvard University in 1927, and in the following year earned a Master of Arts degree in economic mineralogy and geology. Two years later, Fraser obtained his doctorate in economic geology. Between 1930 and 1932, he was hired as an instructor in geology and research assistant at Harvard Engineering School, where he became acquainted with research concerning nickel ores from Sudbury. To his friends from that period he was known as "Dutch." During the 1930s and early 1940s he published extensively, establishing his reputation as an excellent research scientist.

Fraser married and brought his wife to Copper Cliff in 1932, where he had accepted an appointment to the geological staff of Inco. He remained there for three years before returning to the academic world at the California Institute of Technology. His courses in economic geology attracted students from both the United States and Canada. After the United States entered the Second World War, he went to Washington and served from 1942 until 1945 as an assistant divisional chief in charge of ferro-alloys in the Foreign Economics Administration of the United States government. At this time he was heavily involved, through the Office of Economic Warfare, with foreign purchases of nickel and wartime negotiations over nickel stockpile contracts. Recognizing his qualifications, the President of Falconbridge Nickel Mines hired Fraser as mine manager in 1945 to guide the company's postwar development in the Sudbury area as a competitor to Inco. There can be no doubt that his influence and administrative stint with the US government ensured that Falconbridge Nickel Mines would be a major beneficiary of the 1951 and 1953 contracts signed with the United States. Thus, much of the credit for Falconbridge's spectacular progress after the Second World War could be attributed directly to Fraser.

During his sojourn in Sudbury, he was promoted to general manager in 1947. Not afraid to get his hands dirty, Fraser was known to have worn overalls and a miner's hat while crawling in and out of the sixty working slopes at Falconbridge. After being appointed vice-president and a member of the board of directors, he moved in 1950 to the company's head office in Toronto. In 1951, Fraser encouraged D.M. LeBourdais to write a book on Sudbury mining. Titled *Sudbury Basin: The Story of Nickel*, this book was underwritten by Falconbridge Nickel Mines and published in 1953. He likewise led the company to build up its research activities and promote physical metallurgy for the development of new uses for nickel. Fraser was elected president and managing director of Falconbridge in 1957 and Ventures in 1958. He served as the "matchmaker" for the absorption of Ventures by Falconbridge in 1962. His sudden death in 1969, at the early age of 63, occurred just before company officials signed an agreement for Falconbridge to erect a large plant in the Dominican Republic. He left behind his wife, Catherine, and two sons, Ian Bruce and Malcolm. Among his other achievements, he served as the president of the Ontario Mining Association, the Mining Association of Canada, and the Canadian Institute of Mining and Metallurgy. For four years, he also served as Chairman of the Board of Governors for Laurentian University.

---

[a]This biography has been largely derived from "Memorial of Horace John Fraser," *American Mineralogist* 55 (1970): 554–61 and "Falconbridge Head, H.J. Fraser Dies," *Sudbury Star*, February 3, 1969, 1 and 3.

to buy a further 50 million during the same period, and another 50 million in the following five years, subject to certain conditions. What made this deal fabulous was an agreement to pay, in addition to the market prices, a premium of 40 cents a pound on the first 100 million pounds. Given this windfall, Falconbridge expanded production beyond its Falconbridge and McKim Mines to include the development and/or output of eight more company-owned mines: Falconbridge East (1954), Mount Nickel (1954–57), Hardy (1953), Fecunis Lake (1959), Longvack (1955), Boundary (1955), Strathcona, and Onaping (1955), as well as ores from the East Rim Nickel Mines, Milnet Mines, and Nickel Offset Mines companies. As a result of these developments, Falconbridge's employment levels exceeded the 2 000 mark. The Fecunis Lake Mine was unusual, as its location near Levack Mine involved cooperative development work with Inco. Falconbridge initiated a pilot pyrrhotite plant in 1954. In 1959, a noteworthy transition took place; this year marked the first time that the company's production from the North Rim mines exceeded that coming from the Eastern and Southern Ranges.[8]

It was through these means that the US succeeded not only in addressing the critical shortage of nickel for its strategic needs but also in subsidizing Falconbridge Nickel Mines, allowing the company to become a major world producer. The latter was a strategic objective, as it reduced the United States' dependence on Inco as the sole source of nickel for industry and the American military.[9] While these events were taking place, however, Lindsley made a number of tactical errors that resulted in his Ventures Limited holdings, Falconbridge Nickel Company's parent company, being taken over by McIntyre Porcupine in 1957.

The expiry of the DMP-60 contract in 1961, which had accounted for virtually all of the company's sales in the US, necessitated a reorientation of Falconbridge's output back to the civilian market. Concluding that expansion in Europe was not a practical option, the company decided instead to go after a share of Inco's growing US market by cutting the price of nickel. While Inco was stunned by the move, the strong markets of the 1960s deterred it from taking any strong countermeasures. During this decade, Falconbridge Nickel became embroiled in a series of complicated mergers and corporate manoeuvres. In 1962, Ventures Ltd., the vehicle that Lindsley used to control his personal empire, was taken over by Falconbridge Nickel Mines, thereby eliminating his power base. In the meantime, Falconbridge had come under the control of Canadian-owned McIntyre Porcupine. The takeover was seen at the time as a victory for Canadian capitalism, but this proved to be an illusion. Another Canadian upstart, Power Corporation, made some attempts to gain control of McIntyre Porcupine, but its initiative failed. The stage was then set for the takeover of McIntyre Porcupine by two other international giants: Anglo-American and Superior Oil. Anglo-American was first acquired by Superior Oil; Superior Oil, in turn, acquired McIntyre Porcupine in 1967, a step that brought Falconbridge firmly into America's corporate power structure. Falconbridge Nickel was quick to take advantage of this new situation. Following the successful invasion of the Dominican Republic in 1965 by the United States, the company began development work on its Falcondo ferro-nickel operations in that country. The complex, which opened in 1972, confirmed Falconbridge's status as the world's second-largest nickel producer.[10]

On the home front, Falconbridge opened Lockerby Mine in 1976, and by the 1970s, it employed nearly 5 000 workers. Other events then occurred that began to cast a shadow on Falconbridge Nickel's future. These included union–management disputes, increasing competition from other nickel producers around the world, and the establishment of environmental controls on mining pollution by the Province of Ontario in 1970.

Inco (1928–1970s)

The 1930s proved to be a difficult time for Inco, especially 1932, the first time the company did not make a profit. This unusual occurrence was not to occur again until 1982. Inco's fortune was sustained by the fact that it was the sixth largest copper producer in the world and its largest platinum supplier. The steel industry's increasing need for nickel and the growing armaments business in Europe brought renewed prosperity to the company, providing for 80 to 85 per cent of the world's nickel demands. The full extent of Inco's monopoly position is shown by the fact that, aside from the case of nickel oxide, its nickel price never varied between 1928 and 1946—an indication of the company's complete freedom from the normal pressures of competition. As well, there was little threat from any existing deposits found in New Caledonia or at Petsamo in Finland. Competition from Finland had been minimized, since the rights to the Petsamo deposit had been secured by the Mond Nickel Company, Inco's subsidiary in England. Nor did Inco fear Falconbridge Nickel, as its contribution to Canadian production averaged only about 5 per cent of the total. The export of nickel for use in the armaments industry of Europe, which involved a long-term contract with Germany's I.G. Farben in the mid-1930s, again brought calls for the control of nickel exports and even the nationalization of the nickel industry. Inco shrewdly rebuffed these demands by weaving itself into the economic structure of Canada (e.g., by promoting the company's purchase of coal from Nova Scotia) and with a powerful advertising campaign linking the prosperity of the company to "the welfare of Canada."[11]

Inco's expansion in the Sudbury area brought numerous changes to the economic landscape and provided a much-needed boost for construction workers. Many construction projects were created as a result of the merger of International Nickel with Mond Nickel in December of 1928. The new Inco undertook a major expansion program involving mines, smelters, refineries, and power plants. By the end of 1930, a concentrator, smelter, copper refinery, and sulphuric acid plant had been erected. Aside from increases in capacity, these features remained virtually unchanged for the next twenty years. New smelting techniques resulted in the construction of three huge smokestacks between 1929 and 1936, with two that were over 152 metres in height, thereby introducing a new dimension to the local skyline and providing a powerful image of Inco to the outside world. For local residents, a major benefit of the new smelter was the elimination of heap roasting and the pollution this produced. The smokestacks not only served to diffuse waste gases into the atmosphere but also provided strong updrafts for smelting operations.[12] Other developments included the construction of a hydroelectric plant at Big Eddy, the linking of Frood Mine to the Frood Extension, the introduction of a safety program to all mining operations (1930), the establishment of a Geological Branch (1931), the construction of a research laboratory (1937), and the opening of the Frood Open Pit (1938).

During the early 1930s, Inco temporarily closed its Creighton, Garson, and Levack Mines; only Frood Mine remained open. Creighton was reopened in 1933, followed by Garson and Levack in 1936 and 1937, respectively. In the meantime, Inco, through its European subsidiary, Mond Nickel, entered into an agreement with the Finnish government to prospect, mine, and treat nickel ore in the Petsamo district of northern Finland. Exploration began there in 1935, and subsequent work on the Petsamo Mine and smelter continued until the property was occupied by Russian forces in November of 1940. In accord with the armistice agreement signed between Russia and Finland in September, 1944, the Petsamo property was ceded to the Russian government and $20 million USD in compensation was awarded to Inco via the Canadian government.[13]

With five operating mines (Frood, Levack, Garson, Creighton, and Stobie) and the Frood Open Pit in full production by the late 1930s, Inco was well positioned to serve the needs of Canada during the Second World War. In fact, the amount of ore mined by Inco during the war years equalled the production of the company and its predecessors during the previous fifty-four years.[14] When the Falconbridge Nickel Mines refinery in Norway was taken over by German forces, Inco was mandated by the federal government to refine its matte locally. Despite having more than 10 000 employees during the early 1940s, Inco found itself hampered by a shortage of labour; thus, the company started hiring women.

Since 1890, Ontario mining legislation had prohibited the employment of women in mines. Later amendments in 1912 and 1930 relaxed the rules somewhat: women were permitted in mining if they worked in a technical, clerical, or domestic capacity. Due to wartime circumstances and the severe shortage of male labourers, the federal government, using its powers under the War Measures Act, issued an Order-in-Council on August 13, 1942, allowing Inco to employ women in its surface operations in Sudbury.[15] Within two weeks, seventeen women had been hired to work in the Copper Cliff mill and smelter, the first women in Ontario to work in mining production. Over the course of the war, Inco hired over 1 400 women to fill mostly unskilled production and maintenance jobs. In November of 1945, after the war ended and men returned from service, the federal government rescinded its Order-in-Council, arguing that the supply of male labour rendered the continued use of women unnecessary. All women were immediately laid off.[16] It was not until 1970 that the province again made changes to the Ontario Mining Act eliminating the prohibition against women working at surface jobs; the ban against underground work stayed on the law books until 1978.

Following a brief slump after the Second World War, Inco's production returned to normal. The company still used open-pit mining, which accounted in some years for up to 40 per cent of their output. When it became apparent that the ore mined underground was of a lower grade, and that the ore mineable by open-pit methods currently being used at Creighton and Frood (later, at Stobie as well) was becoming depleted, the company introduced a major program to expand underground mining and to develop lower-cost methods of extracting ore. To meet its substantial timber requirements, the company acquired the Geo. Gordon & Co. lumber company in 1948. Another mine, the Murray, was brought into production two years later, as was the Stobie section of the Frood-Stobie Mine. The company's chief endeavour, however, was the mining and concentration of lower-grade ores associated with Creighton Mine. As the pressure for nickel continued unabated, Inco promoted increased production by constructing a concentrator at Creighton as well as an oxygen flash smelting furnace and oxygen generating plant at Copper Cliff in 1951. Flash smelting was introduced as a more efficient and environmentally friendly way of processing ores containing sulphur and also produced liquid sulphur dioxide. Another innovation, introduced in 1950, was the use of aerial electromagnetic surveys to search for possible ore-bearing deposits. Such surveys, with their "bombs" towed beneath the plane, became a common sight in the Sudbury area during the 1950s.[17]

In 1946, Inco had a scare when the United States Department of Justice filed an antitrust suit against the company's subsidiary, charging it with monopolistic practices. The Department sought to force Inco to separate its rolling mills from the parent company. While Inco won after a protracted court battle in 1948, it was nonetheless required to modify its marketing and pricing practices drastically, setting the stage for the end of its monopoly position over the next few decades.[18]

The outbreak of the Korean War and the ongoing Cold War spurred the international demand for nickel, one that exceeded the available supply. In response to these demands, Inco signed a five-year contract with the Defense Materials Procurement Agency (DMPA) in 1954 for the "quick delivery" of nickel and copper outside of the company's normal market deliveries. The contract ushered in a decade of prosperity for Inco, one that witnessed a substantial increase in the company's employment rolls from slightly more than 12 000 in 1950 to more than 16 000 by the turn of the 1960s. As was the case with Falconbridge Nickel, the US contract was a "sweetheart" arrangement, since the price for nickel and copper under the contract reflected the existing market price, plus an allowance for amortization and additional costs of production.[19] Sudbury's skyline changed again in 1954 when two small stacks at the Coniston smelter were replaced by a new 122-metre chimney. In 1955, the life of the Frood Open Pit was extended and the underground mine at Creighton expanded to become the world's deepest mining operation. A major innovation in these mines was the use of rock bolting for ground support, a technique that eliminated the need for the traditional use of timbering.

Another important event was the opening of the company's nickeliferous pyrrhotite treatment plant in 1955–56 for enhanced recovery of nickel, which allowed Inco to engage in the commercial production of iron pellets. These pellets were then shipped to Canadian and US steel mills around the Great Lakes. While Canadian shipments were made by rail, US shipments were first transported by rail to the company's storage area on Goat Island near Little Current, and then put into ore carriers.[20] The completion of this iron ore recovery plant was key to allowing Inco to meet the requirements of the DMPA contract, as it freed up existing smelter capacity and made it possible to treat lower grades of ore hitherto considered uneconomical.[21] The nickeliferous pyrrhotite treatment plant's smelter chimney, rising 194 metres above ground level, was the largest in the world.

Until this plant opened, much of the iron ore had been rejected in the form of tailings from the concentrating process, or as slag after smelting. While the slag piles had a nickel content of 35 to 40 per cent, the iron was of little value because it was linked chemically to silica.[22] The new recovery process resulted in iron ore pellets that contained 68 per cent natural iron. Inco advertising claimed that the product was higher in grade than any being produced in North America. Thus, iron became the fourteenth element mined from the Sudbury ores, the others being nickel, copper, the five platinum group metals (platinum, palladium, rhodium, ruthenium, and iridium), gold, silver, cobalt, selenium, tellurium, and sulphur. The economic impact of the plant was so great that in 1960, Inco introduced plans to triple the plant's capacity.[23] This expansion was completed in 1963. Another outcome of the expansion was that fewer pots of molten slag needed to be transported to the slag dumps, resulting in less frequent cascades of fiery lava visible to Sudburians and tourists. Yet a third expansion phase was undertaken in 1972. While the plant was viewed at that time as the economic "gem" underlying Inco's profitability, this perception proved to be illusory.

In the meantime, Inco's sintering plant at Copper Cliff, dubbed by workers as the "Black Hole of Calcutta," was closed in 1963, due to the high number of employees from that plant who had contracted and died of cancer.[24] Other significant events during the 1960s included the expansion of Creighton Mine, which incorporated the deepest continuous mine shaft in the Western Hemisphere, the opening of new mines at Little Stobie, Totten, McClelland, Kirkwood, Coleman, Copper Cliff North, and Copper Cliff South, the construction of a new mill at Frood Stobie, and the introduction of computers to mine administration by James Grassby, the company's manager of special projects.[25]

While these events were taking place in Sudbury, Inco's attention shifted to other Canadian and global mining sites.[26] These initiatives were initially tied to the United States' wish in 1956 to expand the supply of nickel to meet its military objective of "combating Soviet-dominated communism." Then there was a major discovery at Thompson in northern Manitoba. In quick response to Ralph Parker's claim that "it looks like we've hit the jackpot," Inco signed an agreement with the Province of Manitoba to develop the ore body and to build a new town. At the same time, an understanding was reached with the Office of Defense Mobilization of the United States to provide it with nickel from Thompson by 1961. When the latter understanding was cancelled by the US in 1957, Inco responded by developing new markets for Manitoba nickel in a manner similar to that undertaken by Stanley in the 1920s.[27] Inco's mining and refining enterprise at Thompson opened in 1961 and continued to operate there until 2010, when Vale announced the phasing out of its mill and smelter operations due to declining reserves.

This major development in Canada notwithstanding, in the 1960s Inco decided that its future lay in the tropics. The conventional wisdom at the time was that 80 per cent of the world's nickel supplies were in the tropics in the form of lateritic ores.[28] Inco first focused its attention on the lateritic nickel deposits of Guatemala. In 1960, the company gained control of the American-owned Hanna Mining Company's concession in that country and formed a new company called Eximbal. With the political support of the United States government, Inco acquired political concessions favourable to Eximbal, such as a forty-year mining lease renewable for twenty years. The project did not get the formal agreement of the Guatemalan government until 1971, and it was 1977 before Eximbal started strip-mining nickel. Gauging that the social and political situation was favourable, Inco in 1968 obtained exclusive exploration rights to a 64 750-square kilometre tract of land on the island of Sulawesi, Indonesia. Another factor behind this decision was the proximity of the Sulawesi deposits to the huge nickel market in Japan. Although it was long known that Indonesia had sizeable deposits of lateritic nickel, the prevailing political situation in the country had deterred investors until President Sukarno was deposed in 1967, and replaced by General Suharto. Ten years later, Inco's nickel plant at Soroako was opened, and its first shipments of nickel matte were sent to Japan in 1978. In recognition of these changes and its long-term goal of diversifying itself, the company's name was officially changed to Inco Limited in 1976. Both "nickel" and "of Canada" were conspicuously absent from the company's new name.

The decision made by Inco to borrow heavily for the first time in its history to fund two laterite projects in Guatemala and Indonesia had serious long-term consequences. The horrendous debt acquired by the company required it to take out loans that would come back to haunt its Canadian operations. Whereas previously Inco had sold nickel to maintain its position, it was now forced to sell nickel to pay off its debts. Other chickens came home to roost. The nickel market "fell out of bed" in 1971–72, and Inco was burdened with unsold inventory just as the debt payments required by the 1960s expansion programs peaked. At the same time, Inco's decision to concentrate on mining high-grade, relatively inexpensive nickel and its reluctance to sell a cheaper form of ferronickel allowed Falconbridge Nickel to increase its market share from 9 to 14 per cent. Relying on its traditional monopoly role, Inco likewise failed to negotiate any long-term contracts for the simple reason that "it never had to."[29] The company did not realize that its traditional attitude, "if you don't like it, take it or leave it," no longer had the same impact on its customers.[30] Finally, the oil crisis of 1973 caused costs for lateritic nickel to skyrocket because it required more energy for smelting and refining than Sudbury and Manitoba ores. Even Inco's acquisition of ESB Inc. in 1974, a leading manufacturer of large storage batteries using nickel, came to naught. The company believed that ESB's sales, fuelled by the growing demand for batteries in a world running out of oil, would help balance cyclical downturns in nickel demand. However, this did not come to fruition, and the ill-fated venture was written off in the early 1980s. As the trends in table 10.1 show, the stage was set for an end to the mining boom years, and the arrival of several decades of economic downturn and population decline for the Sudbury area.

Global Changes and Absorption by Xstrata Nickel and CVRD (1970s to the Present)

The turbulent years that followed the 1970s resulted in a lessening of demand for nickel and brought with it other changes, notably a dramatic decline in employment levels for both mining companies. Inco and Falconbridge Nickel Mines slowly began to lose their positions as the world price leaders for nickel; as well, there was a breakdown of the former producer pricing system and movement toward a free market pricing system. In 1979, a turning point in the history of nickel marketing was reached when it began to be traded on the London Metal Exchange. Up to this time, Canadian mining companies had established the official price of nickel; after 1979, nickel prices were no longer set by producers; instead the exchange established the price in terms of world supply and demand.[31] The global supply and demand position of nickel was likewise altered. Whereas in 1950 only four countries in the world produced nickel, by the beginning of the 1980s, there were twenty-six producers, many of them in Third World countries with low wages and minimal environmental regulation.[32] Of these producers, Russia's development of the Norilsk deposits was especially important as they exceeded Sudbury in terms of their grade and resources of nickel, copper, and the platinum group of elements.[33] By 1982, Sudbury producers managed to account for only 10 per cent of global production, a drastic change from the 90 per cent figure that they had prior to the Second World War. An industry historically dominated by relatively few players operating under a stable system of producer prices thus began to fragment.

While these events were taking place, the Sudbury area was changing. Concerned about the impact of the first oil crisis, the loss of market share, and payroll expenses, L. Edward Grub, the new chairman of Inco, trimmed labour costs and sought to improve productivity. Under his direction, a process of downsizing was initiated that lasted for another three decades. By 1984, the number of Inco employees had been reduced by more than one half; two years later, William James, the president of Falconbridge Nickel Mines, accomplished the same feat. During this interval, Falconbridge Nickel Mines underwent several ownership changes that resulted in the Superior Oil Company of Houston, Texas, eventually gaining control. A signal of the economic problems of the time for Falconbridge Nickel Mines came in the form of its first corporate losses in the years 1977 and 1981–83. In 1982, the company's name was changed to Falconbridge Limited to more accurately reflect its diversified nature. This diversification trend continued in 1985 when the company acquired Kidd Creek Mines Limited, a major silver, copper, zinc, and gold producer based in Timmins, Ontario.

Both mining companies attempted to adopt new technological approaches to mining in Sudbury, and efficiency became their byword. Innovations in underground mining, begun in the 1960s, were expanded. These changes involved more mechanization, equipment upgrades, and the adoption of high-volume technologies, such as trackless mining machinery (e.g., scoop trams and jumbo drills) and increased blast-hole mining. Surface operations were impacted as well by the move from mechanized to automated production. This was most evident in Inco's new Clarabelle mill (concentrator) opened in 1971, and its Copper Cliff Nickel refinery that began production in 1973.[34] The latter eventually led to Inco's nickel refinery at Port Colborne closing in 1984, and its conversion to refining cobalt and precious metals. Technological change, mechanization, and automation continued unabated throughout the 1980s; in 1989, Inco was awarded the Silver Medal in the productivity category at the Canadian Business Excellence Awards Ceremony for its improved performance and for becoming one of the lowest-cost nickel producers in the world.[35] A key part in this search for enhanced productivity revolved around reduced energy costs. Nickel production was an energy-intensive activity, on average accounting for 20 to 25 per cent of total production costs, so both Inco and Falconbridge Limited sought to reduce their costs by substituting labour and exploiting economies of scale by using more efficient machinery.[36]

Another stimulus came from the lessons learned from the Inco strike of 1978–79, which marked an important milestone leading to less labour-intensive production. By this time, it was clear that market realities were forcing companies worldwide to further streamline, automate, mechanize, and computerize their operations in order to survive.[37] Technology likewise came to the fore with respect to environmental issues. In response to increased public and governmental concerns regarding environmental pollution, Inco erected a new 381-metre Superstack at the Copper Cliff smelter, and closed its Coniston smelter in 1972. Falconbridge also underwent change, closing its pyrrhotite plant in 1972 and its sintering plant in 1978. As was the case with Inco, Falconbridge adopted numerous innovations throughout this period. It was the first company in Canada to own and operate scoop trams underground, a development that revolutionized mining at the Strathcona Mine.[38] Since the Falconbridge operations were too close to the Sudbury

Airport to build a Superstack similar to the one erected by Inco, they adopted a new smelting process in 1978 that permitted sulphur dioxide gases to be captured in an acid plant. The process proved to be successful, as the acid sales came close to recuperating the total costs for the smelting and acid plant project.[39]

These improvements notwithstanding, the advent of the 1980s and another oil crisis (1979–82) brought a series of "gloom and doom" articles from major newspapers such as *The New York Times,* concluding that the Sudbury nickel mines had hit bottom. The recession, which began in 1981, caused a marked drop in nickel demand; for the first time since the late 1940s, demand declined for two consecutive years. Nothing in Sudbury's history compares to the despair of this period.[40] This pessimism could be attributed to the existence of enormous nickel stockpiles, the announcement of major layoffs in 1977, the strike against Inco in 1978–79, another round of massive layoffs in 1982, and the ensuing strike and company shutdowns. Sudbury thus acquired the reputation of being the "unemployment capital of Canada." It was not until 1987 that the market began to change. Due to an unforeseen rise in the demand for stainless steel, the closure of some nickel producers around the world, and the decreased availability of stainless steel scrap, Inco and Falconbridge began to recoup on their investments in technological change; by 1989, their combined profits were in excess of $1 billion CAD. The nickel industry then experienced yet another competitive era following the breakup of the Soviet Union in 1991, and the privatization of the largest nickel producer in the country, Norilsk Nickel. With the downsizing of the Russia's military-industrial complex, the country's internal consumption plummeted, which led to a surge of nickel exports worldwide and reduced prices. These events led to Norilsk becoming the world's largest nickel producer. Foreshadowing ownership events that were to follow, Falconbridge Limited, after a prolonged struggle lasting more than a year, was acquired in 1989 by Noranda Inc. and Trelleberg AB. Following a rise in nickel prices, Trelleberg later sold off its Falconbridge shares in 1994 for a considerable profit.

As the 1990s dawned, environmental issues became an increasing concern for both Inco and Falconbridge Limited. Responding to a 1985 Ontario government regulation requiring the company to reduce its emissions, and the public relations fallout from serving as the largest single source of sulphur dioxide pollution on the continent, Inco launched an abatement program in 1989 to substantially lower its emissions, and reached its target levels in 1993. The company then commenced production in its Lower Coleman mine and reopened its Garson mine, which had been closed in 1986 due to a serious rockfall. Facing sluggish demand and increased competition from Russian exports in 1992, Inco slashed executive salaries, decreased capital expenditures, and closed its Ontario and Manitoba facilities for several weeks in 1993 and 1994. Both Inco and Falconbridge Limited continued downsizing throughout the decade. To assist in debt reduction and the financing of its capital projects, Inco sold the equity interest in TVX Gold that it had created in 1991. In 1994, the company expanded its research into automated mining and provided more funding for new nickel applications. Two years later, Inco gained control of the Voisey's Bay deposit discovered in 1994 in northern Labrador by outbidding its competitor, Falconbridge Limited. It was a costly venture for the company, and its development was complicated by First Nation claims, the need for an environmental impact study, and the province's insistence that the metal mined in the area be refined there.

Meanwhile, competition from Russia, Australia, and Cuba cut into Inco's profits, as did a strike in Sudbury. Burdened with considerable debt, the company promoted new technologies such as "telemining" (mining from a distance) and developed specialized market niches. While Inco had toyed with the concept of robo-mining in the late 1980s, its research efforts in this direction eventually culminated in the initiation of a Mining Automation Program (MAP) in 1996. In cooperation with other partners, Inco's Stobie mine was used as a laboratory to test revolutionary technology based on the surface control of underground jumbo drills and robotic scoops. These tools were intended to reduce downtimes for miners, who often spent up to two hours travelling to and from their work sites, and to reduce health and safety concerns.[41] Exploration was started on its Victor deposit located at the northeast rim of the Sudbury Basin.

Falconbridge was likewise affected by the 1985 Ontario regulation. The company introduced new capital projects between 1988 and 1993 to conform to the regulation, and took further action to decrease its labour force. Looking to the future, the company commenced production from its Craig mine in 1991. This mine became fully operational in 1995. Steps were taken to upgrade the company's local smelter in anticipation of receiving Raglan concentrate from northern Quebec in 1998.

The new millennium held yet more challenges and opportunities. Both the Inco and Falconbridge smelters were affected by new Control Orders issued in 2002 and 2005 by the Ontario government that mandated reduced ground level concentrations of sulphur dioxide. In 2002, Inco and the Province of Newfoundland and Labrador agreed upon the principles for the development of the Voisey's Bay nickel deposit. For the nickel industry, 2004 was in many ways a year of records, heralding a period of growth and expansion that lasted until 2009. Another major event in that year was the startup of Inco's subsidiary, Voisey's Bay Nickel Company, and the first shipment of its concentrates to the smelter in Sudbury. Most startling was the friendly offer made by Inco to acquire its longtime rival Falconbridge Limited. The motivation of the merger was to avoid a takeover of either company by foreign competitors, and to achieve synergies in their Sudbury operations, which were based on same kinds of extraction and smelting. Since both Inco and Falconbridge had extensive operations in the Sudbury Basin, the "New Inco" was expected to cut costs by US$350 million per year or more by optimizing concentrator and processing capacity, and sharing procurement and capital savings. The offer created a sense of local euphoria. Not only was the proposed merger considered to be a "marriage made in heaven" and a "win-win" situation for Sudburians, it likewise received widespread union acceptance.[42] From a Canadian perspective, it would have made Inco a nickel and copper powerhouse on the global scene, the leading producer of nickel and the fifth-largest producer of copper.[43] This offer was not successful, and the state of euphoria was replaced by one of dismay. One of the issues responsible for the failure was that while the Canadian government granted regulatory approval for the merger in January 2006, similar approvals did not come from Europe and the United States until six months later.

During this interval, other proposals came to the fore. Thus began the greatest mining drama ever seen in Canada: both companies were set upon by hostile takeover bids, a complicated process that was to last another year before its final resolution. The press went into frenzy discovering new plots and closely scrutinizing the changing cast of global corporations. Swiss-based Xstrata submitted a hostile takeover bid for Falconbridge Ltd. in 2005, resulting in a bidding war between Inco and Xstrata. In May, 2006, Teck Cominco submitted a hostile bid to purchase Inco on the condition that it agreed to abandon its acquisition of Falconbridge. Later, in June, Inco and Phelps Dodge proposed a friendly merger of Phelps Dodge, Inco, and Falconbridge. Inco's offer of cash and shares, however, was not as attractive to Falconbridge shareholders as Xstrata's offer; Inco's offer lapsed in July. Xstrada then made a successful bid to acquire Falconbridge in August of 2006. As a result of this event, the historic Falconbridge name disappeared from the local mining scene.

This spectacular development was followed by yet another on August 14, when Brazilian-based CVRD extended an offer to buy Inco. This offer was approved by Inco shareholders on October 23. In January 2007, a new company, CVRD Inco Ltd. (later changed to Vale Inco and then Vale Canada) was created. The purchase marked the end of one of the longest and messiest takeover wars in Canada's corporate history. The saga that included regulatory hurdles, a bidding war, patriotic battle cries, and more plot twists than a soap opera, came to an end in a way that few Sudburians predicted.[44] While the initial reaction of the Steelworkers union to the takeover of Inco was positive, union officials nonetheless cautioned the company that in the long run they could "be their best ally or their worst enemy."[45]

Following a calm hiatus in 2007, dropping nickel prices led to another round of company–union confrontation. Xstrata Nickel announced in 2008 that it would close its Craig and Thayer-Lindsley underground mines, which were nearing the end of their productive lives, place the Fraser complex on care and maintenance status, and shelve its Fraser Morgan development project. This was followed, in 2009, by a cut of 686 permanent jobs in the Sudbury area. These events notwithstanding, a three-year collective agreement was reached between Xstrata Nickel and CAW/Local 598 in February 2010. Shortly thereafter, the company announced plans to bring its Nickel Rim South mine into full production, reactivate its copper excavation at Fraser Mine in partnership with Vale, and make enhancements at Strathcona Mill and the Sudbury Smelter. Xstrata's smelter is now noteworthy in that only 25 per cent of the nickel comes from Sudbury, the rest coming from its Raglan and Australian mines.

The labour situation that arose between Vale Inco and the Steelworkers union was radically different. What emerged was a classic confrontation between a foreign-owned company and a foreign-owned union. Early in 2010, the company suggested that the number of employees at its Sudbury mines was double that of its other "world-class" operations.[46] When talks broke down, a strike began on July 13, 2009. For the first time in its history, the company took the unprecedented step of resuming production during a strike. The sight of smoke rising from the Superstack was an unwelcome sight to the striking miners, who now realized the company's intent to increase production using its own non-striking workforce.[47] The strike ended July 8, 2010, making it the longest in the company's history. In that year, Vale Inco sent another message of its own, disassociating itself from Sudbury's mining history by eliminating the name Inco, thus making the lyrics of Stompin' Tom Connors' 1967 song prophetic: "We'll think no more of Inco on a Sudbury Saturday Night." The intent here was clear: there was a new boss in town—Vale Canada. The company introduced a different management style, and a new approach emerged framed around the belief that "Sudbury does not have the capacity to change organically. It will have to be done by us."[48] These actions caused residents to write negative letters to *The Sudbury Star* stating that Vale's actions were appalling; there was even a proposition that it was "inappropriate to associate Inco's *good* name with a company like Vale."[49]

In 2011, Vale undertook major initiatives estimated at $3.4 billion involving its new Clean AER project, the CORe project at Clarabelle Mill, and the completion of Totten Mine, the first new mine in Sudbury in almost forty years.

Joint Venture Partners in the Greater Sudbury Area

An interesting development in recent decades has been the rise of new and innovative ideas about the Sudbury Igneous Complex (SIC) and, in particular, how minerals over time managed to migrate into the footwall zone surrounding the crater and below the SIC. Due to prevailing geological theory, these areas had never been systematically studied with modern exploration tools. As these ideas gained strength, a number of smaller mining companies acquired properties along the footwall and further outward. One of the leaders in advancing this form of exploration was the Wallbridge Mining Company. Formed in 1996 by Risto Laamanen and others, it was established in the belief that many parts of the Sudbury Igneous Complex remained unexplored, and that more attention should be paid to its footwall deposits. As a result, Wallbridge started to acquire properties in the Sudbury area in 1997.[50] By 2010, it had emerged as the third largest landowner in the greater Sudbury area, with forty-one holdings of more than 66 000 hectares of land.[51] As these properties are still in their exploration phases, there has been little employment impact to date; however, working agreements have been made with both Vale Canada and Xstrata Nickel to ensure that future mining output will be processed locally. In order to assist Wallbridge in developing its North Range properties, a South African company, Lonmin Plc, became a financial backer and shareholder in 2012.

Another new company is KGHM International, a Polish-owned company that assumed ownership of Quadra FNX Mining in 2012. The latter was originally incorporated as Fort Knox Gold Resources in 1984. Its name was changed to FNX Mining Company in

2002. Unlike Wallbridge Mining, FNX focused its attention on acquiring small mining properties from Vale Canada and Xstrata Nickel and processing its ores at Vale's Clarabelle mill and Copper Cliff smelter. In 2002, it acquired five properties from Inco (the McCreedy West, Levack, Norman/Podolsky, Kirkwood, and Victoria mines) on the condition that any future production from these locations would be processed by Inco. Work on these sites began in 2003, and in the following year the company opened its first mine, McCreedy West; production from Levack mine was started in 2006. In May of 2010, FNX teamed up with Quadra to create a larger entity called Quadra FNX Mining Ltd.[52] The company's Victoria Mine, scheduled to open in 2017, is unique in that its development will fall within the framework of an impact and benefits agreement signed with the Sagamok Anishnawbek First Nation.[53] Currently, KGHM is working in Partnership with Vale on the Onaping Depth Project to eventually mine ore bodies below the old Craig mine.[54]

Two other junior companies are First Nickel and Ursa Major Minerals. In 2005, Falconbridge followed Inco's earlier path of divestiture, selling Lockerby Mine in 2005 (that it had run between 1976 and 2004) to First Nickel Inc., a Toronto-based company. Operational from 2006 until 2008, the mine was put on care and maintenance mode until it reopened in 2012.[55] Ores produced at this minesite will be processed at Xstrata's Strathcona mill before being transported to the Falconbridge smelter.[56] Ursa Major Minerals currently owns the Shakespeare open pit mine located near Lake Agnew and has a processing agreement with Xstrata; work at this site, however, was temporarily suspended in 2012. That same year, Ursa Major became amalgamated with Prophecy Platinum, a Canadian-based exploration company. The company is also drilling in the former Nickel Offset mine found in Foy Township.[57]

The location of the mining operations in the Sudbury area in 2011 is illustrated in figure 10.2. At that time, there were five mining companies with active operations in the area: Vale Canada, Xstrata Nickel, Quadra FNX Mining, Ursa Major Minerals, and First Nickel. A listing of the employment figures associated with these mining operations is found in table 10.2. While this data is incomplete, it nonetheless indicates the increasing role played by junior companies and contractors within the mining sector. Given the trend for Vale Canada and Xstrata Nickel to outsource their older minesites to newer companies, the increasing number of properties being acquired by the latter, and assuming a continuing demand for minerals worldwide, it is likely that the employment level at these companies will soon approach that of Xstrata Nickel.

Cliffs Natural Resources Ferrochrome Smelter

A new factor was introduced to the local mining industry in 2012, when Cliffs Natural Resources, a Cleveland-based company, announced its intention to erect a ferrochrome smelter in Capreol. The manufacturing plant will produce ferrochrome for use in the production of stainless steel. While several sites were vying for this facility, it was clear that the Sudbury area had the best chance of successfully developing the project. According

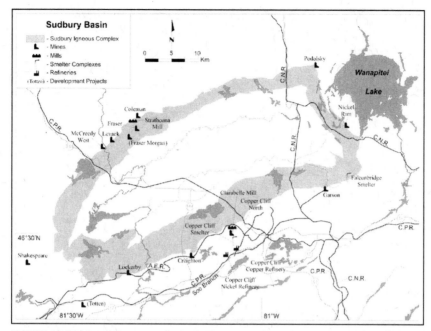

FIGURE 10.2 Mining Operations in the Sudbury Area, 2011. From Ontario Prospectors Association, *Ontario Mining & Exploration Directory 2011* (Thunder Bay, ON: Ontario Prospectors Association, 2011), 27–28.

to one company source, the site was chosen because "Sudbury offers the best combination of transportation logistics, labour, a long mining tradition, community support and access to power."[58] This proposed expansion of the mining sector could be considered historically significant, as it will solidify Greater Sudbury's position as a global mining and mineral processing centre. Requiring a proposed construction force of 450 and another 450 permanent positions when completed, the project has the potential to ensure a stable environment for growth in the mining sector for years to come.

Fundamental Questions

Two fundamental questions arise from the mining drama that occurred in the Sudbury area in 2005 and 2006. The first pertains to the failure of Inco and Falconbridge to merge their operations during the latter part of the twentieth century. The second asks how it was possible that two Canadian-based mining icons such as Inco and Falconbridge Limited, with lengthy histories extending back to 1902 and 1928 respectively, could be taken over by Brazilian and Swiss interests. The two questions, of course, are intimately related. Michael Atkins, President of *Northern Life*, has gone so far as to state bluntly that the "Canadian management of our ... mining assets has been mediocre and incompetent

**TABLE 10.2   Mining employment by company
sector in the Sudbury Area, 2011**

Company	Employment
First Nickel	
Lockerby Mine	120 (projected)
Ursa Major Minerals	
Shakespeare Mine	30 (mainly contractors)
KGHM	
Levack/Morrison Mine	224 (194 contractors)
McCreedy West Mine	98
Podolsky Mine	132
Victoria Mine (under development)	
Vale Canada	
Copper Cliff North Mine	
Creighton Mine	
Garson Mine	
McCreedy East/Coleman Mine	3 649
Stobie Mine	
Clarabelle Mill/Copper Cliff Smelter, Nickel & Silver Refinery,	
Sulphuric Acid Plant & Liquid Sulphur Dioxide Plant	
Totten Mine Project	150 (projected)
Xstrata Nickel	
Nickel Rim South Mine, Fraser Mine, Strathcona Mill,	
Falconbridge Smelter & Sulphuric Acid Plant	1 211 (299 at Smelter)

Source: Ontario Prospectors Association, *Ontario Mining & Exploration Directory 2011*, 27–28.

at best." Atkins made another assertion: "The failure of Inco to buy Falconbridge at an early stage from Noranda, when it was for sale at bargain basement prices, will go down in history as perhaps one of the most stunning management failures in Canadian corporate history, followed closely by its inability to buy Falconbridge, when it was expensive, a few years later."[59] Along the same lines, many locals have long perceived that the location of the Inco and Falconbridge Limited mines, processing plants and undeveloped properties in the Sudbury area made a joint venture seem like a no-brainer. Alas, it was not to be!

The history of antagonism between the two local mining companies was an important factor behind this merger fiasco. From the outset, Inco viewed Falconbridge simply as an upstart to be tolerated, initially because of the threat of anti-monopoly legislation in the United States, and later because of the subsequent impact of the DMPA contracts of the 1950s. Despite the obvious savings possible from uniting their side-by-side operations in the Sudbury area, the two companies maintained their acrimonious relationship into the 1990s. Even the Honourable Judy Erola P.C. from Sudbury, former federal Member of Parliament for the Nickel Belt riding who was appointed to the Board of Directors of Inco in 1987, found the existing conflicts between the two companies hard to understand.[60] According to John Fera, President of the United Steelworkers Local 6500, the two companies "hated each other" so much that joint venture projects commonly found in other Canadian mining camps were rarely attempted.[61] While the nickel discovery at Voisey's Bay in 1996 provided another opportunity for some form of cooperative endeavour between the two companies, this prospect fell by the wayside, at considerable cost to Inco.

An insightful view of Inco's historical relationship with Falconbridge was made by Val Ross in a devastating critique of the company published in *Canadian Business* magazine in 1979.[62] In this article, Ross traced the unique role that Inco played during its hegemony of the nickel market. After its formation in 1902, the company forged a belief in Canada and throughout the world that Inco *was* nickel and that it was simply "the king." This strategy was so effective that it pervaded the mindset of politicians in Ottawa, Toronto, Sudbury, and Washington. In Canada, the company showed distain to its relations with governments and other nickel companies such as British America Nickel Corporation and Falconbridge Nickel. Inco also largely ignored its environmental record, and looked down upon unions. Whenever provincial or federal legislators suggested economic policies inimical to its interests, the company responded by threatening to close down its plants and move its operations elsewhere if they were implemented.[63] This raised the spectre of Sudbury becoming a ghost town. What made this approach possible for so long was the fact that Inco had a monopoly on nickel, one of the world's most crucial minerals, especially during times of conflict such as the First and Second World Wars, and the Korean and Vietnam Wars.

Born rich, Inco relished its role as being one of Canada's largest companies. Its impressive headquarters erected at One New York Plaza and First Canadian Place in Toronto gave tangible expression to its imperial role. This setting allowed Inco to acquire a sense of arrogance and hubris—an exaggerated sense of its own importance. Backed by its monopoly position, Inco proved capable of eliminating its early competitors, and would most likely have undercut Falconbridge had it not been for the US Department of Justice and its threats of trust-busting legislation in 1928, 1946, and 1975. Inco's corporate state of mind and arrogance toward Falconbridge existed up to the 1990s, blinding senior company officials to the shifting nature of the mining industry that would ultimately leave Inco in its wake. The company's attitude was thus an apt demonstration of the renowned economist Peter Drucker's assertion that hegemonies eventually collapse, and that "a power that has hegemony always becomes arrogant ... and overweened.... It becomes defensive and a defender of yesterday. It destroys itself."[64] The historical events that took place around the turn of the century showed that while Inco may have been a multinational corporation in terms of its structure, it was not global in attitude and competitive ability.

Changing forces of demand and supply and an outdated management style left Inco and Falconbridge vulnerable to foreign takeover. While one of the greatest achievements of Inco over the years was its scientific prowess in expanding the demand for nickel beyond military uses, and making it an indispensable part of everyday life, this asset also proved to be a liability. The company was often caught short-handed, notably during the Korean and Vietnam Wars. This opened the door for newer producers elsewhere in the world. While it is true that both Inco and Falconbridge made remarkable strides in terms of increased mechanization, automation, and reduction in labour costs following the oil crises of the 1970s and 1980s, this had a long-term downside effect, since their enhanced competitiveness made them increasingly attractive to other mining companies. As well, global competition dramatically increased in the early 1990s following the collapse of the Soviet Union and the flow of metal from Norilsk Nickel, now the world's leading producer of the mineral. By the early 2000s, Canada had fallen to the rank of the world's third largest nickel producer. Both Inco and Falconbridge now had the name "victim" written all over them. The two corporations, as was the case for other Canadian firms such as Dofasco and Hudson's Bay, were seen globally as being reactive rather than proactive, as "the hunted" rather than "hunters."[65] The attempt on the part of Inco to merge with Falconbridge after the millennium was a tactic that proved to be "too little and too late." Inco's management style, seen by many as stodgy, bureaucratic, and slow to make decisions, was part of its undoing as a stand-alone entity in the rapidly changing and consolidating global mining industry.[66]

Given the historically weak policy stances of the Government of Canada and the Province of Ontario with respect to foreign takeovers, it simply became a matter of time before a key Canadian resource like the mines of the Sudbury Structure was allowed to fall into Brazilian and Swiss hands. A sad commentary indeed!

CHAPTER 11

From Company Town Setting to Regional Constellation (1939–1973)

The global situation of the 1940s and 1950s was fortuitous for the Sudbury area, as it set the stage for major economic and population growth. This growth, in turn, allowed the region to shed its former colonial frontier image. The relatively simple town of 1930 had matured into a crowded city, and the faint remnants of the dirty thirties had long disappeared. New waves of immigration resulted in an expanded and more diversified cultural milieu. Due to its strategic importance for the war effort, the area weathered this turbulent period better than most other urban centres in Canada. The outbreaks of the Korean and Cold Wars continued this fortuitous trend. Provincially, Sudbury's position in Northeastern Ontario was strengthened considerably. While the region attained the population level associated with that of a metropolis, however, the reality was that the urban structure lacked an essential feature associated with this status. Rather than displaying a strong central core surrounded by a circular zone of suburbs stretching into the hinterland, the population was dispersed in a pattern influenced by local topography, water bodies, the Sudbury Structure, the distribution of mining ownership and a variety of company-town settlements. The widespread clustering of the population thus came to resemble that of a constellation of stars centred on a larger orb, the City of Sudbury. While this aggregate form had some centripetal or uniting features, the political reality was much different. A multitude of municipal structures and other centrifugal forces came into being that overshadowed these integrating features. Divisiveness rather than unity ruled the political scene. While the core City of Sudbury continued to dominate the region in the postwar years, the ongoing vibrancy exhibited by its suburbs, outlying centres in the Valley, and elsewhere kept its dominance in check. A major casualty for the city was the loss of the downtown's previous centrality and vibrancy.

By the late 1960s, these local issues all came to the forefront; provincial policies related to political restructuring also came into play. These twin forces eventually culminated in the formation of the Regional Municipality of Sudbury in 1973.

The Second World War Era

No better harbinger of Subury's significance to the rest of the world can be found than the royal visit made by King George VI, Queen Elizabeth, and Prime Minister Mackenzie King on Friday, June 2, 1939, just prior to the outbreak of the Second World War. On this special occasion, the King and Queen of Canada were able to see, by means of a trip down into Frood Mine, the enormity of the mining resource that was to play such an important role in determining the outcome of the military conflict.[1] It was perhaps the most festive event in local history. While the Second World War caused many local ripples, the overall effect was that Sudbury "suffered less than most cities and prospered more than most."[2] It was partly for this reason that the municipal political scene remained stable throughout the 1940s under the helm of Sudbury's mayor, Bill Beaton (biography 8).

While Sudbury prospered more than most cities, it nevertheless experienced its own problems. The most pressing was that of crowding. The city's growth rate of 74 per cent in the 1930s was followed by another growth spurt of 32 per cent in the 1940s. This more than doubled the population in two decades. Compounding the crowding issue was the fact that, despite five annexations, the city's boundaries had increased by only some 134 hectares between 1892 and 1951.[3] Most of this expansion occurred in the southern part of the city, where the boundaries were extended to the Lily Creek area. Since much of the

---

Biography 8

WILLIAM "BILL" BEATON
"Mayor of the People" (1895–1956)

Bill Beaton was born on August 19, 1896 in the township of East Gwillimsbury, York County in Ontario. His parents were Scottish immigrants, and he was the youngest of six children. The family moved from the farm first to Huntsville and then to North Bay, where Beaton received his schooling. He worked for the Temiskaming and Northern Ontario Railway in 1912 and was then hired by the CPR in 1915. He married Isla Robertson in 1921, and they raised five children. He transferred to Sudbury as assistant chief clerk of the CPR in 1921, where he remained until 1927. That year he opened his own insurance business, which he operated until his death.[a]

Beaton's political career began in 1936 when he was elected to city council as an alderman. Re-elected as alderman for the next three years, he tried for the mayor's seat in

1940, but lost. He won the mayoralty in 1941, and remained undefeated for the next eleven years. His long tenure symbolized the relative stability that existed at the city level throughout the 1940s. Recognizing that Sudbury was a workers' community, Beaton supported the Mine Mill union's right in 1943 to hold a spring parade that was ultimately denied by the Sudbury Police Commission. While he ran for political office at the provincial level in 1943 and 1948, he was defeated on both occasions. It was clear that Sudbury voters preferred to keep their local politics separate from their workplace and the provincial government. Beaton effectively served as "a mayor of the people" for most of his political career, and supported virtually every proposed civic project, but the citizenry eventually tired of what they perceived as a one-man rule on council and bounced him from the mayoralty in favour of Dan Jessup, a local grocer, in 1951. He was later elected to Sudbury's Board of Control in 1955 and 1956. He likewise served one term as president of the Ontario Municipal Association.[b]

Beaton was an enthusiastic go-getter and very involved in sports. There is no doubt that the prevailing "cult of the sports personality" that existed in the area was a major factor in his political longevity. In 1919, he was one of the founders of the Northern Ontario Hockey Association. He rebuilt what later became known as the Sudbury Canoe Club, and it was under his direction that canoeing in Sudbury became a serious competitive sport. Bill Beaton and Gib McCubbin finished first in the senior tandem race at the Canadian National Exhibition in 1926, beating Olympic paddlers and the world record by four seconds. Beaton and his wife Isla also won nine consecutive mixed tandem canoe races in Northern Ontario.

One of Beaton's main objectives as mayor was to force the province to devise special revenue for mining communities to compensate them for their inability to tax mining companies under the Assessment Act. His lobbying was effective, and such compensation arrived in 1952. Two of Beaton's other accomplishments involved the replacement of the Sudbury-Copper Cliff Suburban Electric Railway Company with City Bus Lines, and the completion of the Sudbury Arena. While the electric railway had been of exceptional value during the Second World War for transporting workers to essential jobs, the rise of the automobile and the advantages of gas-powered buses made the system obsolete. In 1950, the city purchased City Bus Lines and phased out the last of the last of the streetcars. In the same year, he was instrumental in hiring two planning consultants to assess land use issues in the city and its environs. The sporting scene received a major boost in 1951 with the opening of Sudbury Arena, a structure that replaced the old Palace Rink, which had been demolished in the late thirties. The new arena eliminated the city's reliance on Stanley Stadium located in Copper Cliff.

The Beaton Classic, an annual sports marathon in the city, is named in Beaton's honour. He passed away in 1956. In his eulogy, Rev. E. Lautenslager stated that "Bill Beaton gained early renown in a sport that is typically Canadian. He did everything in life the way he paddled—all out from start to finish."[c]

---

[a]"Great Lives Lived in Greater Sudbury," *South Side Story*, January, 2005, 23.
[b]Much of the information for this biography was derived from "Held Record as Mayor, Bill Beaton Dies at 59," *Sudbury Star*, April 2, 1956, 1 and 3.
[c]Cited in "Hundreds Pay Final Tribute to Con. Beaton," *Sudbury Star*, April 6, 1956, 1.

land within the city consisted of rocky highlands or was owned by the CPR, many areas were overcrowded (aerial photograph 11.1). According to a planning study undertaken in 1950, the city's gross density of population was approximately 50 persons per hectare, a higher degree of crowding than that which existed in any Canadian city of the same order. The study concluded that local urban development featured many undesirable living conditions, and that ugliness and disorder were prevalent everywhere; equally disturbing was the fact that many of these characteristics were emerging in the outlying areas. The study correctly predicted rapid growth in the areas beyond the city limits in the next five years, a development that called for immediate regional planning.[4]

Another major issue was that of transportation and parking. This was most evident in the downtown due to its compactness, its inability to expand because of CPR land ownership, the rocky terrain, and the intersecting nature of the road and railway network.

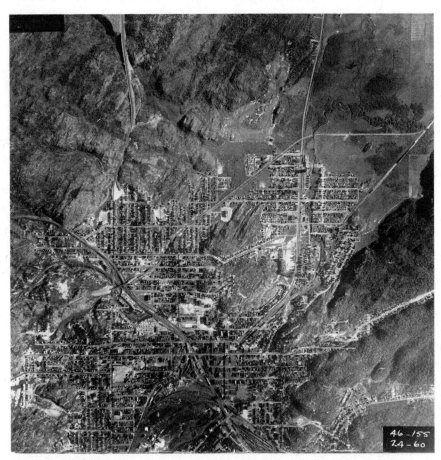

AERIAL PHOTOGRAPH 11.1 Sudbury (1946). *Source*: Ministry of Natural Resources Records, Archives of Ontario, 46-155, 24-60 (RG 1-429).

Since the major roads ran from southwest to northeast, whereas those of the main CPR railway line and its Algoma branch ran from southeast to northwest, a number of dangerous intersections evolved, slowing traffic and causing congestion, notably at the Riverside Drive and Elm Street crossings. Unlike many other cities, Sudbury had no inner or suburban ring roads to divert traffic. Motor traffic through the city, therefore, had to travel through the downtown, and for this reason the intersection of Elm and Durham Streets became the best-known street corner in the downtown. This distinction was architecturally enhanced by the imposing facade of the Federal Post Office and its magnificent clock. While the city was the first in Canada to establish parking meters in the downtown (installing them in 1940) as a way of alleviating its traffic woes, this solution proved to be only partially successful.

The downtown exhibited a liveliness and vitality that is non-existent today. It was a "people's place," where residents often knew one another by name. Since the adjoining neighbourhoods were within easy walking distance, men's and women's clothing shops, movie houses, hotels, and diverse retailers such as Kresge's, Cochrane-Dunlop Hardware, and Morse Credit Jewellers all thrived in the area. Daytime grocery shopping was available since A&P and Loblaws still had their establishments situated in the core. Every weekend, throngs of high school students would flock to the downtown for their weekly dose of French fries at the Chinese-owned Radio Lunch and Manhattan restaurants.

The Second World War resulted in many valiant contributions to the war effort. Thousands of men and women enlisted, and of this total, 489 died in combat. Fundraising efforts were undertaken by many groups, and the 1941 Victory Loan Bond campaign was highly successful. With respect to the latter, the small Chinese community played a huge role.[5] Due to the shortage of labour, women took jobs with the CPR and Inco, and filled other roles traditionally held by men, such as service station attendants.[6]

Numerous issues arose, however, involving the ethnic and language makeup of the population. There were controversies regarding leftist organizations, the imposition of enemy alien status on many individuals because of race, and the conscription issue. In 1940, an Order-in-Council (PC2363) was passed by the Government of Canada banning sixteen organizations with Communist leanings, along with their halls. In the Sudbury area, this order affected the Ukrainian Labour Farmer Temple Association and the Finnish Organization of Canada, both of which had halls located on Applegrove Street. In another Order-in-Council passed in the same year, certain ethnic groups were designated as enemy aliens (along with their Canadian wives) including Austrians, Czechs, Finns, Germans, Hungarians, Italians, Romanians, Slovaks, and Ukrainians. To overcome the stigma of enemy alien status and the need to register and be fingerprinted by the local Registrar of Enemy Aliens, many ethnic organizations took extraordinary actions to clearly indicate that they were not Communist or fascist, and showed public support for the war effort. Conservative elements within the Ukrainian and the Finnish communities were among the most vocal in showing their support for the Allied cause.[7] While the Orders-in-Council were rescinded at the end of the war, for many residents they left a bitter taste that remained for years.

Another divisive issue was the 1942 referendum asking voters to release Prime Minister Mackenzie King from his promise not to introduce conscription for overseas service. Though Sudbury as a whole voted in the affirmative, its "yes" vote was less substantial than those made by communities elsewhere in the province. This weak support was due in large measure to the "no" vote of French Canadians residing in the Flour Mill neighbourhood, and the Italians living in Little Italy in Copper Cliff.[8]

When the war was over on August 15, 1945, the relief in Sudbury, as in the rest of Canada, was profound. While the impact of the Second World War was muted relative to most other Canadian cities, the end of the war in Sudbury was, for some unexplained reason, marked by the outbreak of Canada's only riot on V-J Day.

While construction of public buildings and homes continued during the last half of the decade, no one expected the great events that occurred in the 1950s, when American intervention from Washington wrought a dramatic transformation in the regional landscape.

American Intervention and Regional Population Shifts

During the 1950s, the region boomed in response to the influence of the American "military-industry complex," and the outbreak of the Korean and Cold Wars.[9] There were three important consequences to this American intervention. First, there was a dramatic increase in mining activity and employment. Second, this economic expansion permitted Sudbury area communities to collectively attain the population size of a metropolis in the eyes of Census Canada. Third, improvements in transportation finally began to diminish Sudbury's remoteness and hinterland status in relation to Northeastern Ontario and the rest of the province. By the advent of the 1960s, Sudbury had acquired many characteristics of a provincial central-place. These new attributes supported the transition of the area toward a more mature, service-oriented economy. Other changes involved the growth of a white-collar class, and new concerns relative to land use planning.

The explosive growth of the region in the 1950s was an outcome of the United States government's decision to encourage international competition for strategic minerals and break up monopolies such as Inco in order to contain Communism and promote global stability. The authors of a landmark report issued in 1952 by the President's Materials Policy Commission (also known as the Paley Report), concluded that among the shortages of strategic materials in the US, none was more critical than nickel.[10] To offset this deficiency, the United States spent US$789 million in the 1950s to acquire nickel via stockpiling and special purchase agreements. Although Inco and Falconbridge Nickel were parties to these arrangements, the latter proved to be the main beneficiary, because of the US need to stimulate a more competitive supply system. Sudbury likewise benefited economically from an agreement signed in 1954 between Eldorado Mining and Refining Ltd. and the US government for the purchase of more than $1 billion worth of uranium ore from Elliot Lake between 1957 and 1962.[11]

These developments spurred the arrival of natural gas to the Sudbury area. In 1956, the city approved a natural gas contract with Northern Ontario Natural Gas; this decision had major political consequences, as the city's mayor, Leo Landreville, was later forced to resign from the Supreme Court of Ontario bench after allegations that he had received stock favours in exchange for the contract.

The effect of the American contracts on employment levels in Sudbury was immediate and profound. The increase in mining that Inco had initiated as a result of the Korean War (1950–53) served as a prelude for more expansion. Falconbridge Nickel grew rapidly, and its economic influence spread to the Northern Rim. Significant population growth occurred, along with a reshaping of the settlement pattern. These changes set into motion Sudbury's first major postwar era of transformation. The population of the area grew a remarkable 62 per cent between 1951 and 1961, from 85 597 to 138 152. In the same interval, 3 000 mining-related jobs were added at Inco and Falconbridge Nickel. The resulting in-migration of people proved to be more than the City of Sudbury could handle. Little usable land remained to be developed, other than rocky escarpments and some low-lying areas. The high density situation was confirmed in 1953 by the Dominion Bureau of Statistics, which reported that with "42 410 jammed in 9 450 units," Sudbury was the most crowded city in Canada.[12]

The scarcity of developable land in the city and increased land prices brought about an outflow of the population into the less crowded outlying areas (aerial photographs 11.2–11.3). Three patterns of urban expansion took place. The first involved the spread in McKim and Neelon-Garson townships and the north part of Broder Township. In Mckim, residential areas sprouted around Lake Ramsey, on the flat plain situated north of the Flour Mill neighbourhood area, and later along LaSalle Boulevard. Veterans' housing projects came into being on Belfry and Atlee Avenues. A powerful symbol of the northeastern area's growing influence was the opening of the New Sudbury Shopping Centre in 1957. Farther east, also encouraged by Veteran's projects, growth appeared beyond Minnow Lake in Neelon Township, while to the south, pockets of settlement emerged in the Lockerby area of Broder. These new subdivisions caused the population of Neelon-Garson and Broder Townships to more than double in the 1950s.

The second distributional change was the rise of new subdivisions in the peripheral townships found west and south of the city. The western thrust occurred in Drury, Denison, Graham, and Waters Townships, whereas the southern penetration reached into the Long Lake area in unorganized Broder Township. Unlike the more urban settlements in the areas closer to the city, the outer townships had lifestyles more recreational and rural in nature, many of them reflecting a Finnish flavour.

The third trend was the rapid conversion of agricultural land in the Valley into residential subdivisions, notably along Highway 69 north from McCrea Heights to the Town of Capreol, and along Highway 541 from Azilda to the Town of Levack. This agricultural land became available for settlement purposes because milk and potato production was not profitable there, and farmers desired to supplement their incomes by working in the mines or subdividing their lands. During the fifties, the townships of Dowling, Balfour, Blezard, Rayside, and Hanmer all recorded significant population increases, ranging from 152 to 285 per cent. A regional view of this reshaping process is provided in figure 11.1, which illustrates the sprawling nature of the existing and proposed subdivision activity that occurred from 1946 to 1959. This important map shows that the subdivision activity that took place in the postwar years not only consolidated the constellation form of the area but also proved to be a forecast of the geographical shape of the regional municipality that came into being in 1973.

The sprawling trend in these undeveloped areas had important consequences related to the future planning, servicing, and governance of the Sudbury area. Sprawl notwithstanding, some compact and orderly developments were planned in the City of Sudbury,

AERIAL PHOTOGRAPH 11.2   McKim Township North of Minnow Lake (1946). *Source*: Ministry of Natural Resources Records, Archives of Ontario, 46-155, 24-63 (RG 429). This photograph shows the future site of big box stores and the New Sudbury shopping centre.

notably at the Inco and Falconbridge Nickel townsites. Planned subdivisions were also constructed in the Beaton, Alexandra Park, Lakeview, Northern Heights, Marymount, Holditch, and Park Ridge areas. These residential areas were easily recognizable on maps, as their winding streets formed a sharp contrast to the traditional gridiron face of the earlier city.

Outside the city, company townsites were expanded and new ones created in response to the US stockpiling and special purchase agreements. In 1950, Inco expanded its housing supply at Levack and work was begun on a new townsite known as Lively. Incorporated as a town in 1953, it consisted of 450 homes and soon achieved the standard three thousand population level typical of other company towns in the area.[13] Construction simultaneously began on the Canadian Forces Station to service the Pinetree Line radar

AERIAL PHOTOGRAPH 11.3   The Lockerby Area (1946). *Source*: Ministry of Natural Resources Records, Archives of Ontario, 46-157, 204-9 (RG 1-429). This photograph shows the Idylwylde Golf and Country Club, the start of the Four Corners area, and the future sites of the Moonglo and Maki Avenue subdivisions.

defence station. Another important development was the creation of a new townsite by Falconbridge Nickel in the Improvement District of Onaping in 1956. Built to serve the needs of the Hardy and Levack mines, this attractive community reached a population of more than 1 000 by the end of the decade. While these military and mining townsites were compact and orderly developments, their overall distribution of population within the Sudbury Igneous Complex (SIC) continued to reinforce the constellation setting.

An Evolving Cultural Mosaic

The rapid growth of the postwar economy drew new immigrants from Europe, and the ethnic setting changed considerably (table 11.1). The era was marked by a major influx of Italians, especially after the last restrictions against enemy aliens were lifted in Canada.[14] Significant numbers also came from Germany, Poland, Finland, and the Ukraine. Large-scale immigration continued throughout the 1960s, notably for the Finnish and Italian

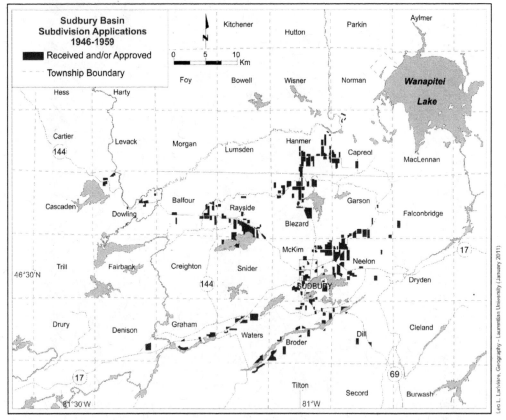

FIGURE 11.1. Subdivision Activity in the Sudbury Area (1946–1959). From the Ontario Department of Municipal Affairs, *The Sudbury Area*, Toronto, 1960, 7.

groupings. One of the distinctive features of this period was the high degree of visibility associated with the immigrant population, as the urban landscape was transformed by the construction of new churches and halls, such as the construction of the Polish Hall in the West End in 1946 and St. Casimir's Church on Paris Street in 1955.[15] There was, however, one exception to this pattern of ethnic visibility. For a variety of reasons related to the Second World War and postwar prejudices, the German population stuck to the privacy of their homes for gatherings and cultural events. This blending into the general population was reinforced by their preference not to live in clusters, and an attempt to "hide their origin as best they could."[16] Another factor was the City of Sudbury's refusal to permit Germans to establish their own clubhouse on Howey Drive in 1958.[17] German newcomers found some consolation through membership in the Lutheran Church, and by establishing their reputation as excellent workers in the mining economy and as *hausfrau* domestics.[18] Beginning in the 1970s, postwar immigration from Europe declined: thereafter, international immigration ceased to be a major element shaping the demography of Sudbury.

As shown in table 11.2, postwar immigration likewise affected the religious composition of the area. Since most of the new arrivals were Roman Catholic, the historical

TABLE 11.1 Ethnic origins for the Sudbury area (1951–2006)

Ethnic Origin	Year						
	1951	**1961**	**1971**	**1981**	**1991**	**2001**	**2006**
Canadian					2685	74955	64345
French	16060	40012	58080	51415	78750	59680	62615
English					53520	30310	35745
Irish	(15502)	(37084)	(56985)	(47690)	27925	24915	30430
Scottish					24030	21305	26595
Aboriginal	69	–	605	1445	7265	10301	14605
Italian	1502	8080	10340	8110	11985	12030	13415
German	845	4215	5005	3285	11750	10185	12140
Ukrainian	2571	4942	5625	3395	6860	7140	7590
Finnish	1478	2994	5470	4490	6670	6870	7280
Polish	1127	2845	2900	1835	4750	4275	4750
Dutch	284	1064	1330	930	3415	2720	2890
Total Pop.	42410	110694	155460	148695	156120	153895	156400

Source: Statistics of Canada, *Censuses of Canada* 1951–2006.
Note: The 1951 figures refer to the City of Sudbury. The figures from 1961 to 1991 refer to the Sudbury Census Metropolitan Area. The years 2001 and 2006 refer to the City of Greater Sudbury. The figures in parentheses refer to the British ethnic group. Prior to the 1981 Census, only the father's (paternal) ancestry was reported. In 1981, multiple ethnic origins were allowed and the paternal requirement was dropped. This resulted in many statistical changes in the ensuing censuses. By 1991, there was an increased response to the use of the term "Canadian" as a form of ethnic response. Thus, there has emerged a less rigorous form of ethnic ancestry data and a greater difficulty in making historical comparisons. As well, there is no longer any correlation between the number of ethnic responses and the total population.

pattern of Catholics comprising around two-thirds of the total population was maintained. Within the Protestant sector, the United Church remained the most dominant; other faiths of importance included the Anglicans, Lutherans, and Presbyterians. Starting in the 1970s, the relative importance of the Protestant churches began to decline. Of the minority groupings, the Pentecostals and Jehovah's Witnesses showed the most growth.

A Growing Central-place in Northeastern Ontario

The expansion of the mining industry was not the only factor bringing about rapid population growth in the fifties. Also important was the area's emergence as a regional central-place for Northeastern Ontario, especially in transportation and communications. Until the Second World War, the Sudbury area had a marginalized location with respect to the province as a whole. The city had no direct road connection to Toronto, so North Bay served as the de facto place of choice in the north for many government facilities and tourist attractions, a trend fostered by the enormous attention given to the Dionne

TABLE 11.2   Population by Religion for the Sudbury Area (1951–2001)

Religion	Year					
	1951	**1961**	**1971**	**1981**	**1991**	**2001**
Anglican	3 210	8 060	10 845	9 100	8 330	8 560
Baptist	747	1 838	2 785	2 450	2 205	1 965
Greek Orthodox	716	1 421	1 545	1 230	1 000	445
Jehovah's Witnesses			1 035	745	1 110	915
Jewish	184	233	245		230	145
Lutheran	1 971	6 193	7 150	6 070	4 660	4 180
Mennonite	7	57	160		210	60
Methodist					135	25
Mormons			215		375	225
Muslim/Islamic					470	510
Pentecostal		394	1 055	1 545	2 080	1 740
Presbyterian	1 687	3 002	3 640	2 800	1 795	1 135
Roman Catholic	25 366	69 022	98 345	96 075	100 235	98 220
Salvation Army			645	465	395	155
Ukrainian Catholic	1 152	2 061	2 240	1 495	910	870
United Church	6 880	16 766	19 630	18 575	14 710	12 865
Protestant						2 775
Christian						1 765
Other/None	490	1 647	5 930	8 140	17 270	16 855
Total Population	42 410	110 694	155 465	148 690	156 120	153 510

Source: Statistics Canada, *Censuses of 1951–2001*.
Note: The figures for 1951 are for the City of Sudbury. The figures for 1961 to 1991 are for the Sudbury Metropolitan Area. The figures for 2001 are for the City of Greater Sudbury.

Quintuplets phenomenon internationally. Indeed, after their birth in 1934, "Quintland" in North Bay for some time served as Canada's biggest tourist attraction.[19] North Bay's regional hold over Sudbury diminished considerably with the opening of the Sudbury–Parry Sound stretch of Highway 69 in 1952, and its extension to Gravenhurst in 1956. For hundreds of university students who went from Sudbury to southern universities each year, this new route was a welcome change. Another advance in transportation was the opening of the Sudbury Airport in 1952, and the beginning of Trans Canada Airline flights in 1954. Internally, the streetcar system came to an end in 1950, replaced in 1951 by Sudbury Bus Lines owned by Paul Desmarais, who later rose to the top rank of the Canadian establishment as the CEO of the Power Corporation of Canada.

The communications scene became more sophisticated with the opening of CKSO-TV in 1953, Canada's first privately owned commercial television station, and in 1957 by the all-French radio station CFBR owned by Baxter Ricard (biography 9). These new facilities, along with the established radio stations CKSO and CHNO, weakened the powerful hold that *The Sudbury Daily Star* had on public opinion. Even so, the newspaper's territorial reach continued to expand into far-flung places like Elliot Lake and Manitoulin Island.

While these events were taking place, the position of Sudbury as a retail, wholesale, entertainment, and service centre took on new life because of the discovery of uranium at Elliot Lake. An agreement signed in 1954 between Eldorado Mining and Refining Limited and the US government calling for the purchase of more than $1 billion worth of uranium ore from 1957 to 1962, and the need for a new highway and a townsite with a population in excess of 25 000 had an immediate impact on many sectors of Sudbury's economy. This link bolstered job prospects for women and brought about some gender diversification of the economic base. Many men from Sudbury also acquired mining jobs at Elliot Lake, often returning to Sudbury on weekends.

Education was another factor bringing about a change in Sudbury's centrality. Most noteworthy was the evolution of Le Collège du Sacré-Coeur into the University of Sudbury. Under rights granted by its original charter of 1914, the directors of the Jesuit College decided in 1956 to transform the institution into a university. The University of Sudbury was thus approved by the Province of Ontario in 1957. In the following year, classes were started at temporary accommodations in Sudbury's downtown. Meanwhile, opposition to a Catholic university grew in the form of the Northern Ontario University Association under the vocal leadership of Rev. E. Lautenslager of the United Church in Sudbury. The purpose of the organization was to create a Protestant University or college within a federated university which would include the University of Sudbury. Heated debates regarding the issue took place, and another group even advocated for the location of the university in North Bay. As a result of this debate, and due to the province's policy of not funding denominational universities, an accommodation was reached that culminated in the formation of a federated bilingual university known as Laurentian University in 1960.[20] When the enlarged university opened in this year, it consisted of a non-denominational college as well as two federated religious colleges—the University of Sudbury (Roman

Biography 9

BAXTER RICARD
Proud Supporter of the Francophone Community (1905–1993)

Baxter Ricard was born in Verner, Ontario, in 1905. As a youngster, he moved with his family to various places such as Victoria Mines, Gowganda, and finally to Sudbury, where his father ran Ricard's Hardware on Lisgar Street.[a] The store had the first gas pump in town. Ricard later purchased the store from his father and ran the business until the end of the Second World War, when it was closed due to the decline in lumbering and increased competition from other wholesalers and retailers.

Earlier, however, he had been approached by Senator Hurtubise, James Cooper, Leo Gauthier, and Joe Samson to join with them in acquiring a French-language radio station licence in the Sudbury area from the CBC board. They were successful, and in 1947 CHNO began broadcasting on the air, the first bilingual radio station in Canada outside of Quebec. Ten years later, in 1957, Ricard created a precedent when he became the first broadcaster in Canada licensed to operate two AM stations in the same city. CHNO became an all-English station, and CFBR served as a French outlet.

A measure of Ricard's dedication to the importance of francophone programming is shown by the fact that the French radio station was never a profitable operation, and had to be continuously subsidized by his English radio station. By 1984, Ricard calculated that the French radio station had cost him more than $1 million. With a listening audience of only around 700 at the time, this was not a surprising development. When asked why he still kept the radio station alive, he mentioned the deep influence of his father, who had earlier fought against Ontario's Regulation 17 prohibiting the teaching of French in provincial schools.[b]

In 1972, Ricard launched Northern Cable, a cable television service for Northern Ontario. Later, in 1980, Northern Cable established Mid-Canada Communications to oversee Ricard's television and radio holdings in Northeastern Ontario.[c] Under this company, he became the major shareholder in MCTV, itself a product of the merger of Cambrian Broadcasting (which owned CTV affiliates in Sudbury, North Bay, and Timmins, and another station in Pembroke), and J. Conrad Lavigne's CBC holdings. His operations in Ontario expanded in 1985 with the creation of Mid-Canada Radio, a group of fifteen radio stations, ranging from Wawa in the west to Pembroke in the east. Ricard's holdings were sold off in 1990 in a series of transactions: MCTV was acquired by Baton Broadcasting, Mid-Canada Radio was sold to Pelmorex, and Northern Cable was purchased by a cable division later acquired by Persona.

Ricard was a strong community supporter. He was Chairman of the Sudbury General Hospital Board and the Borgia Street Area Redevelopment Project. As well, he served as president of Sudbury's Chamber of Commerce, the Northern Ontario Chamber of Commerce, the Kiwanis Club, and Club Richelieu.[d] He was the recipient of numerous honours from the Canadian Broadcasters' Association, the Canadian Association of Broadcasters Hall of Fame, and the Northern Ontario Business Awards. He was given an honourary doctorate from Laurentian University in 1987. Ricard died in 1993, leaving behind his wife Alma and no children.

During their lifetimes, Baxter and Alma made large contributions to many building projects in the Sudbury area, including Laurentian University and a residence at Cambrian College. Again reflecting Ricard's deep interest in promoting francophone culture, his wife Alma announced in 1998 the creation of a $23 million educational charity called *Fondation Baxter and Alma Ricard*. The foundation provides generous financing—up to $50 000 a year—for francophones outside of Quebec who have completed an undergraduate degree, and are pursuing a master's degree or a doctorate. The first award was granted in 1999. Another legacy was the naming of a school by the Sudbury District Roman Catholic School Board in honour of Ricard's father, Felix. Ricard's wife, Alma, received the Order of Canada in 2000, and was the recipient of honourary degrees from Laurentian University and the University of Ottawa. She passed away in 2003.[e]

[a]"Baxter Ricard," in *Greater Sudbury 1883–2008*, ed.Vicki Gilhula (Sudbury, ON: Laurentian Media Magazine Group, 2008), 65.
[b]Baxter Ricard, "Too Busy to Retire," *Northern Ontario Business*, December 1984, 25.
[c]*Wikipedia*, s.v. "Mid-Canada Communications," accessed December 9, 2009, http://en.wikipedia.org/wiki/Mid-Canada Communications.
[d]Ricard, "Too Busy to Retire," 26.
[e]"Sudbury Loses Its Greatest Philanthropist; Ricard Dies at 96," *Sudbury Star*, June 4, 2003, A1 and A4; and Alma Ricard, *Fondation Baxter & Alma Ricard* (Sudbury, ON: Fondation Baxter & Alma Ricard, 1998).

Catholic) and Huntington University (United Church). In 1963, Thornloe University (Anglican) became part of the federation; the following year, the university moved from its temporary downtown location to a new campus adjacent to Ramsey Lake Road. Three colleges—Algoma University College in Sault Ste. Marie, Nipissing University College in North Bay, and Le Collège Universitaire de Hearst—became affiliated with Laurentian in 1965, 1967, and 1970, respectively. In 1974, the French-speaking Teacher's College known as l'Ecole des sciences de l'éducation was integrated into the Laurentian system.[21] In these early years, the Jesuit influence within the university was impressive; by the turn of the century, some eighty-three Jesuits were known to have had some association with the institution. Sudbury's involvement with post-secondary education continued with the formation of Cambrian College of Arts and Technology in 1966. The college began classes in 1967 in the Sacred Heart College building left behind by the Jesuits; by 1972, most classes moved to its new campus in New Sudbury. At the time, Cambrian had campuses in North Bay and Sault Ste. Marie, but these became independent entities in 1972. In 1973 the Northern Ontario Regional Health Science School and residence located on Regent Street became the college's third campus.[22]

Sudbury's regional strength extended to include the field of medicine. There were three hospitals constructed in the city during the 1950s: the Sudbury General Hospital of the Immaculate Heart of Mary in 1950 (at the time the city's highest building), the Sudbury-Algoma Sanatorium in 1952, and the Sudbury Memorial Hospital in 1954. Owned by the

Sisters of St. Joseph of Sault Ste. Marie since 1944, Sudbury General occupied an attractive site adjacent to Lake Ramsey unencumbered by the Bell property covenant.[23] Unlike the Sudbury General Hospital, Sudbury Memorial had no direct affiliation with a religious order. The Marymount School of Nursing opened in 1951, complementing the St. Elizabeth School which had moved into its imposing new nine-storey building on Mount St. Joseph in the same year.[24] The Marymount students were housed in a variety of locations, including the hospital itself, in the nearby Stafford and Silverman residences, and finally, in 1953, in the Mason residence. These two schools were amalgamated in 1968 under the umbrella of the Sudbury Regional School of Nursing. A five-storey residence was subsequently built for the regional student nurses near Memorial Hospital in 1971. In 1973, the school became part of Cambrian College. Another advance occurred in 1967 when Laurentian University admitted the first students to its School of Nursing. The university did this to accommodate a provincial directive that the teaching of nursing would henceforth be incorporated within post-secondary institutions as either diploma or degree programs, independent from any hospital and religious domination. In addition to these nursing facilities, a training school (later known as a Regional School for Nursing Assistants after 1977) for nursing assistants was established in 1954. Other important events included the formation of the regionally based Sudbury and District Health Unit in 1968, and the designation of Sudbury by the province as the regional health centre for Northeastern Ontario in 1968.[25] These moves consolidated Sudbury's medical hegemony in this part of the province.

Changes in the Urban Landscape

The postwar transformation of Sudbury introduced numerous changes to the urban landscape. Old landmarks disappeared, among them Central Public School across from the CPR station in 1950, the Kingsway and Copper Cliff arches in 1950 and 1952, the streetcars in 1950, the Jodouin Icehouse on Lake Nepahwin in 1951, the meat section of the City Market in 1951, the Balmoral Hotel in 1957, and regrettably, the federal post office building at the corner of Elm and Durham Streets in 1959. Other changes involved the social and cultural milieu, such as the opening of the Sudbury Arena in 1951 on the site of the former Central Public School, the construction of the Sudbury Public Library in 1952, the Pioneer Manor Home for the Aged, and the Community YMCA in 1953. The construction of the Sudbury Arena was an important event for sports in the area, as it allowed the community to continue a hockey tradition pioneered by Max Silverman (biography 10). The construction of two elevated Horton-style water towers along Pine Street in 1949 and Pearl Street in 1956 served as high-rise silos defining the western and eastern boundaries of the inner core of the city.[26]

The burgeoning growth of the student population both within the city and the outlying areas necessitated the expansion of existing schools and the construction of new ones. In 1951, Sudbury High School and St. Charles College moved into new quarters. Chelmsford High School opened in 1955, Nickel District Collegiate and Vocational

Biography 10

MAX SILVERMAN
"The Silver Fox" (1906–1966)

Max Silverman was born in New York City on August 25, 1906.[a] He arrived in Sudbury when he was three months old with his parents, who had come from Russia and Poland. After graduating from Central Public and Sudbury High School, he became active in the local hockey scene. Inspired by Fred Donegan, the organizer of the first junior team in Sudbury, Max embarked on a hockey career that lasted throughout his life. After the Northern Ontario Hockey Association (NOHA) was formed in 1919, he travelled with Sudbury's team to every game it played in Canada and the United States. The Palace Rink served as his second home. Starting as a water boy, Silverman worked his way through the ranks.

Calling himself "a poor excuse for a hockey player," he focused on becoming a hockey coach as his dream. Around 1930, he started coaching and managing in the old Nickel Belt League. Knowledgeable about the rulebook, he acquired a reputation for spotting every loophole. In 1931, he coached a mixed junior and senior all-star team that won the NOHA tournament, but lost the all-Ontario final. In the following year, the Sudbury Cubs, with Sam Rothschild (the first Jewish player in the NHL) as coach and Silverman as manager, won the Memorial Cup. He coached the Sudbury Cubs in 1935 and led them to the Richardson Cup and the Eastern Canada championships. He then brought the Copper Cliff Redmen to the Canadian finals in 1937. In the following year, he assembled a team under the Sudbury Wolves banner and took them to Czechoslovakia, where the team won the world hockey title.

Thus, in a span of six years, Silverman had served as manager to a Memorial Cup winner and coached a team to a world hockey title. This left the Allan Cup as the only amateur trophy remaining. In 1954, he almost attained this goal when he coached the Sudbury Wolves to the Allan Cup finals, where they eventually lost a controversial seven-game series to the Penticton Vees, along with the right to represent Canada at the World Hockey Championships. Clearly the dominant team, the reasons for Sudbury's inexplicable loss became legendary in terms of claims and counterclaims, and these remain to the present day. In 1956, he sold the Sudbury Wolves franchise to Harry Smith and his associates.

No man dominated the hockey scene in Sudbury as did Silverman. According to one account, "if there is any man more devoted to hockey than Max Silverman of Sudbury, he would be as hard to find as the proverbial needle in a haystack."[b] However, Silverman made enemies along the way. Elected president of the NOHA between 1946 and 1948, he was later unceremoniously tossed out by the organization. Many fans filled the Sudbury Arena after 1951 simply to boo him, and his fights with spectators became the stuff of legend. Yet, over the thirty years or more that he spent in the NOHA, it was universally accepted that he, more than anyone else, was responsible for the league's greatest era in the 1940s and 1950s. In recognition of Silverman's work, the NOHA gave him a lifetime membership in 1966. While living in Sudbury, he held court regularly at his hotel suite at the Nickel Range Hotel, and later at his apartment on Cedar Street.

At the urging of friends, Silverman decided to enter politics. His first stab at becoming an alderman for the new City of Sudbury in 1960 failed; in his next foray in 1961, he was ahead in the polls for Board of Control, and served as deputy mayor. Due to the illness of Mayor Bill Ellis at the time, he served as the acting mayor for five months. After losing a run for mayor to Joe Fabbro in the next election, he won the top office in 1965. In municipal office as in his hockey career, Silverman was a contentious figure. He adopted a "watchdog" role, and was a prime proponent for having a study of the city administration done by an outside management consultant. He was a strong supporter of the city's need to redevelop the Borgia district. Being a visible public figure, his walks through the downtown were continually interrupted by his admirers. Provincially, Silverman was recognized as one of the most colourful characters in municipal politics. His political life was cut short, as he died from abdominal cancer in 1966. He never married. His funeral was marked by a memorial held in the Sudbury Arena with the pallbearers consisting of many of his former players. After the funeral service, he was buried in the family plot in the Montreal Jewish Cemetery.

[a]For a review of Max Silverman's life, refer to Greater Sudbury Public Library, *Biographical Materials for Max Silverman (August 25, 1899–October 5, 1966)* (Sudbury, ON: Greater Sudbury Public Library).
[b]"Never a Hockey Fan Like Maxie," *Toronto Weekly Star*, January 3, 1953, 9. See also "Rather Good Name Than Great Wealth: Theme of Eulogy Spoken for Mayor," *Sudbury Star*, October 7, 1966, 10.

Institute and Marymount College in 1956–57, Lively High School in 1957, Levack High School in1958, and Lockerby Composite School in 1959. The same decade saw the formation of the Sudbury Male Chorus in 1951, the Sudbury Symphony Orchestra in 1953, and both the Sudbury Philharmonic Society and the Karl Pukara Accordion Orchestra in 1957.

The face of retail likewise diversified. In the downtown, chain stores were enlarged (Kresge's in 1950 and 1959), and new stores added, such as Canadian Tire in 1953, and Zellers in 1958. The Empire, Lasalle, and Plaza cinemas were built in the 1950s, as was a new Sudbury Hydro building (1957) and a federal post office (1958). The habits of grocery shoppers then changed because of the increased use of cars—a trend that brought about the shift of stores from the downtown to inner neighbourhoods where there was more space for parking. Loblaws moved to the intersection of Elm and Frood in 1950, A&P to Elm and Landsdowne that same year, and Dominion to Riverside and Regent in 1951. These shifts paled in comparison to the opening of the New Sudbury Shopping Centre in 1957, the most significant commercial innovation ever seen in Sudbury. Built by Principle Investments of Toronto, it was located on sixteen hectares of land at the corner of Lasalle Boulevard and Barrydowne Road, with parking for more than 2 000 cars. It immediately shifted the centre of the city's commercial and population gravity to the northeast, and showed that "the big city had arrived in Sudbury.[27] The year 1960 also witnessed the birth of one of Sudbury's own icons, when the first "one arch" Deluxe Hamburgers drive-in restaurant opened on Notre Dame Avenue. Others then appeared on Lorne Street (1963), the Kingsway (1966), and Regent Street (1968).[28]

As time progressed, it became clear that the internal changes wrought in the city and urban sprawl in the outlying areas had created serious local and regional planning issues. While tentative planning steps were taken after the Second World War (such as the formation of the joint McKim, Neelon, and Garson Planning Area in 1947, and the the Sudbury and Suburban Planning Area in 1948), they proved to be ineffective because of existing *laissez faire* land use policies and internal wrangling.[29] In 1953, the townships of McKim and Neelon/Garson were designated as planning areas; this step was followed by the City of Sudbury in 1954. Under pressure from the province, the city hired its first planner, Arnold Faintuck, in the following year.[30] Within the city, a temporary bylaw restricting all building outside of the main commercial areas was approved by the Ontario Municipal Board in 1955.[31] This gave the city a measure of planning control until an official plan and zoning bylaw could be approved. Faintuck was replaced by Klemens Dembek in 1957; he retained the position for several decades.[32] Under Dembek's initiative, the city's first official plan and zoning bylaw were approved by the OMB in 1959.

Elsewhere, municipalities in the Valley banded together to form the Chelmsford and Blezard Valleys Planning Board (later renamed the Nickel Basin Planning Board). Other planning initiatives during this period included the formation of the Sudbury and District Health Unit in1956, the establishment of the Sudbury & District Industrial Commission in 1957, and the creation of the Junction Creek and Whitson Valley Conservation Authorities in 1957 and 1959, respectively. The two conservation authorities were established to deal with the costly problems of flooding that occurred each year in the city's downtown and parts of the Valley

While there were many planning issues in the city, the most pressing was the negative effect on the downtown created by the creation of the New Sudbury Shopping Centre. Optimism for the rejuvenation of the downtown's core came in 1958 when the CNoR stated its intent to move its station in the Borgia Street area to a new site in Neelon Township. This spurred the creation of the Sudbury Urban Renewal Joint Committee in 1959, and proposals to rationalize the city's internal transportation network through the construction of an inner bypass around the downtown, and a freeway linking Highway 17 East to Brady Street over the rocky Kingsway plateau. This new highway would have bypassed the Kingsway, which by that time had already emerged as one of the more poorly planned areas of the city. While the transportation projects never materialized, the plans for downtown renewal forged ahead.

The city's rejuvenation began with the release of a general urban renewal report in 1963.[33] Its authors recommended the redevelopment of the Borgia Street area located in the city's downtown district. In 1966, three follow-up reports (Survey, Plan, and Economic Feasibility) were completed, and the two senior levels of government agreed to undertake the Borgia Urban Renewal Scheme. They took on the project because of urban renewal legislation that allowed the federal and provincial governments to cover 50 and 25 per cent of the total costs respectively. An innovative aspect of the process was the selection of Marchland Realties Limited as the developer without going through the competitive bidding phase.

The project, involving some twenty-four hectares of land, got underway in 1967 with the start of land acquisition and demolition. The effect on the landscape was profound. Buildings such as the Queen's Hotel and the CNoR freight sheds were demolished in 1968, and the levelled land was soon covered with deep steel pilings that pushed through the treacherous overburden into the bedrock itself. A new road network was established. In 1970, the first phase of the project was completed, with the opening of 177 Ontario Housing Corporation units to house residents who had formerly lived in the neighbourhood affected by the redevelopment project and the completion of the Holiday Inn Hotel on the site of the former St. Mary's Ukrainian Catholic Church. A new mall was opened in 1971 that included three movie theatres, forty storefronts for retail and services, and later an Eaton's department store as its core attraction.[34] High-rise apartments, a co-operative housing complex, and the relocated domed Ukrainian church were added later.

While the project itself was in many ways one of the most successful in Canada, it still did not deter the majority of shoppers from going to the New Sudbury Shopping Centre, where free parking was a powerful draw. One of the unfortunate legacies of the downtown renewal project was city council's rejection in 1968 of another commercial development on a large parcel of land at the corner of Elm and Elgin Streets. Submitted by Marathon Realty, the real estate arm of the CPR, this project was considered a threat to the Borgia development to which council had been committed via a profit-sharing agreement with Marchland Realties.[35] This short-sighted view on the part of city council not only perpetuated the existence of a huge eyesore but also virtually guaranteed that the downtown would never acquire the vitality possessed by those in other big cities. While some visual aspects of the site were improved by the construction of a Market Square, it never reached its full potential, and the property was taken over by Laurentian University in 2011 for its future downtown School of Architecture.

The Vexing Mining Taxation Issue

Another major issue in the 1950s was one that affected not only the financial stability of the area but also the relationship that existed between the mining companies and their surrounding municipal entities—the manner in which the operations of Inco and Falconbridge Nickel Mines were assessed and taxed. It involved one of the most unfair pieces of taxation and assessment legislation ever enacted in Ontario. Given its importance to Sudbury and other mining communities in Northern Ontario, it is useful to provide some background to show how this inequitable legislation came to be.

Starting with the first regulation of mining activity in 1845, the Ontario government regarded mining as a special industry that could not be compared to other industries for taxation and assessment purposes.[36] While provisions were made for royalty payments from mining companies, none were collected by the province until the Mining Act of 1906 was implemented, at which time a provincial tax on mining profits was introduced in lieu of royalties. Later, under the Assessment Act of 1914, it was deemed that "the buildings, plant and machinery in, on or under mineral land and used mainly for obtaining minerals from the ground or storing the same, and concentrators and sampling plant and the minerals in or under such land are declared not assessable."[37] Despite ongoing protests from mining communities, this unequal assessment continued for decades. Following a concerted effort on the part of the City of Sudbury starting in 1943, the province finally relented and introduced a grant in lieu of the mining tax in 1949.[38] The cheque of $100 000 that the city received that year was significant in that it represented the first time that the municipality had gained some direct benefit from the mines situated outside of its boundary. The grant was later changed to one based on the number of miners who resided in the city. City officials, however, contended that the grant in lieu of the tax system was inadequate, especially considering the fact that the mining profit taxes paid by Inco provided more revenue to the province than any other corporation.[39]

This principle of mine exemption from municipal property taxation was broadened in 1957, when the Ontario Supreme Court ruled that Inco's iron ore buildings west of Copper Cliff were not assessable for purposes of municipal taxation.[40] The judgment thus declared that all smelters and refineries were exempt from assessment. In submissions made to the province in 1964, both the City of Sudbury and the International Union of Mine, Mill and Smelter Workers continued their attacks on this ruling by making three major assertions: (1) despite producing more wealth than any other community of comparable size, the mining tax revenues received by the municipality were far less than those received by other industrial cities in Ontario; (2) the average Sudbury citizen paid more taxes than those living in other Ontario cities, yet had access to fewer municipal services; and (3) of the total amount of tax monies received by the province, only about one-third was returned to mining municipalities.[41]

While the province made no direct response to this submission, it did establish the Ontario Committee on Taxation which, in its 1967 report, made recommendations regarding the taxation and assessment of mining municipalities. These recommendations, however, were regarded by Inco and the City of Sudbury as being even more punitive as they would not provide sufficient revenues for municipal needs. They were also at odds with the taxing principles applied to other manufacturing processors in the province. Inco went so far as to claim that some of the tax reforms presented were a form of "outright discrimination."[42] A similar claim was made by the City of Sudbury in its response to the 1967 report and Inco's submission. Inco's submission was noteworthy, as it marked a significant shift in the company's attitude toward the taxation and assessment issue and the possible implementation of regional government. Inco then made the radical proposal that the Assessment Act should be changed to permit the municipal property assessment of all surface processing facilities, including smelters, concentrators, and refineries.[43] The proposal was striking because Inco finally acknowledged that the future absorption of the Town of Copper Cliff into an enlarged City of Sudbury was simply a part of existing regionalization tendencies that had already affected conservation authorities, social agencies, and school boards.

These events served as the backdrop to the contentious boundary disputes that took place between the City of Sudbury and its outlying municipalities throughout the 1950s.

Formation of the Regional Municipality of Sudbury

While the method whereby Inco and Falconbridge Nickel Mines were taxed and assessed was technically a separate legal issue, it nonetheless remained in the minds of local politicians as the battle raged over municipal boundaries and regional planning matters. The postwar rationalization of boundaries in the region began in a subdued fashion in January 1951, when the OMB granted the city permission to annex sixty-nine hectares of land from McKim Township (fig. 11.2).[44] The affected land was situated mainly south of a line joining the existing Sudbury General Hospital and École St. Denis elementary school. About the same time, the city requested permission to annex all of McKim Township, except the Towns of Copper Cliff and Frood. While the city's position rested mainly on the need to install municipal sewers in this area to avoid polluting Lake Ramsey, Inco argued that this was simply a guise to apply taxes to its Murray and McKim mines. The request set into motion a convoluted series of amalgamation and annexation requests to the OMB.[45] The only decision the OMB made was to grant the city permission to annex thirty-seven hectares for its Northern Heights land assembly scheme on the site of the municipality's former dump in 1957.[46] It was clear that the OMB was hesitant to deal with these boundary problems, as any decision would have important provincial policy implications regarding the future of other mining communities.

Notwithstanding the board's reluctance, serious planning issues forced the OMB to finally deal with the city's amalgamation and annexation requests. Following a series of hearings in 1958, the OMB ruled in November 1959 that as of January 1, 1960, the Town of Frood and the Township of McKim would both be amalgamated with the city, along with the west half of Neelon Township, thereby creating Ontario's sixth-largest city, with a population of about 75 000.[47] Specifically excluded from the amalgamation were the Towns of Copper Cliff and Coniston, and the Townships of Creighton, Snider, Falconbridge, Waters, Broder, and Dill.

Since the ruling made no attempt to reconcile the problem of mining taxation, one of its effects was to spread Sudbury's inadequate municipal services thinner. Thus, instead of resolving the city's taxation problems, amalgamation intensified them, leaving Sudbury in a more serious financial situation than any other municipality of similar size in Ontario.[48] Rather than being a city reasonably compact in area with a high population density trending toward a metropolitan status, the municipality was transformed into a lower-density settlement with several ribbons of urban sprawl.

The ineffectiveness of the OMB's 1960 decision was made clear in the ensuing decade. The financial benefits that were supposed to accrue from amalgamation never did occur; studies later revealed that the city was still experiencing severe financial problems. In 1964, a brief was submitted to the Ontario government concluding that the imposition

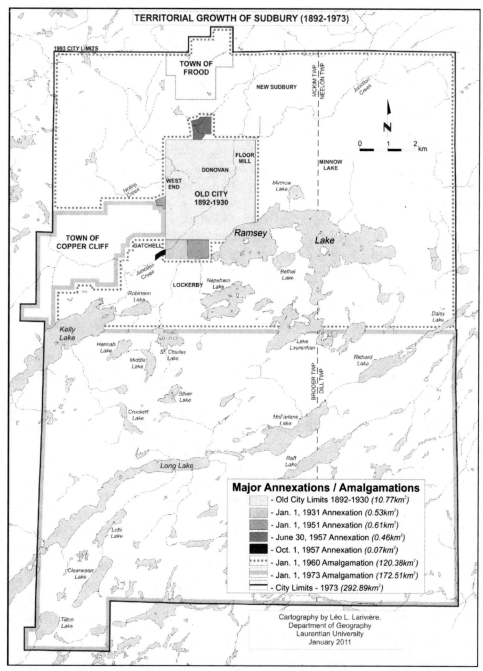

TERRITORIAL GROWTH OF SUDBURY (1892-1973)

Major Annexations / Amalgamations
- Old City Limits 1892-1930 *(10.77km²)*
- Jan. 1, 1931 Annexation *(0.53km²)*
- Jan. 1, 1951 Annexation *(0.61km²)*
- June 30, 1957 Annexation *(0.46km²)*
- Oct. 1, 1957 Annexation *(0.07km²)*
- Jan. 1, 1960 Amalgamation *(120.38km²)*
- Jan. 1, 1973 Amalgamation *(172.51km²)*
- City Limits - 1973 *(292.89km²)*

Cartography by Léo L. Larivière,
Department of Geography
Laurentian University
January 2011

FIGURE 11.2   Municipal Boundary Changes for Town/City of Sudbury (1892–1973). From Carl M. Wallace, "Sudbury: The Northern Experiment with Regional Government," *Laurentian University Review* 17, no. 2 (1985): 88.

of partial amalgamation had left the city a hostage, with inadequate financial resources and unresolved problems.[49] An economic study in 1967 affirmed that the economic tax base was inadequate because the region's largest industry, mining, was exempt, and the mining grants in lieu remained inequitable. The grants were made more intolerable in later years by the decisions of the Assessment Review Board to further reduce the value of mining properties.[50] The same study affirmed that the Sudbury region was economically and sociologically a single unit and that it should have been made a single unit politically. Finally, it claimed that the continued existence of company towns such as Copper Cliff and Lively were a glaring anomaly in mid-twentieth-century Canada.[51] Authors of another 1967 study, known as the *Nickel Basin Planning Study,* likewise concluded that the Sudbury Basin communities constituted a regional entity that required an official plan for the entire area.[52]

While these concerns were being expressed locally, a number of important developments were taking place at the provincial level concerning regionalization issues. These actions led to a reconsideration of the Sudbury area's future. Among them was the province's *Design for Development* (1966) initiative, and the regionalization of health care units (1967) and education (1969). The Ontario Committee on Taxation also recommended that municipalities should be given the right to tax mining properties. It was within this context that the province, through an Order-in-Council in 1969, appointed OMB Chairman J.A. Kennedy to report to the Minister of Municipal Affairs upon the structure, organization, and method of operation of the municipalities in the Sudbury area, and to suggest what reorganization of local government would be required for anticipated future development.[53] The process of 1960 was thus started all over again.

After a lengthy series of local hearings, Kennedy's report, titled the *Sudbury Area Study,* was released in May 1970. An indictment of existing conditions in the region, it called for the consolidation of the area municipalities into one city and five townships, to be administered by a two-tier system of government, with a weak upper tier. This recommendation ran counter to most local submissions, which stated, "give us the additional revenue and leave us as we are."[54] Given the high degree of parochialism and distrust between the City of Sudbury, the Town of Copper Cliff, and the Valley municipalities, it was inevitable that the report would be repudiated by all concerned. The issue polarized the political scene. As Wallace noted, "Copper Cliff was apoplectic, Capreol rebellious, and Sudbury grim."[55] When it became clear that local consensus could not be attained, the provincial government reacted by presenting its own solution in a publication titled *Sudbury: Local Government Reform Proposals* in March of 1971. In this report, many of Kennedy's proposals were rejected; however, the principle of regional government was reinforced through its recommendation of a two-tiered structure, with a powerful upper tier. Other recommendations included a council of seventeen, with eight councillors from the City of Sudbury, nine from the outlying seven area municipalities, and a regional chairman elected by all members of the regional council. The most important functions related to planning, the provision of water, sewer, and garbage services, arterial roads and highways, and public transit were allocated to the upper tier.[56] A year later, on June 29, 1972, the provincial legislature passed an act to establish the Regional Municipality of Sudbury (Bill 164), with some modifications to the earlier reform proposal.[57] Among them was an increase in the size of council to twenty-one members, with ten coming from the City of Sudbury, another ten from the six outlying towns, and a regional chairperson to be elected by the council. A Nickel District Conservation Authority was created, replacing the two in existence. These changes officially came into effect on January 1, 1973 (fig. 11.3).

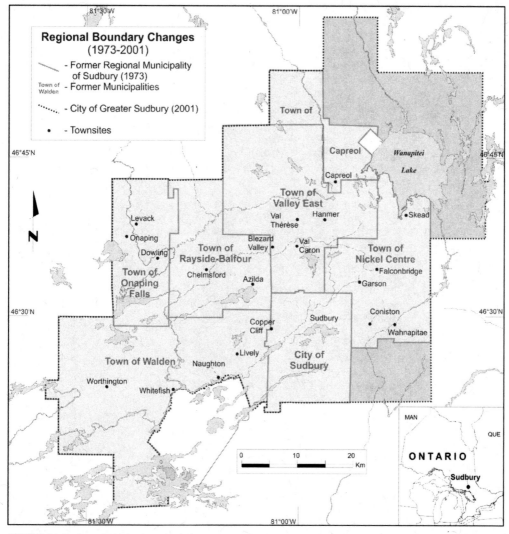

FIGURE 11.3 Regional Municipality of Sudbury (1973). From Léo Larivière, Cartographer and Map Librarian, Laurentian University.

Another important consequence of the new municipality was the shedding of the "company town" status and the attendant baggage with this nomenclature.

By this time, the provincial government had expressed its intention to finally allow municipalities to tax mining properties found on the surface, but not underground facilities. While this was a step in the right direction, its effect was less than anticipated because provincial assessors under-assessed the surface facilities; also, there was a subsequent decline in mining revenue payments.[58] This was the political setting that the Sudbury region found itself as it began to face the turbulent years of the latter part of the century.

CHAPTER 12

From Regional Constellation
to Greater Sudbury (1973–2001+)

While the formation of the regional municipality was a milestone in the political development of the Sudbury area, it proved incapable of resolving many long-standing divisions that existed. The unsettled political and planning conditions were exacerbated
by the turbulence arising out of the decline in the global demand for mineral products
and the concomitant reduction in Sudbury's employment levels. These traumatic events
were followed by one of the most dramatic examples of community change and rejuvenation seen in Canada. Aided by a new regional plan passed in 1978, the creation of the
group Sudbury 2001, and the implementation of the Task Force Reports of the 1980s, the
area began a process of diversification that resulted in numerous changes to Sudbury's
economy and its central-place setting within Northeastern Ontario. This process was
accompanied by a growing "sense of place," a belief among its residents that Sudbury
was now much more than a company town. These changes were of such a significant
nature that Sudbury was selected by the Organization for Economic Co-operation and
Development (OECD) in 1984 as a world model for urban transformation. These changes
notwithstanding, the area continued to experience economic turbulence that continued
until the turn of the twenty-first century. Starting in 1995, the area also came under the
influence of the provincial Harris government's "Common Sense Revolution." As part of
this revolution, the Regional Municipality of Sudbury was transformed into an enlarged
City of Greater Sudbury in 2001. In the decade that followed, a variety of transitional
elements appeared in the landscape that strengthened and diversified the economic and
sociocultural character of Greater Sudbury. These elements were of such strength that they
greatly muted the effects of the 2009–2010 strike by the Steelworker's Union against Vale
Inco. Recent developments since the strike offer considerable promise that the area is now
poised for a period of sustained growth.

Political Uncertainty, Economic Downturn, and Population Decline

The imposition of regional government by the province gave rise to years of political uncertainty and bickering between the City of Sudbury and the other municipalities. The new two-tiered system fostered ongoing parochialism that was exploited by local politicians. Citizens in the outlying areas complained about the remoteness of government and their loss of autonomy. The deep-rooted antagonisms of earlier years were carried into the 1980s with negative consequences. A number of unfortunate political events also occurred. The first regional chairman, Don Collins, a senior Queen's Park bureaucrat, was forced to leave before the end of his term; his successor, Delki Dozzi, died in office (table 12.1). The province's decision to assume control over the creation of a new official plan proved to be an affront to local planners and politicians. The end result of this top-down planning process was an expensive fiasco. After two years of research, the province produced a glossy, computer-generated planning document referred to as the Sudbury Area Planning Study (SAPS). It was quickly shelved by regional council.

Meanwhile, the planning process lost much credibility because of how the planning committee handled the infamous Lily Creek Plaza project in 1972. The proposal involved a rezoning of the Lily Creek area to permit the development of a shopping centre. When the full implications of what this rezoning meant for the area became evident to the public, the stage was set for one of the most complex planning battles ever seen in the province, a process that involved several hearings before the Ontario Municipal Board (OMB), the Lieutenant-Governor in Council, and finally the Ontario Cabinet. The end result was a victory for the Downtown Merchants' Association, merchants from the Lockerby Plaza 69 Shopping Centre, and the powerful and vocal Save the Lily Group, spearheaded by Muriel McLeod and John Rutherford.[1] This event was a milestone for Sudbury. Not only did it illustrate the power of citizen opposition against a variety of vested interests but also it preserved one of the most picturesque and highly visible parcels of land for recreational and environmental purposes.

After locals gained control of the planning process following the SAPS debacle, a new regional official plan (with a grossly over-optimistic population projection of

TABLE 12.1   Regional Council Chairpersons and Mayors of City of Greater Sudbury (1973–2010+)

Regional Municipality of Sudbury Chairmen		Mayors of the City of Greater Sudbury
Don Collins, 1973–75	Tom Davies, 1982–97	James Gordon, 2001–03
Joe Fabbro, 1975–77	Peter Wong, 1997–98	David Courtemanche, 2003–06
Doug Frith, 1977–80	Doug Craig, 1998	John Rodriguez, 2006–10
George Lund, 1980–81	Frank Mazzuca, 1998–2000	Marianne Matichuk, 2010+
Delki Dozzi, 1981–82		

Source: Greater Sudbury Public Library.

228 000 for the near future) came into being in 1978. While the plan was a step in the right direction, it was not until 1987 that the City of Sudbury Official Plan of 1962 was made compatible with the regional plan. In 1981, regional council petitioned the legislature to abolish the "bureaucratic nightmare" of the upper tier, a request that was curtly dismissed by the province. These events were complicated by the long-term employment decline in the mining industry after 1973, made worse by the strike of 1975, the layoffs of 1977, the strike of 1978–79, and the strike and shutdown of 1982–83. The inevitable result was a decline in the population of the Regional Municipality of Sudbury that lasted until the middle 1980s (table 12.2).

A Greater Sense of Place

While the economic downtown in the mining industry and the loss of 10 000 residents in the census decade of 1971–1981 provided a gloomy setting, the Sudbury area continued to evolve. New variations appeared in the human, urban, and economic landscape, resulting in yet another transformation, similar to that which occurred in the 1950s.[2] At the same time, the Sudbury area developed a greater "sense of place," and a growing desire to become a more sustainable community. Demographic changes, such as an aging population and an improved male-to-female ratio gave the Sudbury area less of a "mining centre" look. With respect to the age issue, the population of the Sudbury area exhibited one of the highest rates of "urban aging" in Canada between 1971 and 1981.[3] While Sudbury continued to reflect the Canadian pattern of being one-third French, one-third British, and one-third Other, there were hints of ethnic changes to come. Ethnic clustering was still present, but these concentrations showed evidence of weakening as younger generations moved to more upscale neighbourhoods (as occurred with Finns who moved

TABLE 12.2 Population Trends in the Sudbury Area (1971–2011)

Census Year	Regional Municipality of Sudbury	City of Sudbury	City of Sudbury (% of R.M.S.)	City of Greater Sudbury
1971	169 580	90 535	53.4	
1976	167 705	97 604	58.2	
1981	159 779	91 829	57.5	
1986	152 470	88 715	58.2	
1991	161 210	92 884	57.6	
1996	164 049	92 059	56.1	165 336
2001	153 920	85 354	55.5	155 219
2006		88 708	56.2	157 857
2011		88 503	55.2	160 274

Source: Statistics Canada, *Censuses of 1971–2011*.
Note: The overlap between the RMS and the City of Greater Sudbury is for comparison purposes only.

from the Donovan district to the Northern Heights area), leaving former ethnic halls in a state of physical decay.

The broadening of the ethnic base to include new racial and Aboriginal groups was symbolized by the establishment of the Sudbury Multicultural Centre and the Sudbury Folk Arts Council in the seventies. By 1978, the latter comprised nineteen ethnic/racial groups.[4] The increased Aboriginal presence was evidenced by the founding of the Indian Friendship Centre and Laurentian University's offering of Native study courses. The French population became more secularized as new French-speaking secondary schools such as L'École Secondaire Macdonald-Cartier were constructed (1969). An alternative francophone culture linked to groups such as CANO and La Nuit sur L'Étang emerged. Despite these developments, the visibility of French culture remained relatively weak, and the trend of assimilation continued. While Sudburians of British extraction remained dominant, their ties to the homeland also weakened, as shown by the closing of the Orange Hall and Empire Club, and the opening of the Idylwylde Golf and Country Club's membership to the general public.

The region's cultural milieu broadened when managers of the Sudbury Theatre Centre hired an artistic director in 1971 and opened its own building in 1982. Other community members created the Northern Lights Festival Boréal in 1971 and founded the Sudbury Youth Orchestra in 1972. Finally, the Sudbury Symphony Orchestra became incorporated in 1975. This step enabled Sudbury to shed its reputation as "Canada's Pittsburgh without the orchestra."[5] Through these measures, Sudbury acquired a greater sense of urbanity and pride of community.

The urban setting underwent change. The housing situation was improved by an increase in the number of publicly funded co-operative and non-profit residences—from 400 in 1970 to more than 4 000 by 1980. The downtown was rejuvenated following the completion of the second phase of urban renewal. This process included the addition of Northern Ontario's largest single commercial complex, known as Tower 200, in the present-day Rainbow Centre, the erection of several office and bank buildings (IBM, Scotia, and Royal Trust Towers), the opening of the internationally acclaimed St. Andrew's Place in 1973, and the relocation of Eaton's to the City Centre in 1975. A new civic square was constructed in the early 1970s, featuring an attractive city hall and provincial administration building. This setting was complemented by the erection in 1976 of the nearby Bell Telephone and fire hall buildings. While these eye-catching design features were given a new scenic perspective following the opening of the CPR railway overpass in 1974, the view from the bridge continued to illustrate the influence of the CPR lines in the historic core. The new bridge inspired the local poet Robert Dickson in 1976 to write a eulogy for the old "Sudbury Iron Bridge" that had outlived its usefulness.[6]

The outskirts of the urban core underwent alteration as well, manifest in the revamping of the New Sudbury Shopping Centre in 1974, the development of a large shopping mall in the "Four Corners" area in the southern end of the city in 1981, the Supermall in 1980–81, and the opening of the imposing Canada Revenue Agency building in 1982.

Another major development was the opening of Laurentian Hospital and the partial demolition of St. Joseph's Hospital in 1980.

In the rural areas, urban sprawl was slowed through official plan policies that required the provision of services and residential standards more in accord with the median deemed desirable for all Ontarians.[7] Political life was changing as populists who formerly dominated local politics, such as Joe Fabbro and Maurice Lamoureux, were ousted in favour of professional candidates like Peter Wong, a former city engineer. Other changes occurred that were more nostalgic in nature: Kresge's downtown closed in 1983, the refurbished Grand Theatre (billed as the O'Keefe Centre of the North) went into receivership in 1987, the Church of the Epiphany (constructed previously in line with its original look in 1912) burned down in 1987, and the Creighton Mine, Frood, Crean Hill, Murray Village townsites and parts of Garson were all closed by 1988.[8]

Over time, Sudbury broadened its economic base and improved its central-place status in Northeastern Ontario. With a population in excess of 160 000, the Regional Municipality of Sudbury provided economic opportunities for import substitution in areas such as food processing, printing and design, sportswear manufacturing, and restaurant franchising. The population included a sizeable number of retirees which was significant for the local economy, given the fact that more than 90 per cent of the former employees of Inco and Falconbridge Nickel chose to remain in the region. With a secure retirement income, these retirees supported many of the smaller businesses in the area. The foundations for Sudbury's central-place functions, which had been established in the 1950s and 1960s, were strengthened in the 1970s by a growing white-collar class, a more educated labour force coming from the Laurentian University and Cambrian College campuses, and the creation of the Sudbury Regional Development Corporation in 1974. This body was innovative, as it was one of few municipal organizations in Ontario with its own budget and staff that served to promote local economic development and the image of the community.

The spread of Sudbury's influence into places formerly dominated by North Bay was facilitated by the opening of Highway 144 to Timmins in 1970, and its later extension to the Smooth rock Falls/Cochrane district via Highway 655. The local transportation network was improved with the creation of a public transit system in 1972 and the construction of bypasses and ring roads. Among these can be included the southwest bypass, completed in 1973–74, and Highway 17 freeway leading to Sault Ste. Marie, which opened in 1980. These changes were paralleled by improvements in air transportation. Until the 1970s, Sudbury served mainly as a nodal point for a north–south corridor joining Timmins with Toronto; the arrival of jet service and the introduction of Norontair flights in 1971, however, made the community a growing focal point for air traffic in Northeastern Ontario. After a modern air terminal was built in 1974, an east–west network gave Sudbury direct flights to Winnipeg, Ottawa, and Montreal.

Similar advances occurred in the sphere of communications. In 1972, Bell Canada moved its district office from North Bay to Sudbury, and in 1979, Sudbury was designated as Bell's regional headquarters for the territory from Peterborough in the south to Hudson Bay in the north. By this time, Sudbury had been established as the Northeastern Ontario headquarters for CBC radio broadcasting in both English and French. In publishing, Sudbury's position was enhanced by the formation of Laurentian Publishing by a local entrepreneur, Michael Atkins, in 1973. Since then, Atkins has expanded his enterprise to include other community newspapers and business tabloids, including *Northern Life* and *Northern Ontario Business*.

These developments in transportation and communication fostered the growth of tertiary employment. Sudbury's rise as a regional financial centre became evident in the 1970s with the establishment of bank headquarters, trust and loan companies, investment dealers, credit unions, and the Federal Business Development Bank in the downtown. A variety of wholesaling and distribution businesses were attracted to the serviced industrial parks that were created in the 1970s. The success of Walden Industrial Park, opened in 1975, served to stimulate the creation of others throughout the region. These zoned parks provided much-needed space for new manufacturing enterprises that had previously struggled to find suitable property elsewhere in the region. By 1981, the cumulative effect of these changes was such that Sudbury was considered to be both a service and administrative centre.[9]

Sudbury as a World OECD Model

The Sudbury area experienced numerous economic downturns during the late 1970s and early 1980s related to strikes, layoffs, and the reduced demand for minerals due to the second oil crisis. In Sudbury's economic prosperity forecasts, doom and gloom prevailed. Yet by 1984, Sudbury managed to achieve international fame for its efforts to become a more self-reliant community. It was in this year that the OECD, Government of Canada, and Ontario Ministry of Municipal Affairs chose Sudbury as an international case study for a major centre that had made a remarkable urban-economic transition after a period of decline. The case study concluded that Sudbury managed the remarkable feat of overcoming many obstacles inherited from previous years, and that the area was now on the path to a more sustainable future.[10] How did this come about?

While the case study dealt largely with events that had taken place in the previous decade and devoted considerable attention to political factors, there were other long-term factors that its authors emphasized as well. Among them were Sudbury's unique population size for a resource-based economy, its growing role as a central-place, increased white-collar presence, an expanded tertiary employment base, the impetus provided by the Sudbury Regional Development Corporation, and the implementation of a new official plan. Of critical importance was the creation of the group called Sudbury 2001. In 1977 a number of citizens headed by the innovative planner Narasim Katary and entrepreneur Michael Atkins began to meet informally with representatives from the business, academic, mining, and union sectors to discuss ways of alleviating the effects of mining layoffs in the area; the result was the formation of Sudbury 2001 in 1978.[11] Following the 2001 Conference held in 1978, a formal organization was established with the goal of making the Sudbury region a self-sustaining metropolis by the turn of the century. Assisted by venture capital provided by the province, Sudbury 2001 set out to initiate new businesses, expand existing firms, foster innovative community projects, and make long-term strategic plans based on the principle of sustainability. This effort so impressed the Ontario government that it erected a provincial building in the downtown Civic Square complex.

While the 2001 organization was disbanded in 1987 because of a discredited economic project and allegations of impracticality and mismanagement, it nonetheless must be considered historically as a remarkable achievement.[12] Heralding the transition from the old-fashioned type of reactive planning to one that looked ahead, it reflected a different outlook and attitude, regarding Sudbury as a "home" for its residents. As one well-known Canadian planner noted, "Sudbury 2001 marked the first concrete attempt to evolve locally controlled development programs."[13] It likewise started to discard the long-standing tradition of confrontation and divisiveness that existed in the area in favour of cooperation and community consensus. Equally important, Sudbury 2001 provided valuable lessons for tactical moves that were made later.

Another factor leading to change was that of political leadership. Throughout the 1970s and early 1980s, there existed a unique confluence of politicians at the federal, provincial, and municipal levels. At the federal level, representations were made for special assistance by James Jerome (Speaker of the House), Doug Frith, Judy Erola (biography 11), and John Rodriguez. Similar requests were made at the provincial level by heavyweights such as Bud Germa, Floyd Loughren, Elie Martel, and Jim Gordon.[14] These politicians had considerable political leverage, and they all used the powerful argument that no nation is so affluent that it could afford to throw away a major metropolitan centre.[15] Their efforts were supported by expressions of sympathy for the region's plight from the national media.

The role played by Tom Davies, chairman of the Regional Municipality of Sudbury, was decisive (biography 12). After the summer shutdowns at both Inco and Falconbridge and Inco's announcement of huge layoffs effective January 1983, Davies instituted a dynamic strategy to deal with the future of the community.[16] This response consisted initially of an intense networking process involving the entire citizenry. For the first time in Sudbury's history, Davies managed to bring all the major economic and union forces in the area, including USWA Local 6500, IUMMSW, Inco, Falconbridge Limited, and the Chamber of Commerce under one cooperative umbrella. An evaluation stage followed that succeeded in reducing the overall problem to more manageable proportions. This culminated in the formation of a focused mission statement involving eight sector recommendations: mining, government, business, industry, finance, health, agriculture, and education or training. The statement synthesized 300 proposals into 38 concrete recommendations.

Tom Davies headed a "Team Sudbury" approach, and engaged in intense liaising with senior provincial and federal officials about linking proposed programs to existing

Biography 11

THE HONOURABLE JUDY EROLA, P.C.:
Femme Extraordinaire (1934+)

Judy Erola was born Judy Jacobson on January 16, 1934, one of seven daughters of a dairy farmer with Finnish forebears on both sides.[a] Her grandfather, Thomas Jacobson, came to Sudbury in 1885, and was one of the first workers hired by the Canadian Copper Company. Brought up in the Walden area, Erola attended elementary school in Waters Township and completed her secondary education at Sudbury High School. She was known even as a youngster for her enormous energy and boundless enthusiasm. Her career began in radio and television broadcasting. She started in the radio industry as a copywriter with CHNO AM, established in 1947, and over time found herself doing nearly every job at the station, outside of engineering and management. In 1953, Erola moved to CKSO-TV, where she became the first woman employed by a Canadian television station as a weather reporter.[b] She was a founding member of Sudbury's first theatre group, and an accomplished theatrical performer throughout the 1950s. She married Vic Erola and they had two daughters, Laura and Kelly. While raising the girls, she helped Vic run Erola Panache Landing. Operating the marina with 200 mooring slips gave her valuable insights into the world of private entrepreneurship.

After her husband's death in 1977, Erola knew that she wanted to take a new direction, and she decided to enter the world of politics. In 1979, she won the Liberal nomination for the federal Nickel Belt riding. She lost the election that year to the incumbent New Democrat John Rodriguez. Undeterred, she ran again in the 1980 election, and this time defeated Rodriguez, a victory that surprised many. She was named to Pierre Trudeau's Cabinet, eventually holding the positions of Minister of State for Mines, Minister of State for Social Development, Minister Responsible for the Status of Women, and Minister of Consumer and Corporate Affairs. In Ottawa, Erola developed a lifelong relationship with Monique Bégin, another female member of Trudeau's cabinet.

Among Erola's political achievements was the expansion of the criteria whereby Native women could claim Aboriginal status. She was the first woman appointed to the Cabinet's Priorities and Planning Committee, which debated and decided the direction of government policy. *Chatelaine's* readers voted her Woman of the Year in 1983 for her numerous ground-breaking achievements.

After the end of her parliamentary career, Erola served as president of the Pharmaceutical Manufacturers' Association of Canada, and as a director of the International Federation of Pharmaceutical Manufacturers and Associations based in Geneva, Switzerland, positions that she held until her retirement in 1998. Erola was also the first woman to be appointed to the Board of Directors of Inco Ltd. In 1994 she was appointed to the Board of Directors of the Canadian Medical Hall of Fame. Among her other accomplishments can be included a term as the first fundraising chair of the Canadian Institute of Child Health, and her selection as co-chair of the Legal Education and Action Fund (LEAF) National Endowment Campaign. In

2006, Erola was *ex officio* delegate to the Liberal Party of Canada leadership convention. She currently serves on the board of Equal Voice, an organization which seeks to assist Canadian women in running for political office.

Erola never forgot her Sudbury roots. As a federal cabinet minister, she was a major force in the reshaping of Sudbury, assisting in the creation of Science North and the funding of Laurentian University's Centre in Mining and Exploration Research (CIMER). She likewise worked in concert with the Sudbury Regional Development Corporation to promote Sudbury as a centre of pharmaceutical research. In 1996 she was granted an Honourary Doctorate of Laws from Laurentian University. Her contributions to Sudbury were aptly summarized by Baxter Ricard in 1984: "she surprised us all.... She got more for this area than the last ten members we had here. She worked very hard."[c] Proud of her heritage, Erola was a solid supporter of the local Finnish community, especially the Sudbury Finnish Rest Home Society and the Knights & Ladies of Kaleva.[d]

[a]For an extensive review of Erola's background, see "Judy Erola: The One-woman Energy Resource," 52, 146–47 and 150–52.
[b]"Personalities & Biographies: Judy Erola," CKSO.com, accessed December 14, 2009, http://www.ckso.com/personalities_judyerola.html.
[c]"Baxter Ricard, "Too Busy to Retire," 28.
[d]Saarinen, *Between a Rock and a Hard Place*, 275.

sources of funding. Whenever "steamroller Tom" appeared in Toronto or Ottawa government offices for assistance, everyone knew he represented the entire interests of the Sudbury community. In addition, the regional municipality a Short-term Job Creation Task Force, which did little to diversify the economy, but succeeded in cushioning the financial and psychological effects of the layoffs. As a backup to these economic measures, the municipality also formed a Help Committee to ensure that social services were provided to those who needed them most. To ensure some form of ongoing strategy after the task forces had fulfilled their mandates, the Sudbury Community Adjustment Project (SCAP) was conceived in 1986. It was a private sector non-profit corporation with a mandate to develop new labour adjustment and economic development programs. In many respects it could be considered an offshoot of Sudbury 2001. Considered the only entity of its kind in Canada, it ran successfully until its term ended in 1989.[17] Another aspect of the strategic planning process was the updating of the City of Sudbury's Official Plan of 1962 to meet current conditions, and to assure its compatibility as a secondary plan with the 1978 Regional Municipality of Sudbury's Official Plan. In its draft form, the plan consisted of land use policies linked to infill, recycling, and a Jane Jacobs type of downtown, as well as an innovative corporate plan incorporating human, social, organizational, and economic development issues. While the province approved the land use policies in 1987, it rejected the corporate plan because it promoted socio-cultural and economic development, rather than simply being a land use instrument.[18] Despite this provincial rejection,

Biography 12

THOMAS M. DAVIES
Fervent Sudbury Booster (1934–1997)

Tom Davies was born at Creighton Mine on February 4, 1934.[a] As a youngster, he was a key member of the Creighton Mine Athletic Association juvenile, junior, and senior baseball teams in the late 1940s and early 1950s. He was a graduate of Scollard Hall in North Bay. Davies took his baseball skills to the United States in 1953, where he played ball while studying at Rhode Island University. There he received a Bachelor of Science degree in business administration in 1957. Along the way, Davies became a talented piano player and a fine singer who loved to belt out Irish lyrics; he used the appropriate Finnish accent to sing his famous "Creighton Song." He married Sally Zanier. Davies had employment stints with the Guaranty Trust Company (1958) and a division of the Westinghouse Company (1959–64) in the US before going into private business from 1964 to 1972. Dissatisfied with how the Town of Lively was handling an issue that affected his business, Keller Davies Garage, he decided to enter the field of politics.

Davies' pursuit of politics led to a remarkable career spanning twenty-nine years, and involved him serving as the Mayor of the Town of Walden (1973–81), council member of the Regional Municipality of Sudbury, and Chair of the Regional Municipality for sixteen years (1982–97). His selection as regional chairman was attributed to his remarkable personality and ability to bridge the gap that existed on council between the City of Sudbury and the outlying municipalities. He believed "that building a bridge was always more productive than building a barrier."[b] When Davies assumed the position of regional chairman in 1981, he took charge of laying the groundwork for transforming the community from a mining-based town into the diversified regional capital of Northeastern Ontario. When the local economy experienced a severe downturn in 1982, he developed a strategy to bring the community together, and prepared to make a full frontal assault on Parliament Hill in Ottawa and Queen's Park in Toronto. Armed with a strategic flowchart, Davies had task forces established in 1983 to develop proposals for economic development. This effort culminated in the formation of a focused mission statement linked to existing programs. Since the majority of the proposals were tied to available funding, it was easy for senior government officials to accede to virtually every request he made. Davies used headlines in the *Toronto Star* and *The New York Times* stating that Sudbury had "hit bottom" and was "struggling to stay alive," to great advantage. This was the start of a long list of projects that would flood into Sudbury during the late 1980s and early 1990s. Davies was especially proud of one of his birthplace's legacies: the creation of the Sudbury Neutrino Observatory (SNO) in the depths of Creighton Mine. It was Davies who pulled everything together for this project to succeed.

It was a sad day in the fall of 1997 when Davies informed regional council that he was stepping down from his position because of a terminal illness. Shortly thereafter, a Tom Davies Retirement Concert was held in the Walden Arena, attended by 2 100 people. Other tributes followed, including an honourary Doctorate of Laws from Laurentian University; Civic

Square and the Walden Community Centre were renamed in his honour.[c] On December 11, 1997, Davies passed away. His passing sparked an outpouring of tributes on a scale seldom seen in the Sudbury area. Another lasting tribute was the renaming of his home township to Creighton-Davies Township. He left behind five children and the establishment of the Tom Davies Millennium Fund.

[a]For a retrospective of Tom Davies' life, see "A Tribute to Tom Davies," *Northern Ontario Business*, December 1997, 9–16B.
[b]Cited in "Davies' Estate is Enormous," *Sudbury Star*, December 13, 1997, A4.
[c]"Regional Councillors Say Good-Bye to Their Boss," *Northern Life*, November 28, 1997, 3.

regional politicians and bureaucrats continued to use the corporate plan as a useful internal document.

The task force strategy resulted in a number of success stories. In 1981, the province provided financial support for Science North in 1979, a tourist attraction championed by Edwin Carter and John McCreedy of Inco Ltd. With the assistance of Sudbury representatives James Jerome, Judy Erola, and Doug Frith, the federal government provided added financial assistance in 1982.[19] The end result was an outstanding architectural and tourist feature that fostered a new community image and served as a stimulus for the convention and hospitality sector.

In 1982, the federal government aided the local economy with its decision to construct the stunning, multilayered structure known as the Sudbury Taxation Centre. This relocation did much to lessen the long-standing problem of unemployment and underemployment for women. In addition, the task forces laid the framework for longer-term planning initiatives undertaken by the Sudbury Regional Development Corporation (SRDC) and the city's innovative corporate plan. Another successful initiative was the expansion of the land reclamation program started in 1978. The cumulative effect of these efforts was such that Peter Newman, writing for *Maclean's* in 1991 was able to write an article outlining "Sudbury's Sunny Renaissance."[20] Along the same vein, *Chatelaine* magazine cited Sudbury as one of the top ten places to live in Canada in 1991. In his insightful assessment of Sudbury's history from 1973 to 1985, Katary has gone so far as to declare this era as being the "golden age" of Sudbury—a period during which the mining town became a mindful town.[21]

The Struggle Continues

Despite Sudbury's so-called sunny renaissance of the 1980s, the region did not attain the goal of becoming fully "sustainable." This fact is made clear in table 12.3, which illustrates the local pattern of net migration from 1980 to 2010. The data show the number of people who moved into or out of the municipality annually, using postal codes collected from previous income tax returns. Net migration was negative from 1980 to 1988, and then it was positive from 1988 to 1993, only to be reversed between 1993 and 2002. The back-and-forth cycle was repeated again from 2002 to 2008, and from 2008 to 2010. If the period between 1980 and 2010 is taken as a whole, the overall result was a negative net migration (–12 730). Thus, it was clear that the cycle of booms and busts had not disappeared from the Sudbury scene.

These figures serve as a useful measure of the area's development path since 1980. Between 1980 and 1988, there was a net outflow of 9 545 people to other areas, mainly to the burgeoning areas of Southern Ontario. One of the push factors for this migratory

TABLE 12.3   Migration Estimates for the Sudbury Area (1980–2010)

Year	In-migration	Out-migration	Net Migration
1980–1981	6 421	7 476	–1 055
1981–1982	5 234	6 299	–1 065
1982–1983	4 193	6 463	–2 270
1983–1984	4 373	5 736	–1 363
1984–1985	4 230	5 493	–1 263
1985–1986	4 601	6 394	–1 793
1986–1987	5 519	6 243	–724
1987–1988	6 149	6 161	–12
1988–1989	7 244	5 529	1 715
1989–1990	7 599	5 556	2 043
1990–1991	6 843	4 892	1 951
1991–1992	6 615	5 043	1 572
1992–1993	5 619	5 551	68
1993–1994	5 033	6 125	–1 092
1994–1995	5 107	6 069	–962
1995–1996	5 376	6 099	–723
1996–1997	4 831	6 427	–1 596
1997–1998	4 588	6 879	–2 291
1998–1999	4 252	7 175	–2 923
1999–2000	5 359	6 917	–1 558
2000–2001	4 919	5 657	–738
2001–2002	5 509	6 010	–501
2002–2003	5 739	5 380	359
2003–2004	5 575	5 261	314
2004–2005	5 730	5 206	524
2005–2006	5 716	4 874	842
2006–2007	5 106	4 874	407
2007–2008	5 308	4 798	514
2008–2009	4 458	4 708	–250
2009–2010	4 245	5 105	–860

Source: Statistics Canada, *Migration Estimates*, Cat. No. 91C0025.
Note: The figures from 1980 to 2000 are for the Sudbury Regional Municipality Census Division, and the figures from 2001 are for the Greater Sudbury Census Division.

trend was the impact of new technologies and mechanization adopted by the mining industry, and the ensuing loss of jobs. From 1988 until 1993, the situation reversed itself as the efforts made by the task force in the early 1980s began to take hold. The result was a net in-migration of 7 349 persons from 1988 to 1993, giving rise to new developments and an upsurge of confidence in the local economy. This confidence was bolstered in 1988, when Sudbury hosted the World Junior Track and Field Championships and an attractive Ukrainian Seniors' Centre was erected in the downtown.[22] This was followed by the creation of Sudbury's first International Film Festival (known as Cinéfest) in 1989. This event has proven to be so successful that it continues to the present day.

In 1989, a decision was made to transfer Ontario's Ministry of Northern Development and Mines to Sudbury as part of the province's decentralization strategy. In 1991, the ministry opened its new headquarters in downtown Sudbury, and in the following year another striking building was completed on the campus of Laurentian University to house the Ontario Geological Survey (OGS), the Miners Health and Safety Centre, and the Mining Research Directorate, a division of the Ontario Mining Association.[23] Another significant development in 1991 was the opening of the Northeastern Ontario Regional Cancer Centre and its 70-bed Daffodil Terrace Lodge. The latter development happened only after a twenty-year planning delay because of resistance from the provincial health care bureaucracy and entrenched cancer clinic interests located in Toronto.[24]

The decade of the 1990s brought about a return to net out-migration, with younger people leaving at an alarming rate. Between 1993 and 2002, a total of 12 384 people left the area for other parts. Unfortunately, unlike many southern communities in the province, this loss was not offset by the arrival of international immigrants. Cutbacks undertaken by Inco and Falconbridge Nickel continued to take their toll, as did the effects of increased nickel production in other parts of the world, especially Russia. The downsizing of public sector employment after the defeat of the New Democratic Party in Ontario, and the arrival of the "Common Sense Revolution" in 1995 left its mark, as did the closing of Eaton's in 1998. Following the turn of the twenty-first century, new developments and the lingering euphoria of earlier economic planning efforts were sufficient to bring about another period of net in-migration from 2002 to 2008. As expected, the strike of 2009–10 also led to some out-migration.

While it is true that the Sudbury area did not experience the severity of economic decline faced by many other northern resource communities, the souring of the local economy and rising unemployment levels nonetheless affected confidence in prospects for the immediate future. As a result of this concern, the Regional Municipality of Sudbury instituted another community economic development planning process, called the Next Ten Years Project. Replicating what had been done in the 1980s, sixteen sector committees were created, which prepared 385 proposals and initiatives. These were reviewed by more than 370 delegates who attended the Next Ten Years Conference in November of 1992. At this conference, an integrated project listing was developed and a vision established to make the Sudbury region a balanced and sustainable Total Quality Community.[25]

Unfortunately, this planning initiative was minimally effective, and it did not achieve many successes.

Two other well-intentioned initiatives met with a similar fate. The first was the 2020 Focus on the Future vision, undertaken in 1999, that sought to provide a direction for the region's future official plan.[26] The second effort was The New Way project. In 1999–2000, this strategy defined ten clusters framed within the context of Sudbury evolving as a "smart, sustainable and socially responsible community."[27] To some degree, the lack of success by these initiatives was a result of bad timing—there was no crisis, nor were the two senior levels of government enthused about providing funding for Sudbury megaprojects. Ongoing political divisions within the regional municipality took their toll. There were tactical issues as well, such as the dearth of attention given to the importance of locally based initiatives.[28] Perhaps most damaging was the lack of attention paid to mechanisms for implementing the good ideas brought forth. This was a factor that Tom Davies clearly understood and dealt with effectively in the 1980s.

Despite the mini-slump, the region's economic setting continued to evolve. Inbound and outbound call centres appeared in the late 1980s that provided expanding employment opportunities. This call centre setting encouraged Canadian Blood Services to locate its national blood centre in Sudbury in 2004.[29] In 1993, IMAX came to Science North, strengthening the facility's position as Sudbury's premier tourist attraction. Sudbury's importance in education grew with the opening of Collège Boréal in 1995, a francophone college of applied arts and technology. Located adjacent to the city-owned Terry Fox Complex, with seven campuses and forty-four service centres, it contributed greatly to the economic and social positioning of francophones in Sudbury and elsewhere in the province.[30] The college's opening reinforced Sudbury's reputation of being Northeastern Ontario's premier post-secondary destination.

Spurred by Mayor Jim Gordon (biography 13), the City of Sudbury became one of the first "wired" Canadian cities with the construction of a fibre optic network. By 2003, the network had 2 400 kilometres of cable, providing the digital capability for attracting more call centres to the area.[31] In 1997, a Strategic Energy Plan was approved, linking key municipal facilities and reducing energy costs and carbon dioxide emissions.[32] The area's image got an international boost with the opening of the Sudbury Neutrino Observatory (SNO) in 1999. Situated two kilometres below the ground deep in Creighton Mine, the lowest background radiation particle detector in the world is designed to sense solar neutrinos through their interactions with a large tank of heavy water. Experiments that ran from 1999 to 2006 at the SNO have had a major impact in the field of physics. For example, the SNO was one of the first laboratories to prove that neutrinos have mass. Since 2006, the observatory has been expanded to undertake further experiments related to projects known as SNOLAB and SNO+.[33]

The most striking change in the city's economic landscape was caused by the big box store phenomenon, which began in 1999. This new retail orientation, strongly promoted by Mayor Jim Gordon, came about as a result of the regional municipality's concerted

Biography 13

JIM GORDON
"Red" Tory and Visionary Mayor (1937+)

Born in Noranda, Quebec, in 1937, Jim Gordon moved to Sudbury at the age of two. He was in the first graduating class of St. Charles College, and later acquired a degree from Assumption University (now the University of Windsor). After marrying his wife Donna, the family moved to Espanola, where he was elected to the town council in 1965. Gordon and Donna eventually had six daughters. In 1966, he returned to Sudbury, where he taught at St. Charles College, Espanola High School, Confederation Secondary School, and Nickel District. His political career in Sudbury started in 1967, when he ran for the provincial Tories, losing to the incumbent Elmer Sopha. Later in the year, he was elected to Sudbury's city council as an alderman, followed by a term on the Board of Control. His instincts as a "political animal" soon came to the fore, as shown by his ability to out-manoeuvre Andy Roy in 1976 to succeed as city mayor, following Joe Fabbro's appointment as regional chairman.[a] In line with his political beliefs, Gordon promoted "fiscal responsibility," which he would support throughout his career.[b] In 1977, his tenure was marked by a major controversy centred on his intervention with the Minister of State for Amateur Sport that resulted in the 1981 Summer Games being awarded to Thunder Bay rather than to Sudbury. Gordon managed to avoid council censure over the loss of these games.[c] He was re-elected as mayor in 1978 and 1980. In 1979, he stirred the local scene by criticizing Inco's stance during the six-month strike of 1978, a step previously considered tantamount to political suicide.[d]

Gordon again revealed his political smarts when he outfoxed George Lund to win the Progressive Conservative nomination in 1981 for Sudbury's riding. He went on to defeat Bud Germa, thereby becoming the first Progressive Conservative member to win the Sudbury seat since 1959. Gordon remained the MPP for Sudbury until 1987. His call for a public inquiry into Inco Ltd. and Falconbridge Ltd. and for the possible public takeover of the two mining giants shocked the members of his own party, and won him the reputation of being a red Tory.[e] During his stint in Toronto, he served as the parliamentary assistant to the Minister of Health, and for a short period of time, as Minister of Government Services. He used his political clout to secure the Ontario Government's commitment to build the Northeastern Ontario Regional Cancer Centre in Sudbury. This was achieved despite the strong objections of the medical establishment in Toronto.[f] He served in the legislature until 1978, when he was defeated by Sterling Campbell in the Liberal sweep of David Peterson. Undeterred by this loss, Gordon ran for mayor of the City of Sudbury, defeating Peter Wong in 1991. This set the stage for his long tenure as mayor of Sudbury. In 2001, he became mayor of the new City of Greater Sudbury, a position he held until his retirement in 2003.

As Michael Atkins, President of *Northern Life*, succinctly stated, "there is no easy way to sum up Jim Gordon,"[g] While he was in many ways a controversial figure, for decades Gordon was at the centre of just about everything that happened in Sudbury. According to *Canadian Reader's Digest* in 2000, he was considered one of the three most innovative mayors in the country. Despite presiding over some of Sudbury's most difficult times—massive layoffs in the mining industry, vitriolic strikes, population decline, and the steady erosion of the tax base, he introduced visionary measures to diversify the region's economy. Among these can be included the initiation of a "big box" strategy to make Sudbury the service centre for Northeastern Ontario, the installation of telecommunication and electronic companies (with over 400 kilometres of cable), and getting the province to commit to turning Highway 69 into a four-lane highway within ten years at a cost of $1 billion.[h] Gordon considered his main achievement to be the role he played in the founding of the Northern Ontario School of Medicine.[i] As well, Gordon never broke from his long-standing belief in the need for solid fiscal management. Throughout the 1990s, Sudbury was largely debt-free, and didn't have a tax increase in nine years.[j] While there were many who lauded his consistency in this respect, there remain critics who maintain that one of his legacies was a huge backlog of infrastructure projects for his successors.[k] In 2003, his legacy was honoured by the renaming of the popular Ramsey Lake Boardwalk as the Jim Gordon Boardwalk.

---

[a]"Cohesive Leadership is Key in Difficult Municipal Year," *Sudbury Star*, January 8, 1976, 4.
[b]"New City Mayor to Seek Mandate," *Sudbury Star*, January 7, 1976, 3.
[c]"Mayor Gordon Strikes Back at Games Critics," *Sudbury Star*, October 19, 1977, 1.
[d]"Gordon Blasts Mother Inco," *Sudbury Star*, March 14, 1979, 1.
[e]"Gordon 'Left His Mark' on Sudbury," *Sudbury Star*, June 14, 2003, A3; and "Riding Profile for Sudbury," Election prediction project, accessed August 13, 2011, http://www.electionprediction.org/1999_ontario/north/sudbury.html.
[f]"Appreciating the Perspiration & Inspiration of Jim Gordon," *South Side Story*, July 2003, 3.
[g]"No Easy Way to Sum Up Jim Gordon," *Northern Life*, June 26, 2003, 8.
[h]Claudia Cornwall, "These Mayors Mean Business," *Canadian Reader's Digest*, August 2000, 92–93; "End of an Era," *Sudbury Star*, June 14, 2003, B6; and "Power Broker," *Sudbury Star*, July 19, 2003, B8.
[i]Sarah Lashbrook, "James Gordon," *Sudbury Star*, February 26, 2011, 13.
[j]Cornwall, "These Mayors Mean Business," 93.
[k]"End of an Era," *Sudbury Star*, B6.

policy to make Sudbury "the retail destination" of choice for Northeastern Ontario. At the time, few appreciated what kind of impact this policy would have on the community. With attractions such as Costco, Home Depot, Chapters, Lowe's, Best Buy, Toys R Us, and the Silver City Complex, all conveniently located on one site at the RioCan Centre, customers were drawn from existing and new market areas including Sault Ste. Marie, North Bay, Timmins, and northeastern Quebec. This development was significant in that it had "a life of its own" that was not tied to the mining industry.[34]

The big box initiative has had a major effect on Sudbury's downtown. The core area has exhibited difficulty in adapting itself to the presence of the new stores. Despite several naval-gazing studies, the repurposing of many buildings, and the emergence of numerous bars and nightclubs, the downtown still fails to leave a mark on Sudbury's cultural mentality. Focused more on sports and recreation issues rather than "bohemian" pursuits,

Sudburians have not adopted this area in their hearts and minds.[35] This attitude, however, has shown some signs of change, evident in the decision to use a downtown location for Laurentian's new School of Architecture, and similar plans to move the location of the Franklin Carmichael Art Gallery, Place des Arts, and perhaps a casino into the core area. The renewed emphasis on the downtown is likewise apparent in the proposed Downtown Master Plan, presented in 2012, which seeks to give the area a revamped image and cultural focus: the plan includes a new arts and culture district, a conference centre and four-star hotel complex, a technological park, new greenspace, and an emphasis on public art, beautification, and design innovations.[36]

Common Sense Revolution

From 1995 to 2002, all municipal structures in Ontario were profoundly influenced by the Common Sense Revolution policies the Conservatives implemented after their victory over Bob Rae's New Democratic Party. Harris' platform, which called for significant spending cuts, large tax cuts, and the elimination of the province's huge deficit, had major implications for the Sudbury area. One consequence was the passage of The Savings and Restructuring Act in 1996. This impacted the province in several ways, notably through the amalgamation of local municipalities into larger entities requiring fewer elected officials, the reshaping of hospital structures, and the downloading of provincial responsibilities to municipalities. The rationalization of the municipal system in the Sudbury area began in August 1999, when the province announced its plan to restructure four regions, including the Regional Municipality of Sudbury. A special advisor, Hugh J. Thomas, was appointed by the province to make recommendations concerning the municipal restructuring of the area. After consulting with various interests, Thomas' report was issued in November 1999.[37] In it, he noted the ongoing debates and dissatisfaction in the area regarding municipal governance. On the basis of these considerations, he recommended that the Regional Municipality of Sudbury and the existing area municipalities be dissolved, and a single-tier government known as the City of Greater Sudbury be established. The proposed entity consisted of the existing municipalities, plus the unincorporated Townships of Fraleck, Parkin, Aylmer, Mackelcan, Rathbun, and Scadding in the north, and the unincorporated Townships of Cleland, Dill, and Dryden in the south (figure 12.1). The inclusion of the townships to the north was designed to ensure the protection of Lake Wanapitei as the main water supply for the Sudbury area. These boundary changes came into effect on January 1, 2001.

With a population of 153 920 and an area encompassing 3 354 square kilometres, Sudbury in 2001 became the largest city by size in Ontario. For the first time in its history, a single political unit existed that roughly paralleled the physical boundaries of the Sudbury Structure. Thomas introduced a council of twelve part-time councillors and a full-time mayor, a decision that reduced the number of local politicians by thirty-five. The new single-tier of governance did not satisfy everyone, and calls later came for

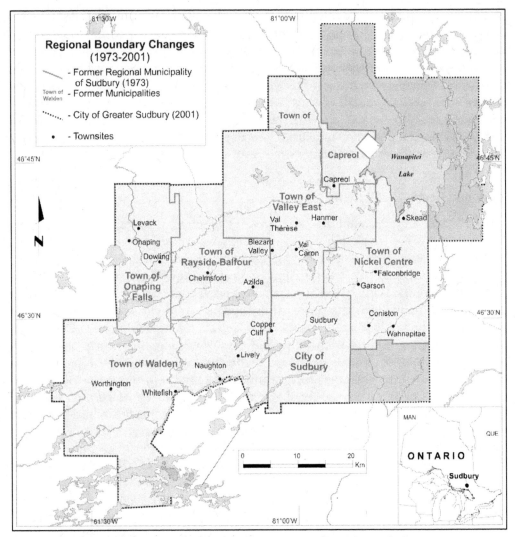

FIGURE 12.1 Regional Boundary Changes (1973–2001). From the City of Greater Sudbury.

de-amalgamation. To meet these concerns, former MPP Floyd Laughren was asked to make recommendations to improve the quality of city services to outlying communities. Laughren's report was completed in 2007, and it made recommendations about the city's ward structure, communications, transportation, recreation, and transit services.[38] While several of the report's recommendations were implemented, the study showed that the historical division between Sudbury's urban core and its outlying communities were still a political fact of life.

Another aspect of the Common Sense Revolution was the Health Services Restructuring Commission's (HSRC) 1996 directions to hospitals and recommendations to the Ministry of Health for restructuring hospital and other health services in the Sudbury area. Under the restructuring plan, the three existing hospitals (Laurentian, Memorial, and Sudbury General) were to be combined under a new Sudbury Regional Hospital Corporation and all acute, chronic, and rehabilitation services as well as mental health inpatient and out-patient services would be provided at a renovated and expanded Laurentian Hospital site. Provincial government officials calculated that this move would result in annual savings of more than $40 million. While the amalgamation went ahead as planned in 1997, the renovation to Laurentian Hospital did not. The construction of Phase 1 that began in 1998 was to be completed in 1999. This did not occur; rather, it was the start of a long and trou-bled history that saw Phase 1 halted in 2002, after massive cost overruns. The unfinished hospital served as a visual embarrassment to all concerned until 2007, when the project was restarted, and a new Phase 2 approved. The expanded hospital opened in 2010, eleven years after its original completion date, following construction cost overruns in the order of $185 million. It was a memorable occasion when the last patient was moved out of the St. Joseph's Health Centre on March 28, 2010.

There were other concerns related to the hospital restructuring process, notably the fact that the new facility had fewer beds than the previous one. On top of that, alter-native level of care (ALC) patients who should have been housed elsewhere were now occupying those fewer beds. Another issue was the lack of parking, a situation that forced many hospital workers to seek spaces in the more distant Bell Park area. A decision was made in 2010 to allow the Memorial Hospital site to remain open until the ALC situation was resolved. The restructuring process involving General Hospital (also known as the St. Joseph's Health Centre), led to a political squabble in 2010 which saw the Sisters of St. Joseph sell the facility to a private developer rather than the City of Greater Sudbury. This led to considerable outrage in the community, as the hospital was located on a prime site adjacent to Lake Ramsey and Bell Park.[39]

The final aspect involving the Harris government's changes to the province was the convening of the "Who Does What" Advisory Panel to counsel the government on ways to deal with the blurred responsibility for the delivery of local and provincial services.[40] The panel recommended shifting responsibility for funding hard services (such as road maintenance and sewers) to municipalities, and soft services (for example, social assistance and education) to the province. The province responded by imposing its Local Services Realignment (LSR) initiative in 1998, which uploaded the costs for education to the prov-ince while downloading full or partial responsibility for subsidized housing, social assistance, public transit, child care, public health, and land ambulance services to municipalities. The government promised that the process would be revenue neutral, but later abandoned the term when it became clear that the process was more about provincial fiscal savings rather than improving local service delivery and accountability. In the same decade, the province transferred major assets, including water treatment plants, highways, bridges, and social housing units to municipalities. These actions shifted the funding of many services from income tax to property tax, a process that caused fiscal problems for all municipalities in Ontario, especially those in the Sudbury area, where the mining tax base was shrinking.

In response to this situation, the Liberal Government of Ontario that came into power after the Conservatives in 2002 set out a new direction for provincial–municipal relationships in 2006. They issued the *Provincial–Municipal Fiscal and Service Delivery Review* in 2008, and made other changes as well, to offset some of the inequities inherent in the policies of the Common Sense Revolution.

From Blue Collar to White Collar

The creation of the City of Greater Sudbury in 2001 was followed by several years of net in-migration, and a modest increase in population. From 2002 until 2007, the city attracted almost 2 500 new residents. For the most part, the attraction of Sudbury was not due to employment in mining, but rather job opportunities in other sectors, notably the white-collar sector, and other factors such as retirement and the availability of social services not easily available in other parts of Northeastern Ontario. As shown in table 12.4, the shift from blue-collar to white-collar employment has been an ongoing trend since the 1950s. These figures must be used with some caution for statistical purposes, but the overall trend is clear—while blue-collar employment fell by almost one half during this interval, the rise in white-collar employment (43.4 per cent) was in the same order of magnitude.

Table 12.4. Sudbury Labour Force by Industry (1951–2001) (%)

Industry	1951	1961	1971	1981	1991	2001	1951–2001
Mining	28.0	32.9	25.3	15.6	10.9	6.4	-21.6
Agriculture & Forestry (Other Primary)		9.4	0.7	0.4	0.8	0.7	-8.7
Manufacturing	20.5	13.2	14.8	12.0	7.9	6.4	-14.1
Construction	7.3	6.1	10.2	5.1	7.2	5.8	-1.5
Transportation, Communication, Utilities	9.7	6.6	6.5	6.6	6.5	7.8	-1.9
Trade (Wholesale & Retail)	9.7	14.7	13.9	18.3	18.1	16.6	+6.9
Finance, Insurance & Real Estate	1.3	2.6	2.9	4.3	4.5	4.1	+2.8
Public & Private Services	14.1	18.8	21.6	30.3	34.7	44.0	+29.9
Public Administration		4.4	4.3	6.9	10.1	8.2	+3.8

Source: David Leadbeater, *Mining Town Crisis; Globalization, Labour and Resistance in Sudbury* (Halifax, NS: Fernwood, 2008), 24.
Note: The manufacturing category includes smelting and refining. The services category includes public sector-funded health, education, and social services, as well as private sector business and personal services. In 1951, public administration is included in services. In 1991, other primary is included in mining. Data for 1951 are from the Sudbury Census Division, and for 1961 to 2001 from the Sudbury Census Metropolitan Area.

Blue-collar jobs showed a long-term pattern of decline that began in the 1970s, but this decline has recently shown signs of recovery. This rebound has been spurred in part by the creation of the Sudbury Area Mining Supply & Services Association (SAMSSA) in 2003, a key manufacturing sector comprised of mining supply and service companies. This organization has attempted to reach, in cooperation with other groups, the long-term goal of making the Sudbury area the "best mining and supply service area in the world." This laudable goal of shifting the future emphasis on mining from that of production to mining technology and related services has been strongly supported by both city council and the Greater Sudbury Development Corporation.

One authority estimated in 2011 that this sector employed about 13 800 people in the region, and injected some $4 billion into the mining economy.[41] What is important to note here is that the mining service and supply side has now become more than twice as important as the mineral extraction sector in terms of its employment impact.[42] The intent of SAMSSA is to make the supply cluster an economic "growth engine" of Sudbury's future.[43] A number of developing clusters of new knowledge focused on post-secondary mining education, research, and development are supporting this endeavour, among them Laurentian University's Mining, Innovation, Rehabilitation and Applied Research Corporation (MIRARCO), the Centre for Excellence in Mining Innovation (CEMI), the Centre for Integrated Mining Technology (CIMTEC), and Cambrian College's Northern Centre for Advanced Technology (NORCAT). The last centre was the creation of Darryl Lake, who Michael Atkins regards as being "one of the most extraordinary social entrepreneurs in northern Ontario's history."[44] Under Lake's leadership, NORCAT has linked mining companies to small- and medium-sized businesses capable of providing creative solutions; some have signed multi-million-dollar contracts with NASA and the Canadian Space Agency to address the material challenges of travel to Mars.[45] Unfortunately, this important shift toward mining-related technology and the creation of a distinctive research cluster in the city has met with little support from the provincial and federal governments.

During the first decade of the 2000s, the building sector provided considerable employment. Much of this activity was influenced by the indefatigable Rick Bartolucci, who has served as the MPP for Sudbury riding since 1995. During his tenure, which included the posts of Minister of Northern Development and Mines, Community Safety and Correctional Services, and Municipal Affairs and Housing, he drove home an agenda that greatly benefited the Sudbury area.[46] Blue-collar trades were kept busy by record-breaking home construction, the expansion of Highway 69 and the Southwest bypass to four lanes, expansions at Cambrian College, Laurentian University, new big box retail facilities, the opening of a second Wal-Mart in the South End, and developments in the health field (Sudbury Regional Hospital and St. Joseph's Villa). This activity, in turn, created white-collar opportunities related to the opening of Dynamic Earth in 2003 and other expanded exhibits at Science North, the completion of Collège Boréal, the Northern Ontario School of Medicine in 2005, the growth of the film and television sector, and the completion of four signature hotels by the Marriott and Hilton chains. The latter phenomenon was stimulated by Sudbury's multi-year hotel occupancy rates, which are among the highest in Canada. While the number of employees working in the arts and culture sector increased significantly, this area of activity still ranked well below the Canadian and Ontario norm.

The cumulative effect of the above activities was sufficient for *Canadian Business* magazine to rank Greater Sudbury in 2008 as one of the top forty Canadian cities in which to do business, a ranking that rose from twenty-fourth to thirteenth in one year. This accolade notwithstanding, concerns remained regarding Sudbury's economic future, especially with reference to the area's Talent Index—a measure of the proportion of the population with a bachelor's degree or higher. The figure for Greater Sudbury at 14.3 per cent still ranked well below the provincial average of 22.8 per cent.[47] To meet this concern, the Greater Sudbury Learning City initiative was created in 2011 to make the community "a city of learning."[48]

The area took another major step forward in 2006, when the City of Greater Sudbury passed its new official plan. Approved by the Ontario Municipal Board in 2008, the plan simplified land use planning by integrating thirteen planning documents covering the former Regional Municipality of Sudbury into one document. Policies were established to facilitate objectives laid out in the city's long-term economic development plan.[49]

Many of the positive developments that occurred in the early 2000s were only minimally affected by the Steelworkers strike of 2009–10. Unlike past strikes, the work stoppage had less of an economic and psychological effect on the citizenry. One indication of this muted impact was that no member of SAMSSA filed for bankruptcy through the labour disruption.[50] The ambivalence local businesses showed regarding the actions of the union compelled Leo Gerard, International President of the Steelworkers union, to issue an ultimatum: "You're either on our side or Vale's."[51] As the strike dragged on into its eleventh month, a growing weariness with the actions taken by both the union and Vale developed locally and provincially. The end of the strike in July of 2010 was welcomed by the entire community; many concluded that the weak effects of the strike provided solid evidence of the success of the diversification efforts that had previously taken place. Rather than being seen solely as a mining and union town, there was optimism that Sudbury could now be a more diversified community, one based on education, health, retailing, mining-based research, retirement living, and tourism.

Other Transitions

Recent years have witnessed other transitional phases with respect to Sudbury's socio-cultural setting. Among these can be included the aging of European postwar immigrants, the emerging dynamism associated with the francophone population, the increasing importance of Aboriginal migrants, and the rising status of women within the urban milieu.

Mention was made earlier regarding the vitality exhibited by the European immigrants who came in the 1950s and 1960s. Since then, immigration from Europe has rapidly declined. While table 12.1 suggests the ongoing numerical importance of these ethnic groups, the figures are deceptive, as they include people of multiple origins whose associations with European ethnicity can be considered more "symbolic" than real. The virtual disappearance of the once-vibrant hallscape and the graying membership of ethnic-based churches give proof of this aging phenomenon. European ethnicity has now come to be identified with senior's homes and complexes, such as those that have been built by the Finns, Ukrainians, Poles, Italians, and Ukrainians. Despite this aging process, the ethnic factor remains persistent in the form of annual celebrations, such as the Canadian Garlic Festival held at the Ukrainian Seniors' Centre, the Italian Festival at the Caruso Club, the Greater Sudbury Celtic Festival, the Greek Festival held at the Hellenic Centre, and Finlandia Village's Mayfair.

Another recent trend has been a surge in francophone-related cultural organizations, and associated entrepreneurial activity. One source goes so far as to assert that "Francophone cultural institutions are, by any account, the most active, coherent and dynamic in the city of Greater Sudbury" and that Sudbury has "become something of a centre of excellence for *Les Arts d'expression.*"[52] In September 2010, for instance, there were celebrations for the 35th anniversary of the Franco-Ontario flag's creation in Sudbury, as well as anniversaries for other francophone groups such as ACFO, Carrefour Francophone, Club Richelieu, Les Patriotes, Centre Victoria Pour Femmes, and la Galerie du Nouvel-Ontario.[53] Another example was the Canadian Francophone Games (Jeux de la Francophonie Canadienne), held in July of 2011.[54] Other institutions embracing francophone culture include CBON-FM (the French language arm of the CBC), le Théâtre du Nouvel-Ontario, Librairie Grand Ciel Blue, la Bouquinerie du Moulin, Le Salon du Livre, and Éditions Prise de Parole. Eight francophone organizations have also proposed the creation of a Place des Arts or performing arts centre in the downtown housing a theatre, restaurant, and other related venues.[55] This expanding institutional framework has been an important factor providing employment opportunities for francophone graduates from Sudbury's post-secondary institutions.

In like fashion, Aboriginals have broadened their cultural imprint on the human landscape. Examples of this phenomenon include the creation of the N'Swakamok Friendship Centre, which assists Aboriginals migrating to or already living in Sudbury, the construction of the Shkagamik-Kwe Health Centre, which provides primary health care and traditional healing methods, and Sudbury's hosting of Northern Aboriginal Festivals. Aboriginal students are making their presence felt, not only in the elementary and secondary school systems but also in the realm of post-secondary education with the establishment of a Native Studies program at Laurentian University, the opening of a Metis and Aboriginal Centre at Collège Boréal, and the First Nations College Experience Program at Cambrian College.[56]

The unfortunate reality, however, is that outside of the downtown core, the Aboriginal population has remained largely invisible.[57] This is surprising, given the fact that the 2006 Canadian Census found the number of persons reporting some form of Aboriginal ancestry (14 605) exceeded those recorded for Italians, Germans, Ukrainians, or Finns.[58] Given the recent tendency for Aboriginals to flee reserves for cities (some Aboriginal leaders now estimate that as many as 70 per cent now reside in urban settings), and the fact that there are approximately twenty reserves situated within 300 kilometres of Sudbury's radius, many predict that the number of Aboriginals living there will grow in the future.[59] This group, therefore, will have to be seriously considered when dealing with the city's cultural and economic future.

Another contemporary manifestation of Sudbury's transitional nature concerns the greater role being played by women in many parts of the city's setting. This increased influence has been enhanced by the Greater Sudbury Business and Professional Women's Club. Created in 1945, the club has expanded its visibility in recent decades, mainly by stressing female entrepreneurship and political involvement. Another important institution is LEAF Sudbury (Women's Legal Education and Action Fund), which is one of the country's most active branches. Among other contemporary developments can be included the elections of Marianne Matichuk as Sudbury's first female mayor, Diana Colilli as the first woman to head the influential Caruso Club, and Joanne Caouette as the first female president of the Sudbury and District Home Builders Association.[60] Women have also served as the presidents, at one time or another, of all three post-secondary institutions in Sudbury. While these advancements may be commonplace in other Canadian communities, they warrant special attention in a resource-based community such as Sudbury.

CHAPTER 13

A Union Town?

While Sudbury's history has been intimately associated with the corporate aspect of resource extraction, this linkage also brought with it another aspect of the mining spectrum—unionism. Indeed, Sudbury has long had the reputation of being a union town. While most Sudburians have traditionally taken pride in this image, for others it has been regarded as a dubious distinction. The latter view, for instance, is explicit in the book *For the Years to Come*, a history of International Nickel of Canada written by one of the company's chairmen in 1960, where the existence of Mine Mill did not even warrant mention in the book's index.[1] When viewed in the context of Inco's traditional hegemony in Sudbury and its influence in the corridors of power in Toronto and Ottawa, and the lack of interest shown by other Canadians to Sudbury's woes related to hazardous working conditions, mining assessments, and environmental issues, it was inevitable that some counterforce to this capitalism would appear. This resistance came in the form of the only option available to workers: unionism, notably via the International Union of Mine, Mill and Smelter Workers (IUMMSW), known locally as Mine Mill. For three decades, Mine Mill had an honourable tradition of supporting its union members and the wider community through cultural programs and fundraising activities. Its presence was sufficiently strong in the 1950s to encourage the rise of unions in other sectors of the community. Throughout its early years, Mine Mill successfully faced many challenges, some of them internal, and others from a variety of external forces elsewhere in Canada and the United States. While these forces ultimately combined to bring about the demise of the union at Inco's operations in 1962, Mine Mill managed to reign supreme at Falconbridge's operations. The takeover of Mine Mill at Inco by the United Steelworkers of America (USWA or the Steelworkers) in 1962 is legendary in the annals of union history, and many miners and retirees today still remember the bitterness that accompanied this epic battle.

As subsequent events revealed, the victory of the Steelworkers' union over Mine Mill turned out to be bittersweet for the two factions and the community at large. Beginning in

1966, the Sudbury area witnessed a turbulent pattern of recurring strikes against Inco/Vale Canada and Falconbridge Ltd./Xstrata Nickel that has continued into the present day. The strike involving Vale Canada and the Steelworkers that took place in 2009 and 2010, however, was a watershed in terms of the history of labour and management relationships in the Sudbury area. The outcome of the strike not only represented a major victory for company management but also demonstrated the futility of the traditional adversarial bargaining approach, and illustrated the need for more flexible stances on both sides to ensure the long-term viability of the mining company and the union. The muted response on the part of the community to the strike raised the question as to whether Sudbury could any longer be considered a hardcore union town.

Early Beginnings

The beginning of unionism in the Sudbury area started with the IUMMSW.[2] This union can trace its roots back to 1893, when the Western Federation of Miners (WFM) was formed at Butte, Montana. The first local of the WFM in Canada was established at Rossland, British Columbia in 1895. It was 1906 before the first miners' union was formed by the WFM in Ontario, at Cobalt. Two more locals were then formed at Garson Mine and in Sudbury in 1913. At these sites, Finnish immigrant workers were especially supportive due to their previous exposure to socialist thought in Finland. As I have written elsewhere,

> working class radicalism in Sudbury expressed itself most strongly in the mining industry, where the eighty-four-hour work week and dangerous conditions were the norm. Hazardous conditions in the mines, and the feeling among work- ers that coroners' investigations always absolved the mining companies in the death of miners, encouraged Finns to turn to socialism and unionism as a match to corporate power. The first strikes at Canadian Copper in 1899, 1903 and 1904 were led by Italians; it was not until 1909 that Finnish activism started at the Mond Mine in Garson. When the mine supervisor demanded that a Finn by the name of Gus Viitasaari pay for a damaged tool, he was fired for his refusal to do so. Despite the lack of a union, the next shift went on strike demanding that he be rehired and not forced to pay for the drill. The Mond Company reluctantly agreed to this first show of strength among the Finns. From this point on, the trail of the Finnish mine worker could be seen in every subsequent strike.[3]

When the two unions showed signs of growth, mining companies responded by hiring the Pinkerton spy agency and ransacking union offices in Garson and Sudbury for membership lists. These lists were then used to blacklist union members. The mining com- panies' practice of blacklisting was made public at the hearings of the Royal Commission on Industrial Relations, held in Sudbury in 1919, where a miner gave the following testi- mony: "I know that immediately a man takes an active part in labour organization he is discharged from his position."[4] Particularly irksome to miners was the fact that employment could be brought to an end on the simple whim of a boss, or saved by means of "whisky seniority."[5]

Accidents and the lack of compensation were other serious concerns. While the Ontario Bureau of Mines' 1912 report noted that there were 43 deaths and 341 serious accidents at the mines and workplaces regulated by the Mining Act of Ontario, ensuing investigations resulted in the prosecution of four workmen and the fining of only one company for $100. As the Bureau's 1913 report blandly states, "the majority of non-fatal accidents were due to carelessness and incompetence of the injured workers."[6] Judgments such as this were the norm, so mining companies found it easy to fire workers and limit compensation awards.

As a result of continued company threats, both of the existing unions disappeared in 1916. In the same year, the WFM disbanded and was renamed the IUMMSW. Another local was formed at Coniston in 1919, but it too floundered. These failures notwithstanding, the province was sufficiently worried about the level of labour unrest to pass the no-fault Workmen's Compensation Act in 1914, and additional legislation in 1918, limiting the hours of work in the mines to eight in any one day.

Following the First World War, the advance of unionism in North America was stalled by a severe economic depression, internal divisions involving the International Workers of the World (IWW) in the United States and the One Big Union (OBU) movement in Canada, and campaign against unions by both government and industry. Government attacks were fostered by fears of the Russian Revolution spreading to North America. The mining shutdowns that occurred in Sudbury between 1920 and 1922 were sufficient to extinguish any spark of unionism that might have prevailed. After 1925, the IUMMSW ceased to function in Canada. It was not until the election of Franklin D. Roosevelt as President of the United States in 1933, and the passage of the Wagner Act in the same year giving workers the right to organize, that Mine Mill came back to life. In 1935, Mine Mill became one of the founding unions behind the formation of the Committee (later Congress) for Industrial Organization (CIO) in the United States.

The events in the United States spurred Mine Mill's reentry into the mining camps of Northern Ontario in 1936, especially in Sudbury where organizers thought that victory would allow the union to sweep into other areas. To this end, two locals were formed in 1936 and 1937, in Sudbury and Garson respectively. Due to continued company resistance, the destruction of a union office in collusion with law enforcement, and a lack of support from the international headquarters in the United States, the two unions were forced to dissolve before the outbreak of the Second World War. Order-in-Council 2685, issued by the federal Mackenzie King government in 1941, stated that "workers should be free to organize in trade unions, free from control by employers or their agents." It looked fine on paper, but its power was illusionary because the government failed to introduce any implementation measures. This weakness was apparent when the gold mine owners at Kirkland Lake, in 1941–42, successfully defied federal legislation by refusing to recognize or negotiate with a Mine Mill local. While the three-month strike that occurred during the winter was ostensibly a huge defeat for the union, its long-term effects were the opposite. When the Kirkland Lake miners left the area to seek employment in other industrial centres of the province, including Sudbury, they were powerful forces encouraging the spread of unionism.[7]

In Sudbury, the situation was complicated by the industry's refusal to relocate experienced miners from Kirkland Lake. The plight of the Kirkland Lake workers gained considerable support from the public and many church pulpits. Even more ominous in the eyes of the provincial and federal governments was the rise of the Co-operative Commonwealth Federation (CCF) political party. Standing for economic reform, the party won a crucial federal by-election in 1942 that prevented the Conservative leader, former prime minister Arthur Meighen, from entering the House of Commons. In the following year, the CCF became the official opposition in Ontario. The growing strength of the labour movement and the CCF's popularity in the provincial polls resulted in the passage of the Trade Union Act by the Ontario Liberals in 1943. In response to these events, the federal government enacted Order-in-Council PC 1003 establishing Wartime Labour Relations Regulations in 1944: these guaranteed the right of workers to choose the union they wanted to represent them without outside interference, and directed employers to bargain in good faith. These legislative enactments were potent legal forces that cleared the way for a new era of unions in Sudbury, one that lasted until the election of the Harris government in Ontario.

The stage was thus set for the rise of unionism in Sudbury. The new era started when Mine Mill Local 598 was formed, in considerable secrecy, in 1941. When the union openly established an office in downtown Sudbury early in 1942, there was a brutal attack on union officials that was condoned by the police and the Sudbury media, notably *The Sudbury Star* (referred to by mine workers as the *Inco Star*). Due to this opposition, Mine Mill found it necessary to initiate a secret sign-up campaign and created its own weekly paper known as the *Sudbury Beacon*. Despite the success of Local 598 in signing up members, Inco steadfastly refused to recognize Mine Mill's bargaining authority and asserted that it had an agreement with its own company union known as the United Copper-Nickel Workers (UCNW) established in 1942 (this union was popularly called the "Nickel Rash"). A similar company union was formed at Falconbridge Nickel.

When Bob Carlin, running for the CCF in Sudbury, won a landslide victory, the die was finally cast in the union's favour. Following certification as a union in December 1943, Mine Mill became the bargaining agent for employees at both Inco and Falconbridge Nickel Mines, with Mel Withers serving as the first president to more than 10 100 employees.[8] Once certification was announced, the bargaining process moved with surprising speed. The first contract between Mine Mill and Inco was signed on March 10, 1944; a similar agreement was reached between Mine Mill and Falconbridge Nickel Mines on April 19 of the same year. Local 598 was then certified at the Canadian Industries (CIL) sulphuric acid plant and negotiated a contract for workers there by June 15, 1944. History was made without a single man-day being lost in production. The signing of the contracts set a pattern, whereby negotiations with Falconbridge Nickel Mines and CIL would follow the precedent set by Inco. With the agreements in place, Local 598 not only acquired the honour of being the largest local chartered by Mine Mill, but also the largest in Canada. From this point in time, union presidents became part of the power structure in the Sudbury area (table 13.1).

One notable gain in these first contracts was the replacement of the earlier "face-to-face" interpretation of the eight-hour day with the "collar-to-collar" provision. Previously, the eight-hour day had been calculated from the time miners arrived at their exact locations in the mines until they left at the end of the shift. The new provision meant that their hours of work would begin as soon as they entered the mine collar to be lowered by cage into the mine. The time clock would then run until miners reached the surface. For miners, this often meant the working day was shortened by more than an hour with no loss of pay.[9] Another important aspect of these contracts was a provision inherited from the WFM constitution forbidding discrimination by the company or the union against any employee on the basis of sex, race, creed, colour, nationality, ancestry, place of origin, or *political opinion*. The clause regarding freedom of political expression later became the dominant

TABLE 13.1 Presidents of Mine Mill/CAW and Steelworkers Locals in Sudbury

Local 598 at Inco and Falconbridge

Mel Withers (1942–45)
James Kidd (1946–47)
Nels Thibault (1948–51)
Mike Solski (1952–58)
Don Gillis (1959–62)

Steelworkers Local 6500 (Inco/Vale)	Mine Mill/CAW Local 598 (Falconbridge)
Don McNabb (1963–65)	Tom Taylor (1962–64)
Tony Soden (1965–67)	Nels Thibault (1965)
Homer Seguin (1967–70)	Robert McArthur (1966–69)
Mickey McGuire (1970–76)	Jim Tester (1969–74)
Dave Patterson (1976–81)	Emile Prudhomme (1974–77)
Ron MacDonald (1981–87)	Joe MacDonald (1977)
Dave Campbell (1987–98)	Jack Gignac (1977–80)
Gary Patterson (1998–2000)	Emile Prudhomme (1980–82)
Dan O'Reilly (2000–01)	Ed Leger (1982–84)
Jim Gosselin (2001–03)	Rick Briggs (1984–95)
John Fera (2003–10)	Rolly Gauthier (1995–2001)
Rick Bertrand (2010+)	Rick Grylls (2001–09)
	Dwight Harper (2009)
	Richard Paquin (2009+)

Source: Office of the President, United Steelworkers of America Local 6500, and Office of the President, Mine Mill Local 598/CAW.

issue in the battle for labour supremacy in the United States and Canada. Other important gains were made with respect to seniority rights in all departments, and the introduction of grievance procedures.[10] These accomplishments were significant. For the first time in Sudbury's mining history, workers had finally gained some dignity in the workplace.

The Heyday of Mine Mill (1944–1958)

In line with the original mandate of the WFM, Local 598 turned its attention to spreading the cause of unionism into the service industries. While this mission was done in part to prevent the Canadian Congress of Labour (CCL) unions from organizing the service industries in Mine Mill strongholds, the broader desire to raise minimum wages that existed throughout Sudbury at the time was equally important. To face this situation, Local 902 was chartered as a General Workers' Union in 1949. Existing CCL unions consisting mainly of bartenders quickly signed up. The ambitious campaign by Mine Mill to organize the remaining service workers in the area caused considerable consternation and resentment among Sudbury's merchant class.

Despite several setbacks, Local 902 was able to boast twenty-four contracts (mainly at hotels) by the end of 1950. Among these contracts was one signed with the Sudbury Brewing and Malting Company. Another union achievement was its organization of grocery chain stories that were making their appearance in Sudbury. In a rapid-fire campaign, all the clerks at Dominion Stores were unionized by 1952. In 1954, Mine Mill became the first *bona fide* union at Loblaws in Ontario. Organizational drives continued, so that numerous bakeries, dairies, laundries, downtown shops, hardware stores, and a few minor industries were brought into the fold. Certification of the 5- and 10-cent chain stories proved to be more elusive. Metropolitan Stores decided to close its operations in Sudbury rather than dealing with the union. The opposition shown by Kresge's (now known as Kmart) was so strong that a union was never formed in its store. Nonetheless, by 1956, Local 598 held fifty contracts in the Sudbury area. Although overshadowed by the size of Local 598, General Workers' Union Local 902 managed to significantly contribute to the working-class history of the area, especially for its French-speaking members.[11]

Another Local 598 aim was to serve as a working-class force within the social and cultural fabric of the community—a phenomenon previously unknown in the Sudbury area. To broaden its community appeal, Mine Mill instituted a welfare plan, participated in fundraising drives such as the Red Feather campaign (one of the forerunners to the present-day United Way Centraide movement), donated an iron lung to the Sudbury General Hospital to assist polio patients, paid for the maintenance of a mobile physiotherapy unit, and actively supported the construction of the Sudbury Community Arena. At the same time, it worked to fill the void that existed in the working-class sporting and recreational sphere, especially in Sudbury itself, which did not have many of the facilities common in the company towns. Entertainment for workers was often centred on the pubs made famous by Stompin' Tom Connors' "Sudbury Saturday Night" lyrics, and the sporting and hall activities developed by ethnic groups such as the Finns, Italians, Poles, Ukrainians, Slovaks, and Serbs.[12]

Local 598 continued the Western Federation of Miners tradition of building halls to serve its members. Using monies from the union's building fund, a new hall was built in Sudbury between 1949 and 1952. Other Mine Mill halls were constructed in Garson (1953), Coniston (1956), Creighton (1957), and Chelmsford (1959). These venues served as meeting places for union members and were often used by the community at large, because they offered a variety of facilities and programs including dances, banquets, movies, babysitting services, bowling alleys, and boxing matches. The union sponsored baseball, hockey, and basketball teams as well. A unique venture Mine Mill took on was the creation of a campground on a 65-hectare property on Richard Lake in 1951. A lodge and several dormitories were added later.

In addition to this sporting and recreational infrastructure, the union ventured into the cultural sphere. Under Weir Reid, Mine Mill's cultural program flourished, and its working-class members began to enjoy activities previously restricted to Sudbury's elite. Unfortunately, Weir's notable legacy was besmirched because of the intense union rivalry that occurred between Mine Mill and the Steelworkers after the strike of 1958 (biography 14).[13]

Another Mine Mill legacy of the WFM was the important role accorded to the wives of the miners through Ladies' Auxiliaries. These groups were formed with the belief that "a union is only half organized if it doesn't have the women." These auxiliaries were a constitutional adjunct and integral part of the union itself, and they participated fully in the life of the organization. As stated in the Canadian Constitution of Mine Mill at the time, "It shall be the aim and purposes of this Union to advance and promote the organization of Ladies' Auxiliaries.... It shall be the right of the Auxiliaries to send delegates to all Conventions of the Union, to attend Local Union meetings and be free to express their opinions and make recommendations to these bodies." During Mine Mill's initial drive, the Ladies' Auxiliary Local 117 was formed in 1943. The women canvassed homes, spoke to the wives of miners, and distributed handbills. In the heyday of Local 598's recreational, cultural, and children's camp programs, the women in this local were prominent in every activity. When the union halls were built in the outlying areas, auxiliaries appeared in these areas as well. During the 1958 strike, they played a leading role, blocking the efforts of local politicians to start a back-to-work movement among local women.[14] They were so influential within the union that one of the first steps taken by the Steelworkers after their victory at Inco in 1962 was to disband the auxiliaries.

Seeds of Dissent

The early years of Mine Mill in Sudbury were relatively peaceful and positive for the entire community; however, while the rank and file members were united in their defiance of the mining companies, there were negative undercurrents that hinted at the power struggle to come. These seeds of dissent had their origin in the period prior to the Second World War.

Biography 14

WEIR REID
"The Last Angry Socialist" (1918–1971)

Branded by *Time* magazine as the "last angry socialist" and the "unofficial, unforgiving conscience of the left in Canada," Weir Reid was one of the most influential and controversial figures in the Sudbury area during the 1950s and 1960s.[a] He was born on March 18, 1918, to a farm family in Georgetown, Ontario. Throughout his life, he was plagued with various illnesses, such as diphtheria, circulatory problems, and dizzy spells. Following high school, he enrolled in courses at McMaster University and the University of Toronto. Reid had solid convictions concerning social justice and the brotherhood of man. This interest was later expanded to include the co-operative movement, low-rental housing, the expansion of libraries, unionism, and socialism. With respect to the latter, he saw himself as a "keeper of the purist socialist flame."[b] While he had a brief flirtation with the Communist Party, he thought of himself as a self-styled Marxist. He believed fiercely in classic socialism rather than social democracy. Reid viewed the latter philosophy in Canada of simply being the "milquetoast of social reform with no deep commitment to giving the means of production to the people."[c]

Reid married Ruth May in 1944. He then began a varied and colourful career involving community housing, flooring, real estate, and the Central YMCA in Toronto. Through these positions he developed managerial skills, honed the ability to speak in public, and gained experience in various elements of the arts. His true love, however, was in developing recreational and cultural programs for the working class.

When the manager of Mine Mill's camp and its hall quit, Reid was hired in 1952 to take his position. Reid's experience fitted well into what Mine Mill had established earlier in the field of sports and recreation. He expanded Mine Mill's efforts by enlarging the summer camp program and creating ballet and theatre groups. His competence in drama and dance did not go unnoticed, and before long, Sudbury's middle-class leaders praised his efforts. His work in creating a ballet school at Sudbury's union hall under the supervision of professional teachers trained at the National Ballet was to have been crowned by the appearance of the Winnipeg Ballet under the union's auspices in 1954; unfortunately, this performance was cancelled due to pressure from the US State Department based on Cold War anxiety. Reid produced and directed plays for the local's theatre group known as the Haywood Players. He brought in popular entertainment acts such as Jamboree shows, circuses, Pete Seeger, and the Travellers. The famous singer Paul Robeson gave his first performance here outside of the United States in 1956, after the travel ban imposed on him in 1952 was lifted. Reid's attempts to run popular programs on Sudbury's two radio stations, however, were banned because they were considered too controversial.

Reid's influence at the Mine Mill camp was pivotal in his attempts to create a working-class culture in Sudbury. He expanded what had been a day camp into a comprehensive program for children under the direction of well-trained YWCA directors and counsellors.

A lodge was constructed on the site as well as dormitories to accommodate overnight campers. A new diving tower, picnic tables, and Sunday church services were added to serve the general public. Under Reid's direction, the camp developed a program that paralleled the YMCA experience.

Reid's influence at the camp was crucial to his role in promoting Mine Mill's worker culture, so it is ironic that this success would bring him to the brink of ruin. Due to his influence in Sudbury's public life at the time, Reid was caught up in the bitter clash over the control of Mine Mill after 1955. When the reform slate was elected in 1959, one of their first acts was to fire Reid without severance. He was also the victim of a trumped-up assault charge. When these two actions did not lead to the desired results, the reformers charged him with other fabricated crimes. While these charges were later dropped, Reid's reputation had been smeared and his livelihood undercut. In the process, the union's cultural and recreational endeavours were destroyed.[d] Within a few years the dance schools were closed, the theatre disbanded, and the remaining programs diminished. Reid was rehired to run the camp in 1964 and 1965, but his victory was a hollow one, as the camp was never able to regain its former glory.

In 1971, Reid passed away, a victim of the circulatory disease that had always affected his health. Besides his wife, he was survived by four sons and one daughter. In a *Sudbury Star* article covering his death, he was referred to being "knowledgeable, cultured, with a brilliant mind and sardonic wit."[e] It is sad, that in the annals of Sudbury's union history, such a gifted individual as Weir Reid could be used as a pawn for so-called "reformist" union members who did not hesitate to put their own ideology ahead of the collective good of unionism.

---

[a]"The Last Angry Socialist," *Time*, May 3, 1971, 9.
[b]"Controversial Union Figure, Weir Reid, 53, Dies Monday," *Sudbury Star*, August 3, 1971, 3.
[c]"The Last Angry Socialist," 9.
[d]Buse, "Weir Reid and Mine Mill," in Steedman, Suschnigg, and Buse, *Hard Lessons*, 280.
[e]"Controversial Union Figure, Weir Reid, 53, Dies Monday," 3.

As in other parts of the world, radicalism in the Sudbury area consisted of many factions and ideologies, ranging from revolutionaries and communists on the far left to evolutionists and socialists on the more conservative side, or, as Jason Miller stated, there was "left-wing union versus right-wing union."[15] An anti-Communist observer, Frank Southern, wrote that "these two factions were the initial people who were going to be active in the union. That was the beginning of the big political fight in Sudbury. It started from day one."[16] This factionalism was later exacerbated by the formation of different labour organizations such as the OBU and the IWW, and competing government parties.

The relationship between the labour movement and political power was brought into sharp focus in the provincial election of 1943, when Bob Carlin was nominated as Sudbury's CCF candidate. At this time, the Canadian Congress of Labour (CCL) endorsed the CCF as the political arm of labour in Canada. Carlin, a powerful union leader in Sudbury and close ally to Reid Robinson, the International President of Mine Mill and a Communist sympathizer, easily won provincial elections held in 1943 and 1945. For his support of Robinson, Carlin was appointed as a representative on the International Executive Board of Mine Mill. In the process, he alienated many in the conservative fold of the union, notably James Kidd, who would later become one of Carlin's strongest opponents.

It was clear that the house of labour was divided. The schism over Communism became the most important issue at Mine Mill's International Convention held in Cleveland in 1946, where the Robinson forces defeated a motion banning Communists from holding office in Mine Mill. Undeterred, Kidd organized a series of heated debates with Carlin for control of local 598. Carlin's support of the Robinson faction and his failure to support motions condemning "Soviet Imperialism" and "Communist totalitarianism" raised the ire of the CCL against him. His voting behaviour, along with the sin of permitting an article critical of another labour leader to be published in *The Union*, Mine Mill's official newspaper, became the reasons that led the CCL to suspend Mine Mill from its ranks in 1949. While local Mine Mill members were not enthused about being suspended from the CCL, they rationalized the situation, believing that even a Communist-dominated union was better than none at all; many of them remembered the old days when no union existed.[17] As well, members acknowledged that Mine Mill had secured many improvements in the collective agreements signed after 1944.

The federal government was also involved with the controversy; it reacted to attempts by Mine Mill to build its strength in Ontario's gold industry using US organizers, including Reid Robinson, by expelling them for their Communist tendencies. In the following year, the IUMMSW was turfed from the Congress for Industrial Organization (CIO) in the United States. This action in the United States was taken following the refusal of Mine Mill to adhere to the Taft-Hartley Act of 1947. The act signified the beginning of the Cold War, requiring union leaders to sign pledges affirming that they were against Communism. Mine Mill's popularity also suffered when its leaders opposed the Marshall Plan, formulated to revitalize war-torn Europe. In the meantime, Carlin's flirtations with Communist elements in Mine Mill aroused the suspicions of his fellow CCF party members; his refusal to assist the party in expelling Robinson and other Communist organizers from the gold fields of Ontario resulted in his loss of the CCF nomination for the provincial election of 1948.

In an action that violated the CCF constitution, he ran as an independent candidate against the CCF nominee in the provincial election of this year. Following the victory of the Conservative candidate, Carlin was summarily expunged from the CCF. Carlin's actions were significant in that they externalized the seeds of dissent regarding Mine Mill. What had previously been a matter of internal division in the union was brought out into the open: the battle now involved powerful state interests in Canada and the United States, as well as the international labour movement. In Sudbury, the fight involved the biased editorials of the influential *Sudbury Star*.

Around the turn of the 1950s, external and internal pressures on Mine Mill intensified. The Korean War brought with it military tensions, ones that were heightened by Cold

War fears and the growth of anti-Communism. New anti-union and civil rights limitations were promulgated in the United States to stifle dissent. The situation can be compared to the paranoia that gripped the United States following the 9/11 attacks in 2001. Among the first targets were the militant unions within the CIO, especially the IUMMSW. Raids on Mine Mill began, based on the belief that it was Communist-controlled. This was the signal for its opponents to attack Mine Mill in Canada and Sudbury.

One of the premises for Mine Mill's expulsion from the CCL in 1949 was that such a move would make it easier for other unions to raid the Sudbury local. This was made evident later in the year, when the CCL approved an application by the Steelworkers to enter Mine Mill's jurisdiction. After this decision, the CCL awarded Mine Mill's jurisdiction to the Steelworkers for the paltry sum of $50 000. Although the decision applied to all of Canada, mining operations in Ontario became the focal point for the Steelworkers attacks on Mine Mill. Shortly thereafter, Mine Mill was decisively defeated in the gold mines of the Timmins area. Early attempts made in 1950 and 1952 by the Steelworkers and conservative dissidents, assisted by the Roman Catholic clergy, to usurp Mine Mill in Sudbury proved to be unsuccessful. It was at this juncture that Nels Thibault and Mike Solski were enlisted to counter the Steelworkers. These two individuals soon gained considerable power within Mine Mill and became formidable opponents to the Steelworkers and others during the 1950s and 1960s. While Mine Mill attempted some forays into the field of municipal and provincial politics to strengthen its position, these moves had little success.

In 1955, Mine Mill made a major administrative decision that was to have serious consequences for Local 598. In this year, Mine Mill became the first international union to be granted autonomy because of its Canadian membership. This movement, started as early as 1943, had been spurred by the "Canadian fact" and nationalistic sentiments that Canadian members should be free to govern their own affairs. The autonomy issue gained momentum in the 1950s because of the difficulties that Mine Mill faced in the United States due to its Communist leanings. This process culminated in 1955, when a Canadian constitution and Canadian executive board were approved by Mine Mill. While this event was welcomed at the time, few realized the problems that this autonomy would cause later for Mine Mill in general, and Local 598 in particular. In 1956, the Ontario Labour Relations Board (OLRB) made a decision that the Canadian Mine Mill was, in fact, a new legal entity. This decision was a strategic loss for Mine Mill in Canada, as it made it virtually impossible for the union to succeed in organizing the uranium mines at Elliot Lake. The ensuing "dogfight" that took place at Elliot Lake made it clear that the battle was less about Communism and more about the struggle for union supremacy.[18] At the same time, James Kidd continued his harassment of Mine Mill in Sudbury. For his actions, he was expelled from Mine Mill. Kidd remained unfazed by this turn of events, as he had by this time become a full-time staff member of the Steelworkers that had established an office in Sudbury to pursue the demise of Mine Mill.

The depression that started in the non-ferrous metals industry in 1957, the negotiations between Mine Mill and Inco in 1958, and the Diefenbaker government's "hold-the-line" dictate all provided a brief hiatus for Sudbury's inter-union battle. The situation changed totally when Mine Mill called the 1958 strike, a step that had tremendous consequences for Mine Mill's future in the Sudbury area.

The 1958 Inco Strike

The first action in what was to become the "year of the strike" was taken by Inco, when it announced on March 15, 1958, that due to the economic recession, it was reducing production and laying off 1 000 employees in Sudbury, and 300 in Port Colborne. This was followed on May 23 by a further layoff of 300 men. On June 17, Inco placed all of its remaining hourly rated workers on a 32-hour week. The fact that the latter two layoffs took place during the negotiating process for a new contract added fuel to the fire. By this time it was clear that Inco, with its substantial stockpile of inventory during a period of reduced demand for nickel, was in a stronger bargaining position; as well, the company had no fear of a production shutdown, as this would allow it time to develop new domestic markets for nickel to replace decreasing military demands.

While negotiations were taking place, a number of wildcat provocations occurred at several plants and mines. Since Local 598 had advocated to its members that they should continue working, suspicions were raised that dissidents within the union were deliberately using these tactics to force Mine Mill into a questionable strike. When further meetings with the company proved unsuccessful, conciliation talks were held. The conciliation board favoured the company position and recommended a one-year contract. Not satisfied with this response, the union went on strike on September 24. For the first time since the chimneys in Copper Cliff were built, the smoke plumes were absent. Thus began a series of mining-related events that were to haunt the Sudbury area for the rest of the century.

As the days grew into weeks, the strike continued. The Roman Catholic Diocese of Sault Ste. Marie, under the direction of Bishop Alexander Carter, then entered the picture. In a pastoral letter issued to every Catholic church in his diocese and read from the pulpits on October 24, the bishop chastised the union for allowing the strike to occur.[19] A company proposal to end the strike on November 27 was summarily rejected by the union. To gather public support, an auto cavalcade went to Queen's Park on December 3; a week later, a meeting of 900 women was held in the Mine Mill hall supporting the twelve-week-long strike. This meeting was held to counter a "back-to-work" movement supported by *The Sudbury Star*. In a front page editorial, the newspaper claimed that the "timing of the strike was catastrophic for the union" and that "it has not been conclusively proven that it is in the interests of the union members to continue with their strike."[20] In response to the meeting held at the Mine Mill hall, Mayor Joe Fabbro, with the support of the area's two MPPs, secured the Sudbury Arena for different gathering of women on December 12, at which time they passed a resolution calling "on husbands to go back to work on a forty-hour week without gains in the first year of a three-year contract." This event severely weakened the financial and moral resolve of the union.

After a new offer was made by Inco on December 22, 1958, one of the most contro-versial strikes in Canada officially came to an end with the new agreement coming into effect on January 2, 1959. While the strike had a major impact on the local community, *The Globe and Mail* reported that there was no indication that the strike had bankrupted any businesses.[21] These events notwithstanding, the strike had a great effect in polarizing the citizenry, the workforce at Inco, and many sectors of the Sudbury community.

The Aftermath

After the ratification of the contract, Mine Mill presented it as a major victory. This senti-ment was not shared by many of the miners who had suffered as a result of the eighty-day strike. Nor was it shared by Inco, which was clearly the financial winner of the battle. When called upon to defend the strike, the Local 598 executive headed by Mike Solski claimed that the union had no other choice, as Inco wanted a strike, and had no inter-est in bargaining in good faith. His latter assertion was certainly correct. With a huge inventory at its disposal, Inco was in a strong position to wait out a lengthy strike. As well, the layoffs and production cutbacks implemented prior to and during negotiations were undertaken to test the resolve of Mine Mill. The company even informed Mine Mill during negotiations that, while it would not resume production during a strike, it would use this period to replace the smokestack in Coniston, upgrade the smokestack in Copper Cliff, and have gas lines installed at its Copper Cliff operations.[22] The company was also aware of the internal divisions that existed within the local, and saw a strike as a good opportunity to weaken Mine Mill. While the threat of outside raiding by another union was not made explicit, Mine Mill had concerns about the Steelworkers lurking in the back-ground. Had a strike not occurred, the Steelworkers would have alleged that Mine Mill had caved in the company's demands. Mine Mill also realized that Sudbury workers had fallen behind the gains reached by the Steelworkers in their Elliot Lake mining contracts.[23] Many of the Inco workers who had temporarily gone to work in Elliot Lake at Denison Mines, where the Steelworkers had a contract, had come back with favourable reports about the union.[24] There were even assertions made by *The Globe and Mail*, the *Canadian Register* (a Catholic newspaper), and Frank Southern that the strike was called to further the cause of the Communist Party of Canada. While there is little evidence to support this claim, this Red-baiting had considerable impact.

Inco's large inventory was the most critical issue that Mine Mill had to face. By calling a strike, Mine Mill went against one of the basic tenets of unionism—never strike against a stockpile; simply wait until that stockpile is gone and the company needs you.[25] Mine Mill failed to heed the advice of Walter Reuther of the United Autoworkers in the United States who, when faced with a similar situation of a huge inventory of cars and trucks, said that it was a useless time to strike. A year later, when market conditions improved considerably, the Autoworkers went on a brief strike, and as a result the automobile com-panies gave them a good contract.[26] In a similar vein, the Mine Mill workers employed by

COMINCO in British Columbia managed to suspend wage negotiations in May of 1958, thereby navigating the union through the economic recession with few scars. It is also worth mentioning that Mine Mill did not seriously pursue the option of a one-year contract recommended in the Conciliation Board's report released on September 9, 1958.

Other criticisms came to fore regarding the handling of the strike itself, which eventually cost the union one million dollars. Since Mine Mill had never experienced a work stoppage before, it had not seriously adopted the idea of having its own strike fund. Local 598's leaders assumed that if the financial situation was strained due to a strike, they could turn to the National Office for help. This did not turn out to be the case. Shortly after negotiations started on April 15, 1958, Local 598 was told by the National Office that it would be unable to provide much financial support in the case of a lengthy strike. This forecast was borne out by later events. When the local's cash reserves became depleted, it turned out that the American and other Canadian sections of the IUMMSW were only able to provide Local 598 with minimal strike relief, leaving the bulk of the strike costs to come from outside donations and the local itself. Given the previous expulsion of Mine Mill from the CIO (Congress for Industrial Organization) in the United States and the CCL (the Canadian Labour Congress, or CLC after 1956), there was no hope of financial support from these national organizations. A strong undercurrent of anger was raised by members, who asked—after contributing dues for fifteen years—why was there such a shortage of funds for strike relief, and why was it not considered to be the right of every union member to collect?[27] Particularly irksome was that the union required strikers to prove with their bankbooks that they had little or no money and needed support. Other sources of resentment were related to the manner in which strike vouchers were issued and used.[28]

It is not surprising, therefore, that many workers at Inco and Falconbridge Nickel, angry at the events of 1958, sought to extract revenge on Solski's former executive board. Unaware of one another's existence at the time, three small workers' groups formed in Falconbridge, Sudbury, and Levack to unseat Solski's slate at the next election on March 10, 1959.[29] These three groups later merged under the Committee for Democratic Leadership and Positive Action, and formed their own slate headed by Don Gillis, Reeve of Neelon-Garson Township. While this reformist group was optimistic about its chances for victory, there remained sufficient concern for them to seek outside assistance. This assistance came in the form of "divine intervention" from the Roman Catholic Church and the University of Sudbury.

During the aftermath of the 1958 strike, it had become clear that the interests of the reformers, the Steelworkers, and that of the Roman Catholic Church were similar: all were opposed to the influence of perceived Communist leadership within the labour movement. By this time, the Steelworkers had been firmly endorsed by the Catholic Church as a model union for the mining industry.[30] To further their cause, the Catholic Church led by the Jesuit Fathers at the University of Sudbury hired Alexandre Boudreau in 1958. Boudreau, a Jesuit-trained economics professor, was given a position as the new Director of Extension Courses. An avowed anti-Communist, he stated that "Mine Mill must be destroyed, and disappear from the map of Canada. This can be achieved only by depriving the Commies of their milch-cow, local 598 of Sudbury."[31] Boudreau initiated the Northern Workers Adult Education Association, and developed courses over the next three years. One of his first courses, titled Leadership Course for Miners, was ostensibly designed to instruct miners who had not completed high school on the true principles of unionism, and teach them to become leaders in their unions.

On closer inspection, it became clear that the Boudreau's real intent was to destroy Mine Mill. For the Steelworkers, the course was an educational vehicle through which they could develop the nucleus of candidates for the upcoming union elections in 1959. Some 140 students enrolled in Boudreau's course, of which the majority were reformist-minded workers whose names had been supplied by James Kidd. In another course titled Northern Ontario Workers Adult Education Association Course on Communism, Boudreau focused on one overarching theme, which was to emphasize the evils of Communism.

While in agreement with the gist of the material provided in Boudreau's courses, Kidd was nonetheless disturbed by his insistence that the main purpose of the trade union movement was the battle against Communism. Kidd suggested that these efforts were likely being subsidized by the Steelworkers.[32] Another of the more contentious aspects of Boudreau's teachings was his assertion that everyone was obligated to take an affirmative stand against Communism, even if it doing so went against democratic principles.[33] As later events demonstrated, this axiom proved to be effective. Taking direct aim at Solski, Boudreau emphasized that even if people like Mike Solski were not Communists, they nevertheless served as effective "front men" who had to be considered as Communist sympathizers.

While these courses were being taught, a reformist slate headed by Don Gillis was busy promoting a program centred on the need to remove Mine Mill's isolationist stance, and bring back into the newly formed CLC and the union mainstream. The slate railed against a central weakness of the Thibault-Solski leadership, emphasizing that opposing forces were not permitted to speak at union meetings. Allegations of financial mismanagement provided another point of attack. The final argument Gillis made was that the time for neutrality had long passed. This aspect of the reformist strategy was designed to attract those members who had not bothered to become involved in union affairs in the past. Such apathy was shown by the fact that in the 1953, 1955, and 1956 elections, the Solski slate had been returned by acclamation. Particular attention was focused on the 2 000 miners who obtained employment in the Elliot Lake uranium mines during the strike, where the Steelworkers were in control. Getting both the uninvolved majority as well as the former uranium miners interested in union affairs paid great dividends; in the election held on March 10, 1959, the Gillis slate won. While the Solski side managed to keep their traditional supporters, it was the votes of formerly uncommitted members that now made the difference. For these workers, the main issue was one of economics rather than loyalty. The only question for many of them was simple: which union could win the highest wages, the most efficient grievance procedure, and the best pension plan?[34]

After the election, a bitter rivalry started between the two factions within the union and the National Office. It became apparent that the majority of the new executive and Boudreau supported the Steelworkers. To start the process of change in this direction, the Gillis supporters (1) declared open war against Communism within Local 598, and undid many of its cultural and recreational achievements; (2) discredited the financial management of the Solski regime; and (3) sought re-affiliation with the CLC. These policies led to intense conflicts with the National Office. The first move was achieved with the sacking of Weir Reid by planting false stories about him in the *Toronto Telegram*. The Roman Catholic School Board assisted in the plan by issuing leaflets to its students reminding their fathers of what should be done. The second step involved an audit of Mine Mill's books by Allistair Stewart, a chartered accountant and well-known CCF member of the House of Commons for fifteen years. While he spent a month examining the records of Local 598, his report failed to produce any serious evidence of previous wrongdoing or stealing. In reality it was not an accounting analysis, but an attack on Mine Mill disguised as a financial audit.[35] There were nonetheless sufficient minor allegations made to cause worry among the members. The final initiative involved the return of Mine Mill to the union mainstream. Many members had come to the realization that Mine Mill's isolation from the CLC had denied the local access to the CLC strike fund and its fundraising apparatus. They also realized that existing differences between the aims of Local 598, the National Office, and the CLC made a return impossible. This conclusion was affirmed at the National Convention held in September, at which time a resolution from Local 598 banning Communists from holding office was soundly rejected.

With another election looming in the fall of 1959, the National Office instigated a plan to regain control of Local 598, and to oust the Gillis administration. A decision was approved establishing two new administrative districts (with District 2 comprising all areas east of Saskatchewan), which had the effect of minimizing Local 598's political clout in Canada. Nels Thibault then announced that he was stepping down as the National President of Mine Mill to contest the presidency of Local 598.

The tone for this election was set not by unionists, but by the Catholic Union Social Life Conference held in Sudbury from October 9 to 11. Mayor Joe Fabbro stated that the reason Sudbury was chosen as the site for the conference was the feeling there and abroad that it held the dubious honour of being the worst hotbed of Communism on all of the North American continent. His message was made clear the next day when *The Sudbury Daily Star* appeared with the headline, "Must Prove Sudbury Not Communist Area." The indefatigable Boudreau was quoted in the newspaper declaring that the election was a "last-ditch fight between Christianity and Communism."[36] The nature of the frenzied attacks at the time was exemplified by charges that Thibault was operating under an alias, and that he could speak Polish perfectly. With Boudreau serving as the driving force behind the Gillis administration, the move to maintain the existing executive proved to be unstoppable. In the end, a record turnout of 81 per cent voted for the entire Gillis slate.

As an added insult to the National Office, union members also elected a reform slate at Inco's Port Colborne operations.

Early in 1960, the CLC rejected Mine Mill's application for affiliation. The CLC considered the situation in Sudbury so volatile that it preferred to play a waiting game. Running battles continued at local membership meetings; the issue of Weir Reid's severance pay almost caused a riot. Local 598 then took a major step by temporarily withholding dues it normally sent to the National Office. This was significant financially, as these funds accounted for half of the National Office's revenues. The National Office countered this action by establishing a District Office in Sudbury. At the Constituent Convention establishing District Two, the pattern of dissention continued, with the old guard forces winning the majority of seats on the District Two Board.

Meanwhile, in Sudbury, Local 598 and the National Office became embroiled in the status of the Mine Mill employees at Falconbridge Nickel. Taking advantage of this uncertainty, Falconbridge Nickel implemented a policy of wearing safety glasses, one that led to an unsuccessful wildcat strike at their operations, and the firing of eleven union workers. This outcome was a bitter blow to unionism in Sudbury. Union rivalry continued when the Gillis administration proceeded to form its own Ladies Auxiliary independent of the National Office. At a raucous joint meeting held between Local 598 and the National Office in Sudbury on December 10–11, the Gillis administration al stated that it intended to pursue every avenue open towards affiliation with the CLC. The incidents described served as a prelude to the near civil war situation about to unfold in the Sudbury area.

The Battle for Inco Begins

The District Two Convention held in Sudbury April 24–29, 1961, served as the scene of the first clash in the all-out battle for union supremacy. In contrast to the local's meetings where it was the old guard leading the confrontation process, the new guard led by Gillis took this opportunity to harangue Chairman Solski and his supporters. Predictably, chaos resulted, and not a single resolution was passed. Since no progress had been achieved, the Gillis executive again did not remit the local's dues to the National Office. This move was supported by the majority of the local's members, as shown by the third election victory of the Gillis slate on June 7, 1961. By this time, the new Local 598 leaders were exploring the option of seceding Local 598 from the National and International Union, and becoming a chartered local of the CLC. When it became clear that the only way Local 598 could get into the CLC was by joining the Steelworkers, *The Sudbury Star* joined in the cause by printing a story under the headline, "Has Steel Begun Drive to Supplant Mine Mill?" Given the threat of losing Local 598's buildings and finances, the National Office succeeded in acquiring a local injunction, allowing William Kennedy, its Secretary-Treasurer, to administer the local on the National's behalf. On August 26, the Union Hall was taken over by Kennedy.

Thus began what became known as the Union Hall Riot, featuring a siege of the Union Hall by hundreds of pro-Gillis members. The hall was barricaded and defended with fire hoses by thirty Mine Mill supporters. Blow-by-blow bulletins from local radio stations ensured a growing crowd to witness the spectacle. The siege continued into the morning, until Sheriff Lamoureux literally read the Riot Act. According to one source, this riot was pivotal, as its public relations impact drove the final nail into Mine Mill's coffin.[37] A later ruling by the Ontario Supreme Court restored control of Local 598 to the elected officers of Local 598. In the meantime, the National Office took the questionable step of signing a mutual assistance pact with the Teamsters in an attempt to bolster its coffers, which had been depleted by Local 598's refusal to pay dues. This decision played into the hands of the Gillis team. They argued that if members had to choose between the Teamsters and the Steelworkers, they should select the Steelworkers. To gauge the feelings of the Local 598 membership, a mass meeting was held at the Sudbury Arena on September 11, 1961. In attendance were representatives of the CLC, the Steelworkers, and the Gillis executive. When Mike Solski and members of the National Office including Nels Thibault were denied entry, the boisterous crowd of some 4 000 began a huge melee, forcing the meeting to end early.

On September 11, 1961, the Steelworkers formally began their campaign, setting the stage for the most complicated application for certification in the history of the Ontario Labour Relations Board (OLRB). At a meeting held later at the Mine Mill hall, the Steelworkers announced news of a grand coup—Bob Carlin had defected to the Steelworkers! Before long, the unsettled situation made it impossible to hold membership meetings. Mine Mill members were angry that while the salaries of the Gillis slate were paid by Mine Mill, these same people were openly working for the Steelworkers. Attempts by Mine Mill to use the courts to deal with this conflict of interest were unsuccessful. Inco took advantage of the divisive situation and refused to bargain. A furious public relations battle ensued, involving the media and a barrage of leaflets. Finally, on November 11, 1961, the Steelworkers made formal application to the OLRB for bargaining rights at the Inco operations in Sudbury, claiming it had the required number of cards signed. Despite a number of seemingly valid objections from Mine Mill, a certification vote was set by the OLRB from February 27 to March 3.

This decision set off another flurry of activity, not only in Sudbury, but elsewhere. At Port Colborne, Mine Mill lost a vote to Steel in December 1961 with disastrous consequences, because it provided psychological support to the Steelworkers efforts in Sudbury.[38] Other assistance to the Gillis faction was given by the deputy commissioner of the RCMP and federal Justice Minister Davie Fulton. Prime Minister Diefenbaker embraced Gillis, appointing him labour representative to a NATO conference in Paris. The Province of Ontario added political support by appointing Gillis as the labour representative on the Ontario Economic Council. Attention too was given to the ruling of the Subversive Activities Control Board in the US that declared Mine Mill to be a Communist organization, thereby linking the Canadian leadership to the Communist Party.[39] In April of 1962, the Steel surge gained momentum with a resounding victory at Thompson, Manitoba that routed Mine Mill as the bargaining agent.

The Sudbury vote took place as ordered. When the results were announced on June 2, 1962, the Steelworkers won by a narrow margin of fifteen votes. This narrow victory, accompanied by the Steelworkers being denied of certification at Falconbridge Nickel because of numerous forgeries, caused many voices, including that of the *Northern Miner*, to call for a new vote. The OLRB refused to permit one, so on October 15, 1962, the history of unionism was radically changed when the Steelworkers were certified as the bargaining agent for all Inco workers in the Sudbury area. By the end of the year, the CIL workers joined the Steelworkers. Local 598, however, continued to be the bargaining agent for the workers at Falconbridge Nickel. In the election held on October 24, 1962, the Gillis slate was defeated, but this was a shallow victory for Mine Mill supporters, given the certification won by the Steelworkers at Inco. The new slate at Falconbridge applied to have the Steelworkers' certification at Inco revoked by the OLRB early in 1963. The OLRB denied the application, and refused to revoke its decision of October 15, 1962, or to order a new vote. It is nevertheless remarkable that Mine Mill, despite all of the forces that had been marshalled against it since 1947, was still able to put up such a strong fight—one that they arguably should have won.

Taking advantage of the chaotic situation, Inco laid off 2 200 employees in 1962, making it difficult for the Steelworkers to negotiate a new contract. Inco took the position that the existing Mine Mill contract was void and that the Steelworkers had to start a new bargaining process. It was another eight months before a new three-year contract was signed in 1963. Undaunted by its loss in 1962, Mine Mill also found some consolation in the fact that Don Gillis, the Progressive Conservative candidate in the federal election that same year, suffered an overwhelming defeat after a controversial campaign involving a riot at the Chelmsford Union Hall with Prime Minister Diefenbaker in attendance. In 1965, Mine Mill tried once more to regain control of the Inco workers without success. The Steelworkers victory at this time was aided by new employees hired in late 1964 and 1965 who had no past loyalty to Mine Mill. Many workers by this time had also become tired of the inter-union rivalry, and voted for the status quo. Other workers concluded that peace was more likely if the Steelworkers remained as the bargaining agent.

The Steelworkers celebrated these events by acquiring the imposing Legion Hall that had been erected in 1947 as its local headquarters. This was a strategic acquisition, as the building was associated with Canada's military history, and commanded patriotism and respect from the public at large. The need for a Steelworkers hall was dictated by the fact that Mine Mill Local 598 still owned all of its previous assets and properties (Mine Mill kept its old home on Regent Street until 1999, when it was sold to the Navy League).[40] After a wildcat strike almost a month long in 1966, and a brief legal strike lasting three days, another three-year agreement was reached between Inco and the Steelworkers.

With the softening of Cold War rhetoric, the discrediting of McCarthyism in the United States, and the growing understanding that a united union movement was more effective than one that was divided, a momentous step was taken in 1967, when the IUMMSW and the USWA in the United States and Canada agreed on a "non-raiding pact" and the merging of two unions. Suddenly, old enemies became new friends. Nels Thibault, for instance, became a successful organizer for the Steelworkers in Canada, and other Mine Mill officers were given staff positions. These actions undercut the validity of previous "anti-Communism" stances that had proved so damaging to Mine Mill.

The Steel Raids at Falconbridge Nickel

The 1967 merger was distinctive in that Local 598 at Falconbridge became the only local of the IUMMSW in North America that voted to remain outside of the agreement. While the events at Inco during the 1950s and 1960s were taking place, the Steelworkers also conducted a parallel campaign at Falconbridge Nickel. It appeared on the surface that there were two campaigns by the Steelworkers in the Sudbury area, but the reality was that the Steelworkers had Inco as its primary focus. The Steelworkers, however, eventually discovered that the organizational culture at Falconbridge was different from that which existed at Inco. Following certification as a Mine Mill local in 1944, the miners at Falconbridge expressed the desire for an autonomous local that would represent their interests first. This did not happen, and until 1962, Mine Mill served as the bargaining agent for both Inco and Falconbridge; during this period, a tradition emerged that the Inco contracts would be first ratified, followed by a similar agreement at Falconbridge. This pattern meant many of the unique concerns at Falconbridge were rarely addressed.

The situation was aggravated after the 1958 strike at Inco; the Falconbridge workers were tied to the meagre gains won there, when in fact they were in an excellent position to bargain with their own company because it had no financial difficulties. In this atmosphere, a referendum was approved by the Falconbridge workers for separate local status. Since many of these workers were loyal to the National Office, its executive board issued a separate charter—Local 1025—to the Falconbridge employees. However, the OLRB dismissed Local 1025's application for certification on May 4, 1960. Two weeks later, many Falconbridge workers staged a wildcat strike lasting almost a month, resulting in numerous penalties and discharges against union workers that were upheld by the OLRB. The National Office then withdrew its charter for Local 1025. The Steelworkers continued their membership drive at Falconbridge, and in late February 1962 made an application for certification to the OLRB. When a number of forged cards were discovered, the Steelworkers realized the weakness of their position, and withdrew their application. The officials of Mine Mill were jubilant over this rare victory. On August 20, 1965, the Steelworkers filed another application before the OLRB that was again rejected due to the lack of signed cards.

The Steelworker's efforts to organize the Falconbridge workers in 1965 were not a complete failure. In that year, Steelworker's Local 6855 was chartered to represent the company's office and technical staff. On June 21, 1966, the Steelworkers filed their third application for certification, only to again concede defeat. It was clear by now that the "raiding stick" would never lead to a successful campaign at Falconbridge; as a result, the Steelworkers decided to "dangle a carrot" before the Falconbridge workers in the form of a merger agreement with Mine Mill. This tactic did not work either. When the Canadian section of Mine Mill voted in favour of the merger in 1967, every local voted in favour with the exception of Local 598. In a surprise concessionary move, the president of Steelworkers Local 6500 made an announcement supporting the position taken by the Mine Mill workers at Falconbridge. While this effectively ended the attempt to merge Local 598 with the Steelworkers, it remained another two years before the legality of Local 598's independence was established. As a result of these events, Local 598 managed to retain ownership of its union hall on Regent Street in Sudbury.

The question is: why were Falconbridge workers so adamant in their rejection of the Steelworkers? One reason for this stance was the higher proportion of miners at Falconbridge. At Inco, miners comprised about half of the labour force, while the comparable figure at Falconbridge was in excess of 60 per cent.[41] The difference was important, as it was the miners who traditionally were the strongest supporters of Mine Mill and its National Office. According to Sheila Arnopoulos, another important but unappreciated factor was the francophone influence: "The French miners did not always figure so prominently in the union, but in the end they contributed an important chapter to Canadian labour history by saving Mine-Mill from absorption by the United Steelworkers of America. Although not always widely recognized, the French were the soul of an historic fight to save one of Canada's most avant-garde and controversial unions from passing into oblivion."[42] In supporting this conclusion, the author brought forward a number of intriguing propositions. First, she asserted that the French miners in Sudbury were traditionally alienated from their elite—the teachers, Jesuits, and professionals.[43] For this reason, they were more loyal to their work and the concept of an independent union than they were to their cultural elite. The union was second only to their families. Second, in the 1950s and 1960s, when the miners were fighting for Mine Mill against the Steelworkers, they received no support from the French elite. Indeed, many French workers felt that the Roman Catholic Church did everything it could to destroy the union. It was asserted by some that French miners could not obtain employment with Inco in the 1930s unless they had the approval of a local priest. Third, the French miners did not care whether their leaders were Communists or not; they judged them on their union record, not their personal politics. Fourth, Mine Mill was important because it made more effort than any other institution to recognize their French-speaking needs. One of the few places where miners could socialize in French was at the Mine Mill halls and summer camp situated on Lake Richard. For these reasons, there was nearly unanimous support among the French-speaking miners for Mine Mill.[44] While francophones played important roles in Mine Mill, it remained until 1974 before a Franco-Ontario miner—Emile Prudhomme—became president of the local.[45] Since then, the French presence within the union has been strong.

Post-Merger Events

In the years following 1967, both unions went their separate ways, each respectful of the other. In 1969, Inco tested the mettle of the Steelworkers, resulting in a 128-day strike. Unlike previous strikes, this one was quiet and orderly. With no nickel stockpile at hand, the Steelworkers outlasted Inco. The strike ended on November 15, 1969, with the union winning major gains in wages and, for the first time, a cost of living allowance (COLA). The union made progress on issues such as the "contracting out" of jobs, training and apprenticeship opportunities, and an evaluation of all job classifications at Inco. The last act resulted in major monetary gains for numerous positions. Falconbridge workers went on strike around the same time and reached a similar settlement, albeit without a contracting out provision. The signing of the 1969 contract set a positive tone for the next three years because of Inco's desire to project a revamped company image. The setting was advantageous for the Steelworkers as well, and its membership rose to a peak of 18 224 in July of 1971. Over the next six months, however, the situation changed as Inco announced cutbacks, layoffs, and the closing of the Coniston smelter. Despite this gloomy setting, the union signed a contract that introduced a new clause allowing workers to retain their seniority throughout any of Inco's operations. Formerly, workers who moved from one department to another lost their seniority. For the first time in mining history, a Joint Occupational Health and Safety Committee (JOHC) was negotiated. During the 1970s, the Steelworkers promoted the concept of mining as a trade, and in cooperation with company officials and Local 598 at Falconbridge, created a "common core" training program for basic underground hard rock mining.

One of the major struggles for the Steelworkers in the 1960s and 1970s involved the status of health and safety legislation in the mining industry. Many Steelworkers continued to support the activist work of Mine Mill member Jean Gagnon, who had worked in Copper Cliff's sintering plant in the 1950s. For his pioneering efforts over the years, Gagnon received the Order of Ontario on January 28, 2010.[46] At the time, responsibility for these issues rested with the Ontario Department of Mines, later the Department of Mines and Forests. One of the serious problems with this administrative authority was that virtually all of the government officials were previous mining executives; former union officials were rarely hired. An intimate relationship thus existed between the provincial ministry and the local mining companies. While such collusion was vehemently denied by government officials, it was well known by union officials and workers that whenever there was a health or safety issue in any of the mines or plants that required investigation, the mining companies were notified in advance about when an inspection would be taking place. This gave mine officials plenty of time to clean up any messes prior to inspection. This collusion was proved in the provincial legislature by Elie Martel, NDP representative for the Sudbury East Riding from 1967 to 1987. In a brilliant manoeuvre, Martel showed that a letter purportedly written by the Ontario Department of Mines and Forests in 1970 concerning serious gas and dust conditions in Inco's Copper Cliff Roaster was actually a rewrite of an Inco letter previously sent to the ministry. This revelation made it perfectly clear that the ministry was working on behalf of the mining companies.[47]

The Steelworkers were likewise concerned with the difficulties that workers had dealing with health issues in both the mining camps of Elliot Lake and Sudbury. An existing problem was that there were several provincial agencies responsible for health and safety. This allowed officials to diffuse blame and create major bureaucratic delays. In many instances, when cases went to court, provincial lawyers did not show up, forcing judges to throw the case out. While Bill 70 was eventually passed to address this issue, its long-term effects were questionable.[48] In the area of workers' compensation, a significant achievement was the establishment of a local branch office of the Workers' Compensation Board, which grew to become the busiest office in Ontario. In line with the provisions of a contract signed in 1975 (following a ten-day strike), and the recommendations of the Ham Commission report issued in 1976, a major epidemiology study was funded by the Joint Occupational Health Committee (JOHC) regarding employees who had worked in the Sintering Plants at Copper Cliff and Coniston.[49] The Sintering Plant at Copper Cliff operated from 1948 to 1963 before it was closed. The Coniston plant was closed in 1972. The study noted the higher incidence of various illnesses, such as silicosis and cancer, experienced by workers in the plants compared to the population as a whole. The report also documented a high number of accidental and violent deaths among Sudbury miners.

Following a massive layoff of 2 800 in 1977 and the accumulation of another large inventory by Inco, the company again assumed a hard bargaining stance in 1978. This resulted in a 261-day strike lasting from September 16, 1978, to June 3, 1979. The strike, the biggest one in Canada's history in lost man-days, proved to be costly for both the union and the company. For Inco, the strike turned out to be a public relations disaster, and a turning point for its corporate image. This was exemplified by a scathing article in *Canadian Business* magazine accusing the company of corporate arrogance. A Laurentian University study published in the same year found considerable community support for the strikers; this support included Mayor Jim Gordon, who claimed that the company was holding the city to ransom.[50] A major gain won by the union in the contract was the "30 years and out provision," which allowed workers to retire with a full pension after three decades of work. This provision proved to be important, as it encouraged union workers to remain in the area and support the Sudbury community.[51] Another layoff, strike, and contract agreement cycle took place in 1982. This time, a one-month strike was followed by a nine-month shutdown due to low demand for nickel. Inco undertook further workforce reductions in 1984 and 1985. This brought the number of Inco's layoffs that had taken place since 1971 to almost 6 000. In 1986, the office workers Local 6855 at Falconbridge staged a successful three-day strike.

Events proceeded well for the Steelworkers, as they signed three-year contract agreements with Inco in 1985, 1988, 1991, and 1994. It required a 26-day strike in 1997, though, before another agreement was reached with Inco, one that the Steelworkers proclaimed provided the "best industrial pension in Canada."[52] Changes in the labour scene then occurred at Falconbridge Ltd. and Mine Mill Local 598.

During the 1990s, two events took place that changed the setting considerably. The first was the merging of the Mine Mill local with the Canadian Autoworkers Union (CAW) in 1993, a step that allowed it to become a member union in the CLC under its new name, Mine Mill/CAW Local 598. The second was the takeover of Falconbridge by Noranda Inc. (owned by Brascan). After its first acquisition of stock in 1989, Noranda's ownership increased to more than 50 per cent in 2000. By this time, Noranda had started to implement a managerial model called Six Sigma, which required the removal of all personal considerations and the use of statistics to guide the company's operations.[53] In line with this new philosophy, Noranda became one of the first companies in Ontario to realize the potential of the Harris government legislation in 1995 that reversed the anti-scab laws contained in the Labour Relations Act passed by Ontario's New Democratic Party government in 1992. With these two considerations in mind, Falconbridge Ltd. introduced a different strategy for its bargaining stance with Mine Mill in 2000. When the company presented a new collective agreement demanding several concessions twelve hours prior to the expiration of the old contract, the union was caught off guard. In the past, there had been a well-understood collective agreement process, based to some degree on personal relationships. It was clear that this was no longer the case.[54]

On August 1, 2000, the company's intent to bring about a strike became a reality. Falconbridge attempted to undermine the local by emphasizing the censure of the CAW by the CLC, and by referring to Mine Mill as a "yellow union."[55] In another dramatic move, Falconbridge elected to scab the legal strike by using nickel concentrate from its Raglan operation in northern Quebec to feed the strikebound smelter in Sudbury. Both management personnel and outside labour were used to run the smelter. The plant's output was then shipped to Norway, raising the ire of the workers at Falconbridge's refinery there and resulting in a five-day support strike. A conciliatory attempt by the union to resolve some outstanding issues was rebuffed by the company. This was a move designed by Falconbridge to prolong the strike beyond six months, at which time the workers would no longer enjoy legal job protection. After the six-month deadline passed, the strike ended on February 20, 2001. While costly for Mine Mill, it emerged from the battle as a still-proud union. In 2004, another strike took place that lasted only three weeks. Events took a more favourable turn in 2007, when, for the first time since 1987, Local 598 managed to negotiate a contract without going on strike.

It was déjà vu in 2009 when more than 3 000 Steelworker workers went on strike against Vale Inco on July 13. Following the pattern set in 2000 by Falconbridge Ltd., Vale Inco opened negotiations by taking an aggressive bargaining stance, demanding pension, nickel bonus, and seniority concessions. Likewise, the company attempted to start bargaining "with a clean sheet." This position was posited by the company because of the existing economic downturn, and its desire to remain competitive even in poor market conditions. In stark contrast to the optimism expressed by company officials at the time Inco was acquired in 2006, Vale Inco management claimed that the Sudbury operations were unsustainable.[56] Adding fuel to the fire, federal Industry Minister Tony Clement uttered the politically damaging quote that without Vale Inco, the Sudbury area would be a "Valley of Death."[57] Other issues of concern for the company were the decline in nickel prices and the realization that perhaps it had paid too much money ($19 billion) to acquire Inco. As the strike carried on, the company announced in August that it would resume production using members of the Steelworkers Local 2020, representing office, clerical, and technical workers, its non-union staff, and some outside workers, mainly for training purposes. A limited production schedule was introduced at two mines (Garson Ramp and Coleman), and the Clarabelle Mill. With no willingness on the part of union negotiators to participate in any form of concessionary bargaining, the stage was set for a prolonged strike. If the union thought Inco was large, it turned out that Vale Inco was massive. The Ontario operations, the be-all and end-all of old Inco, constituted only a small proportion of Vale's global operations. As the chief union negotiator stated, "it was their game. It was their rules."[58] Vale Inco, knowing that it could afford a long strike, simply waited for the mandatory six-month period to pass, at which point the company would no longer be compelled to rehire its former employees.

An interesting aspect of the strike was that many Sudbury residents and the two senior levels of government only showed ambivalence to the workers' situation. So concerned were the Steelworkers about the lack of local support that Leo Gerard, the International President of the United Steelworkers union, felt compelled to deliver an ultimatum to the business community: "you're either on our side, or you're on Vale's side."[59] This threat fell largely on deaf ears. More than eleven months of failed bargaining and mediation attempts passed before the provincial government began to show any real concern. In a press release dated July 2, 2010, the Ontario Labour minister stated that "the impasse ... is not acceptable to the communities nor the government."[60] Spurred in part by this growing provincial concern, the two parties struck a tentative deal. The longest strike in the history of the 108-year-old company officially ended on July 8, 2010. While the contract involved numerous concessions on the part of the union, it found solace in the fact that some improvements were made with respect to Vale Inco's original bargaining stance.

An end result of the strike was a reshaping of Sudbury's image as a union town. Whereas there had traditionally been strong community support for the two unions in their contract negotiations with the mining companies, the situation had now changed considerably. By 2010, support for the union had clearly waned. In many instances, the lack of sympathy was open, with residents expressing little concern for the workers' predicament and even anti-union sentiments. The growing diversity of opinion was especially apparent in the content of bloggers who responded to strike articles published in *The Sudbury Star*. As well, there was what Michael Atkins called a "sea change" within the labour scene itself, illustrated when Steelworkers Local 2020 declined to support its brethren at Local 6500 who had been on strike for nearly nine months. Another indicator of the weakened union position was John Rodriguez's defeat in the 2010 mayoralty race, despite having the support of the local labour council.[61]

The ebbing of union support shown by the community during the Steelworkers' strike notwithstanding, Sudbury's past experience with mining companies clearly dictates the need for a strong and ongoing union presence within the industry. This future, however, will be complicated by challenges related to foreign ownership, globalization trends, and less support from both levels of government. Whether or not the union movement can continue to show the same resiliency in the mining industry that it had in the early and middle years of the twentieth century remains to be seen. Perhaps the time has come for the union movement to return to its earlier philosophy of fully engaging its member's families and being a more active participant in the community, as Mine Mill did so well during the halcyon decades of the 1940s and 1950s. The opportunity to take this approach has been enhanced by the opening of the multipurpose Steelworker's Hall in January of 2012 and its impressive Leo W. Gerard Hall.[62] The symbolic value of this new building to the future image of the Steelworker's union should not be underestimated. Recent actions undertaken by United Steelworkers Local 6500 bode well for this community reorientation.[63]

CHAPTER 14

Healing the Landscape

While Sudbury's past reputation as a union town is undisputed, it has another image that has been around longer and is more pervasive, not only in Canada but internationally as well—as one of the world's most scarified landscapes. Referred to historically as "a bleak landscape of black, scarred and barren rock, denuded forests, and acidified lakes and streams," the area from its inception attracted media attention as an example of the worst effects of the capitalist system.[1] In later years, experts developed a growing understanding of the negative effects that this scarification process had on the people who lived there as well. For almost one hundred years, this degradation of both the human and physical landscape was the stuff of newspaper headlines, magazine articles, and scholarly journals; at the same time, few of these reports demanded any meaningful resolution of the deplorable situation from the mining companies or the two senior levels of government. Not until the emergence of the environmental movement in the late 1960s, and the completion of the "superstack" in 1972 was there any concerted effort by organizations or governments outside of the Sudbury area to deal with the environmental impacts of mining. The reason for this is clear-cut; the rest of the country was not concerned for Sudbury's plight until the pollution became their problem as well.

The new smokestack heralded the start of a radically different and hugely successful transformation in the Sudbury area—a process of healing the landscape unrivalled in any other part of the world. In a rare instance that demonstrated the power of local control, the Regional Municipality of Sudbury, along with its allied institutions, assumed prime responsibility for this ecological renaissance. While much remains to be done, Sudbury's environment today is not only a vastly improved one from that which existed in the past but also it serves as a model for other parts of the world dealing with major ecological damage.

Logging, Forest Fires, and Erosion

The ravaging of the local landscape began prior to the construction of the area's first smelter in 1888. Waves of human activity in the area and elsewhere contributed to the deforestation process. In fact, Chicago's Great Fire of 1871 has been credited with opening the region for logging. Based on the records of the Hudson's Bay Company post in Naughton established in 1824, the history of early lumbering operations, the decision by Jesuits to call their parish Ste. Anne of the Pines, and photos from the early Ramsey Lake tote road CPR line, it is evident that the area at one time was covered by huge stands of red, white, and jack pine intermixed with sugar maple and yellow birch trees.[2] The red and white pine trees were cut and exported to the United States. The effects of this cutting were twofold: first, it resulted in a less attractive succession of birch and poplar trees; and second, it exposed much of the underlying bedrock and topsoil that had hitherto been hidden. The soil, now less protected by the vegetation, was washed away, and the fall frost uprooted the few tree seedlings that managed to catch hold. Throughout the 1800s, there were also many natural forest fires. Following the passage of the CPR mainline, the increasing need for railway ties, locomotive fuel, and pulpwood encouraged clear-cutting of the remaining spruce, balsam fir, and jack pine stands. The slash left behind created ideal conditions for wildfires created by sparks from passing steam trains. When the prospectors arrived, they would sometimes set fire to an area in order to expose what lay beneath and facilitate the staking of claims. After 1888, trees not suitable for lumber or pulpwood were used as roast bed fuel.

The Infamous Roast Yards (1888–1929)

The first smelting processes in the Sudbury area involved the use of roast yards and, after 1920, mechanical roasters. Refining ore meant removing copper and nickel metals from the crushed and sorted host rock made up of sulphur and iron. Iron constituted anywhere from 50 to 75 per cent of the ore itself. The roast yards were used to remove the sulphur content of the ore, while the blast furnaces and Bessemer plants in the smelters burned away the host bedrock and its iron content. The molten iron removed was called "slag." To reduce the amount of material transported to the United States, it was necessary for the roasting yards to be on-site. After a trial roast in December of 1886 proved its practicability, the first roast yard became operational in 1888. Between 1888 and 1929, there were 165 roasting beds at eleven roast yards located in or near Copper Cliff, Coniston, and Creighton (fig. 14.1).[3] Historians estimate that more than 3.3 million cubic metres of wood were burned in these roast yards (the equivalent of seventeen football fields all stacked thirty metres high). Over the forty-year history of the roast yards, researchers estimate that 10 million tons of sulphur dioxide were released from the ores.[4] The locations of the roast yards took minimal account for factors such as the direction of prevailing winds, the impact on surrounding agricultural and forested lands, and plant life; little thought was likewise given to the roast yards' impact on humans.

Once the location of the roast yard had been decided upon, the bed the roast heap was to be situated on was laid out.[5] After levelling the ground, workers dug a perimeter ditch to allow water to drain away. To prevent the loss of valuable minerals by leaching, the surface of the ground was covered by a layer of ore "fines." The construction of rail lines or an aerial tramway to transport the ore to the roast yard and from there to the smelter was the next step in the process. Depending on the amount of ore designated for roasting, a layer of cordwood, usually dead pine, would first be laid out in the form of a rectangle. Coarse ore was then placed on the cordwood, comprising about two-thirds of the whole. A thin layer of medium-sized ore known as "ragging" was added. This was, in turn, covered by a thin layer of fines to prevent open flames. In its final form, the roast heap assumed the shape of an elongated pyramid. The wood underneath the ore was ignited with the assistance of flues built at three-metre intervals. Based on the amount of ore and the size of the heap, the roasting process required anywhere from one to seven months.[6] During this time, the sulphur content of the ore would be reduced to between 6 and 12 per cent. After the fires in the heaps had burnt out, the ore was broken up, first by hand and then by steam shovel, before being delivered to the smelter. The amount of wood required for roast-heaping was enormous. According to one estimate, there were 934 668 cords of wood used while the heaps were in operation.[7] The extent of this usage was particularly noticeable in the Coniston area, where much of the wood cover was removed between 1913 and 1916.

The sulphur dioxide emissions caused widespread destruction. For every one million tons of ore roasted, about 30 per cent was released in the form of sulphur dioxide emissions into the atmosphere. In the process, clouds of sulphur dioxide from the roast yards rolled across the countryside, killing or damaging the surrounding vegetation, adding to the impact created by lumbering and fire. These emissions impaired the growth of vegetation, contaminated the surrounding soil with copper, nickel, and iron particulate matter, and started the process of blackening the rock cover. So long as these fumes had no adverse economic effects, they were not a matter of public concern. However, after new townships were opened for agricultural occupation under the Free Grants Act of 1906, and mining production continued to increase, the sheriff was called upon from 1909 to 1914 to arbitrate awards compensating farmers for damage to their crops from sulphur fumes.

The increased demands for mineral production arising out of the First World War completely changed the situation. Following a number of lawsuits, the Canadian Copper Company (CCC) relocated some of its roast yards to unallocated Crown Lands. In complementary fashion, the province passed Orders-In-Council on October 21 and December 29, 1915 closing all or parts of twenty townships for settlement purposes. These included the townships of Graham, Denison, Fairbank, Creighton, Snider, Water, Broder, and parts of Rayside, Balfour, Dowling, Trill, Drury, and Louise situated in the vicinity of the CCC's roast yards. Mond received the same consideration with respect to townships around Coniston and in the Wanapitei River valley, including Dryden, Dill, Neelon, and specified sections of Garson, Falconbridge, Awrey, and Cleland.[8] To deal with the issue of lands already in private hands, the Ontario government passed the Industrial and Mining Lands

Compensation Act of 1918, authorizing mining companies to buy smoke easements from private landowners, and forbidding existing and future owners from suing mining companies. Falconbridge Nickel followed Inco's example of purchasing smoke easements after 1930. When the 1915 law barring settlement in certain townships began to be repealed in 1921, smoke easements were made mandatory for all new owners. To deal with farmers who refused to sign smoke easements, the province passed the Damage by Fumes Arbitration Acts of 1921 and 1924. One effect of these acts was the establishment of a permanent sulphur fumes arbitrator who had the power to make awards not subject to appeal. The 1924 act remained in force until it was repealed in 1970. By this time, the air pollution problem had become so complicated that it required new legislation, known as the Air Pollution Control Act.

The CCC's first roast yard (1888–1905) was erected southeast of present-day Copper Cliff, now covered by slag piles. A second yard was constructed in 1899 near the site of Copper Cliff's Nickel Park. Possessing the dubious distinction of being in the worst possible location, it remained open only until 1903. In 1899, a third yard was completed north of the townsite. This operation was closed in 1916, and transferred to the O'Donnell site in Graham Township. This yard, the largest of its kind in Canada, was more than 2.3 kilometres long with four parallel tracks running through its length. Open air roasting was

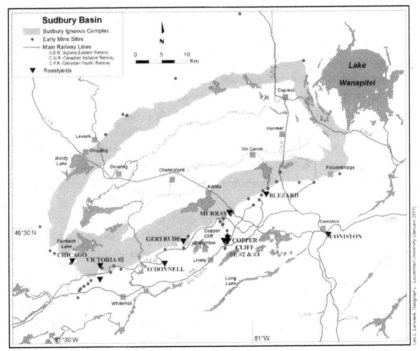

FIGURE 14.1   Early Roast Yards in the Sudbury Area (1888–1929). From Laroche, Sirois, and McIlveen, *Early Roasting and Smelting Operations in the Sudbury Area.*

finally phased out in 1930, when the roaster furnaces in the new smelter at Copper Cliff with its 155-metre smokestack, the largest in the British Empire, were opened in 1930. Though the ore bridges and railroad tracks serving the O'Donnell site were removed in 1938, remnants of the yard remain visible.

In the meantime, Mond Nickel built two roast yards in 1901 at Victoria Mine in Denison Township. The first yard lasted only a few months until a fire forced its relocation to a position between the mine and the smelter. This operation came to an end in 1913 with the opening of the smelter at Coniston. This necessitated another roast yard southeast of the Coniston smelter; that yard was phased out in 1918, after only six years of operation. Smaller roast yards included those located at Blezard Mine, owned by the Dominion Mining Company (1889–94), H.H. Vivian's at the Murray Mine (ca. 1889–96), the Drury Nickel Company's at the Chicago Mine (1892), and the Lake Superior Power Company's at the Gertrude Mine in Creighton Township (1901–03).

To gauge the impact of these roast yards, an 1890 government report mentioned that they were as "unsavoury as a gehenna," that is, a place or state of pain and torture, symbolic of hell.[9] In another geological report dated 1904, the author writes, "a more desolate scene can hardly be imagined than the fine white clay or silt of the flats, through which protrude, at intervals, rough rocky hills, with no trees or even a blade of grass to break the monotony."[10] One area farmer in 1915 was inspired to write,

> Any man being on the road between Sudbury and Sudbury Junction, on the CNR on Monday, August 23rd, could not help smelling the odour of dying vegetation, and on the following day the fields were a rusty dying colour, instead of a living green. Is that not sufficient proof of the damage being done by sulphur smoke[?]... [A]nd I would challenge any man with a head on his shoulders, a face on his head, and a nose on his face to deny this fact.[11]

The roast yards were eliminated after smelters were introduced; this was a significant technological advance, but did not solve the problem. Since the new smokestacks simply diffused the sulphur emissions over a larger area, farmers and others in the more outlying districts began to complain about the damage.

Solid Wastes—Tailings and Slag Dumps

In addition to the vast expanse of land damaged by roast bed effluents, smaller areas of land were rendered barren by the deposition of solid wastes—mill tailings and slag. Both of these forms of waste were significant not only for the large areas that they occupy but also for their environmental and visual effects. These tailings and slag zones cover thousands of hectares of land. While slag dumps are the most visible, they encompass only 12 per cent of the total area; the remaining 88 per cent consists of tailings largely hidden from public view (fig. 14.2).

Tailings

As they are largely invisible to the public eye, there is little appreciation of the geographic extent and environmental impact of the tailing ponds that exist in the Sudbury area. Tailings are the waste product of the milling/concentrating process. At these mills, barren rock is removed from the ore through a system of crushing, pulverizing, and flotation. In earlier times, when the mines were small and only high-grade ores were mined, the rock removal problem was minor and could be easily managed by mills located at the minesite. When the roast yard system was replaced by the new smelter in 1930, a need arose for more finely ground concentrate. While smaller mills continued to serve this concentrate for some time, economies of scale brought most of them to a close. There are now only two—Vale's Clarabelle Mill in Copper Cliff (after 1991), and Xstrata's Strathcona Mill in the Onaping area (after 1988). Since ores now constitute as little as 1 per cent of the mined rock, the crushed rock portion is an enormous volume of waste that must be disposed of elsewhere. Coarser tailings are usually returned underground as mining fill; but because the original ore body expands after removal, not all of the tailings can be put back underground. Thus, the remainder is mixed with water, pumped out as slurry, and brought to disposal areas.

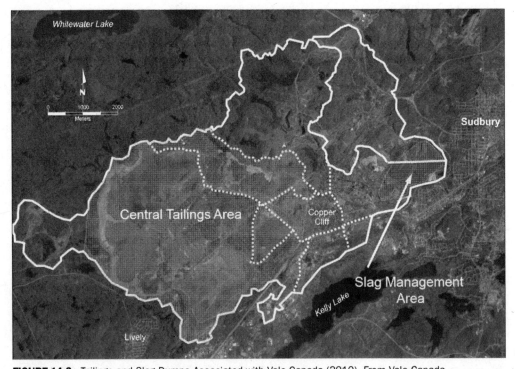

FIGURE 14.2 Tailings and Slag Dumps Associated with Vale Canada (2010). From Vale Canada.

By the 1940s, many of the dammed tailing disposal areas were fifty metres high, equal to or higher than the rocky hills of the local topography. This resulted in the tailings being picked up by the wind and becoming airborne. The blowing dust became a nuisance to local residents, and affected local industrial operations and equipment. These problems led Inco to try to revegetate the tailings basins. The first attempts to stabilize the surface of the tailings in 1947 were failures. This was not surprising, considering that the tailings had no nutritive value, limited water-holding capacity, no soil structure, a low pH, and no organic matter. Stabilizers such as chemical sprays, limestone chips, and water sprays proved ineffective, as were the experimental seeding of grasses.[12] Researchers made a significant breakthrough as a result of experiments in 1956–57 that used limestone, fertilizers, grasses, and companion crops such as rye and mulching. This was the first time in the Western hemisphere that vegetation was successfully grown on tailings.[13] The establishment of a vegetative cover on the tailings resulted in the production of the award-winning film *Rye on the Rocks* in 1969.[14] An unexpected bonus of the tailings program was the creation of a marketable crop of hay for livestock. While Falconbridge Nickel also adopted a vegetative solution in 1957, it differed in that the company planted young Carolina poplar trees in holes filled with loamy soils, which now present a pleasantly green approach to the Falconbridge townsite.[15] The company experienced some difficulty, however, with attempts to rehabilitate its Northern Rim tailings acquired in 1964. Over the years, the Falconbridge operation utilized six different tailings sites, the first being Centennial Park beginning in 1930. Later, Falconbridge experimented with several types of dry and wet covers placed on the tailings before attempting to establish vegetation. Its most visible tailing site, Centennial Park, was reclaimed in 1967.

By the early 1970s, both Inco and Falconbridge Nickel had established their own environmental departments. Through their efforts, an ecosystem began to emerge on the tailings, one that featured both a change in the vegetative cover and local wildlife. The wildlife species changed as the vegetative cover evolved from prairie-like grassland to a scattered tree savanna, and finally to a forest of indigenous trees, such as jack and red pine. In 1974, after consultation with local wildlife clubs, the tailings around Copper Cliff were declared a Wildlife Management Area. Since then, ninety-seven avian species have been identified at the site, which now serves as a stopover point during spring and fall migrations.[16]

While the original objective of establishing vegetation on the tailings areas was merely to stabilize the surface, the quality of the drainage and seeping water from the tailings has evolved into a major concern in recent years. One technique Falconbridge has employed to deal with drainage from tailings basins has been the construction of natural wetlands. These wetlands are located downstream of the tailings basins, and they serve to filter and remove contaminants while passing through the biological and chemical processes integral to that ecosystem.[17] This technique has also been used by Inco.

The geographical extent of the tailings areas around Sudbury is considerable. A total of 22 square kilometres, the Vale repository at Copper Cliff is the largest in Canada, consisting of 10 per cent of all tailings stored in the country.[18] Xstrata Nickel also has extensive tailing areas and ponds. The restoration of these sites since the Second World War is a remarkable achievement, and it constitutes an integral part of what has been called the re-greening of Sudbury.

Slag Dumps

The other form of solid waste from the Sudbury mining operations consists of slag, which is a byproduct of the smelting process. When the concentrate from the mill reaches the smelter, it is melted to form a metal matte and a slag composed of iron silicate. While some slag is used as backfill in the mines, most of it is transported via trains to the slag dumps. Slag dumps can be found around the Falconbridge and former Coniston smelter, but those at Copper Cliff are the most visible. Covering 2.25 square kilometres and up to 61 metres high, this dump is one of the most dramatic features of the regional landscape, due not only to its physical presence but also to its role in Sudbury's imagery and folklore. For many decades, the place to take out-of-towners was the Inco slag dump. The strange, dazzling, volcanic experience was the one spectacle Sudburians could count on to make a lasting impression on visitors.[19]

For younger Sudburians, the slag dumps were attractive for a different reason—it was where teenagers went for a romantic evening to watch the molten slag flowing down the slopes. The Coniston slag dumps were popular among youth for the same reason. When Vale Inco arrived on the scene, however, the company decided that the slag dumps should be a more integral part of Sudbury's reclamation process. Following on the earlier success of Inco's efforts to revegetate and contain the slag dumps with berms along Regional Road 55 between Gatchell and Copper Cliff, the company initiated a program in 2006 to start a re-greening process along a three-kilometre stretch of Big Nickel Road from Highway 55 to Elm Street, making a visible statement to the community that Vale Inco was working on its environmental impact. The project involved grading the slopes, applying clay and peat up to half a metre thick on them, and then hydro-seeding the heaps with a mixture of fescue grasses and clover. By the end of 2008, more than thirty hectares had been covered. The effect has been stunning. Where previously people in Gatchell had to face a black wall of slag dumps, they now see a broad carpet of greenery. As a reporter for *The Sudbury Star* eloquently stated, "the now-green hills are as awesome and commanding an example of human endeavour as Sudbury can hope for."[20] Within a few years, a new phase will begin with the planting of trees and shrubs, many of them from Vale's surface and underground greenhouses.[21]

Acid Plants and Smokestacks

While sulphur fumes from roast yards, tailing ponds, and slag dumps were differing manifestations of environmental damage, they had one thing in common—their effects were localized. With the introduction of larger smokestacks at Copper Cliff, Coniston, and Falconbridge after 1929, the situation changed. Now the waste sulphur dioxide gases resulting from the smelting process were dispersed over wider areas. To some extent, these emissions were reduced by the subsequent construction of sulphuric acid plants at Copper Cliff and Falconbridge. Nonetheless, the remaining volume of waste gases that went up the smokestacks was enormous. Despite the widespread damage caused by these sulphuric dioxide conditions year after year, it was not until 1944 that the first serious study of smoke damage in the Sudbury area was undertaken by mining companies and the provincial government. Released in 1945, the report discovered that "smelling strength" fumigations occurred as far away as 64 kilometres, with "visibility effects" extending even further, to 120 kilometres or more. Evidence of severe tree foliage damage was found in these areas.[22] The Ontario Department of Mines had been monitoring sulphur dioxide emissions through a network of five stations in the area. A series of reports dating from 1952 laid out their findings, but few outside of the ministry were aware of the reports' contents and devastating conclusions.[23] In a more public 1967 report, the situation changed:

> One characteristic feature of the immediate Sudbury area is its apparent endless barrenness.... The airborne pollution emanated from these smelters has destroyed and continues to destroy vegetative growth in varying degrees of intensity within approximately 1 700 square miles [4 400 square kilometres] bordered by Lake Temagami in the northeast, St. Charles in the southeast, Worthington in the southwest and Milnet along the CNR line in the northwest. However, only 40 square miles [100 square kilometres] are classified as severely barren, while a bordering band of approximately 140 square miles [360 square kilometres] supports combinations of herbaceous, shrub and small tree growth.[24]

The first restoration project started in 1969, when Laurentian University and the Ontario Department of Lands and Forests began to develop strategies to revegetate the barren land. Called the Sudbury Environmental Enhancement Program, the department established a mandate to assess the ability of the area's soils to support vegetation. Studies undertaken in the following two years indicated that direct tree planting without any soil amendment was not possible, and led to poor results.[25] The strategy changed in 1971 to introduce different herbaceous species, and to raise the low pH levels through liming. In 1972, the restoration process was given an unintended boost by the completion of the "superstack," and the overwhelming global response to this new icon of the Sudbury landscape. The growing awareness that air pollution was a global rather than a local problem, and a wider appreciation of the impact of acid rain both contributed to pressure for better environmental improvement and re-greening programs. While Inco was quick to tout the new smokestack as a responsible environmental measure, the economic reality was that its older smokestacks needed repairs, and a bigger and taller smokestack was a vital component for allowing the company to increase production.[26] Meanwhile, the growing environmental movement of the late 1960s prompted the province to pass the Air

Pollution Control Act in 1967, and to issue mandatory emission controls for Falconbridge Nickel in 1969 and Inco in 1970; by 1972, these steps yielded significant improvements. Other important measures taken at the time included the closure of Inco's Coniston smelter, reducing emissions at Inco's iron ore plant; Falconbridge Nickel's pyrrhotite plant also closed. Immediately after 1972, these improvements led to a drop in the average concentration of sulphur dioxide emissions by 50 per cent.[27]

The pollution issue was then brought into sharp focus by a secret 1974 federal government report titled, "The Sudbury Pollution Problem: Socio-Economic Background," that was inadvertently made public in 1977. This study stated: "Government has been extremely lenient with Inco and Falconbridge. Historically there have been no prosecutions under applicable environmental legislation, and from 1924–1970, there was a curtailment of a citizen's right to sue for pollution damages, and there has been a lack of government-sponsored research on the damage caused by the copper-nickel smelters."[28] The wheels of change were now set in motion by the public.

Throughout the rest of the century, additional reductions in emissions continued to be achieved in various ways: a new acid plant opened at Falconbridge Nickel's smelter in 1978, more control orders such as the Countdown Acid Rain Program established in 1985, summer shutdowns, labour strikes, periods of reduced production, and new initiatives undertaken by the Regional Municipality of Sudbury.[29] By the year 2002, the average annual emission of sulphur dioxide in the Sudbury area had decreased by a remarkable 90 per cent compared to 1960 levels. It is a testament to the proactive approach taken by Inco and Falconbridge Nickel that both companies actually exceeded the emission reduction objectives laid out by the Countdown Acid Rain Program. The improvements in local air quality have been such that the community now compares favourably with other large urban centres, often ranking above Windsor, Hamilton, and Toronto.[30] In accord with a Ministry of the Environment (MOE) order issued in 2002, the existing limits of emissions after 2007 were to be reduced by one-third again.[31]

In 2005, the Minister of the Environment passed Ontario Regulation 419, marking the biggest improvement in regulating air toxins in Ontario over thirty years. While Vale Canada immediately began preliminary work to meet this goal, it was 2011 before its project to meet the regulation was approved. According to the plan, known as the Atmospheric Emissions Reduction (AER) project, Vale will reduce its sulphur emissions by 70 per cent between 2012 and 2015 through lowering stack emissions, achieving better fume capture around the converter aisle, and reducing emissions in the material handling area. The project will also lessen emissions of dust and metals by 35 to 40 per cent over current levels. An estimated 1 300 workers will be required on site during peak construction. At a cost of $2 billion, this atmospheric emissions reduction project is the largest single environmental investment any company has made in Greater Sudbury's history, and one of the largest ever undertaken in the province.[32] This 70 per cent reduction by 2016, in addition to the 90 per cent reduction realized since 1970, will serve as another historic milestone in the re-greening of Sudbury's landscape.[33]

Land Reclamation Program 1973+

The involvement of the Regional Municipality of Sudbury in the re-greening process began after its formation in 1973. That same year, a Technical Tree Planting Committee was established with a mission to change Sudbury's reputation of being a barren and inhospitable environment. Led by Laurentian University's Biology Department, a study was commissioned to determine the extent of vegetative damage along the regional highway corridors, where the impact was most severe. Other studies led to favourable re-greening results using lime, fertilizer, and seed. While these steps were positive, the reality, however, was that little had actually been done to physically improve the land-scape outside of efforts by the mining companies. The scene changed in 1977, when Inco announced the layoff of 3 500 employees, and the cessation of hiring summer student labour as it had done in the past. This situation led to the creation of a huge federal Young Canada Works program. With the assistance of the province, this program started in May of 1978, beginning one of the world's largest community efforts to reclaim industrially disturbed lands. Sudbury's council passed a Regional Official Plan in 1978, and one of its major objectives was the enhancement of the area's visual quality and image through re-greening techniques.

The projects started in 1978 were focused on two highly visible sites: the airport corridor, and the stretch of highway from Coniston to Wahnapitae. The results quickly demonstrated that dramatic landscape improvement was possible, and that the regional municipality was capable of operating job creation programs of significant size. In 1979, the program was expanded. The focus again was on locations along major transportation corridors. In 1980, a five-year planning program was approved. As the major highway corridors had already been covered with grass, the work in 1981 centred on smaller areas of barren lands. In the following year, a full-time coordinator for the reclamation program was hired. Re-greening efforts then shifted to areas with steep slopes. This resulted in a new innovation—the use of helicopters to deliver raw materials to sites on higher ground, while students worked downhill at labour-intensive tasks.

When Inco announced additional layoffs in 1982, regional officials were success-ful in expanding short-term employment opportunities for unemployed miners under various provincial and federal government programs. Another change in the re-greening process took place in 1983 when, for the first time, a major tree planting component was added. Between 1978 and 1984, re-greening projects employed more than 2 800 workers, improved 2 500 hectares of land, and resulted in the planting of 388 000 trees.[34]

Since 1984, land reclamation has continued unabated (fig. 14.3). By 2010, some 9.2 million trees and more than 50 000 shrubs had been planted by 4 536 workers and 9 981 volunteers. They also limed, fertilized, and seeded more than 3 429 hectares of land.[35] Further innovation was introduced in 2010 with the transplant of forest floor mats containing an array of plants, seed, roots, and organisms that would otherwise take cen-turies to reestablish naturally. These vegetation mats were salvaged from the construction

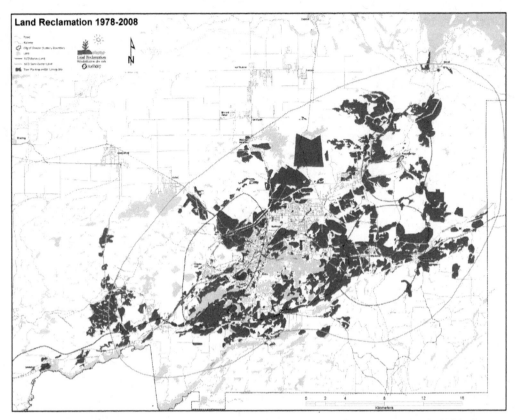

FIGURE 14.3   Geographical Extent of Land Reclamation in the City of Greater Sudbury (1978–2008). From the Planning Services Division, City of Greater Sudbury.

corridor of the new four-lane highway south of the city and laid out elsewhere on barren and semi-barren four-by-four-foot plots of land.[36] An innovative aspect of the tree planting program is the contribution made by Vale's greenhouses. Using a surface greenhouse in Copper Cliff and a greenhouse situated 1 400 metres underground at Creighton Mine, Vale staff grow approximately 250 000 red and jack pine seedlings each year. Of this total, about half are used on Vale's own industrial properties, and the other half are earmarked for Sudbury's Vegetation Enhancement Technical Advisory Committee (VETAC).[37]

The change in Sudbury's image as a result of these efforts has been significant, and well worth the $25 million investment as of 2010.[38] The program has garnered global attention and received more than fourteen awards recognizing its achievements. Among these can be included the Lieutenant Governor's Award, the Government of Canada Environmental Achievement Award, the United States Chevron Conservation Award, and the United Nations Local Government Honors Award.[39] These awards bear testimony to the fact that Sudbury is known internationally as a rare example of a community whose scars are healing rather than deepening. While these examples of the healing of the land-scape are sufficient to make the Sudbury area an ecological success story, the caveat is that, in terms of biodiversity, the area is still in its infancy. Much has been done, but much remains to be done. Every effort, therefore, should be made to ensure that re-greening remains an ongoing process.

Aquatic Issues

Since the City of Greater Sudbury contains more lakes than any other city in Canada, it was also deemed important to deal with the impact of sulphur and acid rain emissions on local aquatic systems, and the alteration of the pH of nearby lakes. Studies have shown that the acidification of the Sudbury lakes began in the 1930s, when the tall smokestacks first appeared. Acidification continued and intensified until the 1970s. By that time, an estimated 7000 lakes within a 17000-square kilometre area encompassing the city were acidified to a pH of 6.0 or lower. Another 12000 lakes beyond the City of Greater Sudbury were impacted as well. Lakes closest to the smelters exhibited the highest levels of acidifi-cation. In accord with general wind directions, lakes found to the northeast and southwest of the smokestacks were the most affected. The acidification process was accompanied by metal contamination from the airborne metallic dust of smokestack emissions. Other complex factors contributing to lake damage included the underlying geology, in-lake chemistry, erosion, storm water discharge, nutrient enrichment, and industrial spills from the large creosote plant that existed on the western edge of the City of Sudbury between 1921 and 1960.[40] Fish populations were also adversely affected by the increased acidity and metal concentrations.

The first attempts to deal with water conditions began in the early 1970s, when scientists added powdered lime to local lakes. In each case, the combination of reduced sulphur dioxide emissions and liming resulted in an amelioration of lake water acid-ity, increased alkalinity, and decreased metal concentrations. While the duration of improvement varied somewhat, researchers discovered that longevity increased if the surrounding catchment areas were also subjected to the liming process. There still remained, however, a time lag with respect to aquatic ecosystems, suggesting that the recovery of biological communities, including fish, may be substantially delayed even after water quality has been improved.[41] In recognition of this ongoing problem, the city's future re-greening program will focus more on increasing tree density within a 100-metre belt around lakes. In time, these elements will add leaves, branches, and trunks to aquatic systems. To keep abreast of aquatic issues, the City of Sudbury has been assisted by a Lake Water Quality Program, organizations such as the Greater Sudbury Lake Improvement Committee, and thirty-one lake stewardship groups scattered throughout the municipality.

In addition to these measures, other projects have been initiated to enhance lake bio-diversity through fish stocking programs. While the Ontario Ministry of Natural Resources

(MNR) has long assisted in local fish stocking, these efforts have recently been aided by Vale's Environmental Department. Following successful efforts to establish a temporary fish farm for baby rainbow trout at the company's surface greenhouse in Copper Cliff, and transport them to the Onaping River in 2011, a pilot project was conceived to create a similar operation underground, using the Creighton Mine greenhouse. The project is based on harnessing pre-existing heat underground (where it is 23 to 24°C year-round), and the utilization of the existing greenhouse. In addition, project organizers plan on filtering the solid waste from fish and recycling it as fertilizer for saplings being grown in the greenhouse to create a uniquely sustainable operation. Such a feat has not been done underground anywhere else in the world.[42]

The Vale Living with Lakes Centre

The history of the Vale Living with Lakes Centre extends back to 1989, when the Cooperative Freshwater Ecology Unit (Co-op Unit) was formed as a partnership includ- ing Laurentian University, the Ontario Ministry of Natural Resources, and the Ontario Ministry of the Environment (OME). For some projects, other partners are also involved. Beginning with an initial focus on the restoration of acid- and metal-damaged waters in Northeastern Ontario, the Co-op Unit has become an internationally renowned research and monitoring team. It recently broadened its research to address other stressors affect- ing the health of aquatic ecosystems such as climate change, trace contaminants, aqua- culture, loss of biodiversity, and excessive exploitation.[43] Scientists working at the centre have helped turn Sudbury's once-polluted waters into a healthy habitat for fish and inver- tebrate communities affected by years of mining and smelting. In some ways, the aquatic changes have been "just as profound and even more so" than those associated with ter- restrial re-greening.[44]

The work of the Co-op Unit was given a significant boost with the opening of the Vale Living with Lakes Centre in August of 2011 on the shore of Lake Ramsey.[45] Originally conceived in 1992 and designed using sustainable materials in 1996, the building acquired the status of being Canada's first "green" institutional building, achieving a Platinum LEED status, the highest certification possible. The design of the facility won the Bronze Holcim Award for Sustainable Construction in 2008, and was named one of Ontario's best and most innovative consulting engineering projects of 2011.[46] Built with Sudbury's future climate in 2050 in mind, the facility conserves water, utilizes solar panels, and relies on geothermal heating and cooling systems.[47] Another unique element associated with the building is the use of blueberries planted on the green roof to absorb heat during the winter and keep the building cool in summer. Supported by a federally funded Canada Research Chair in Stressed Aquatic Systems since 2003 and surrounded by an abundance of boreal lake systems, the centre has added much to Sudbury's international reputation for environmental restoration and innovation.

The Soil Story

As was the case with lake acidification, interest in the metal levels and acidification of soils only emerged after the late 1960s. In the 1970s, studies by local foresters and ecologists showed that soil acidity and concentrations of nickel and copper were elevated in many of the same areas where sulphur dioxide damage had been observed. Researchers discovered that the acidity of the soil, when combined with airborne metals, created an environment toxic to plant growth. To deal with this issue, the MOE passed the Guidelines for Use at Contaminated Sites in Ontario in 1997, which provided regulators and industry with a set of criteria to use when monitoring data. In September 2001, the MOE released a summary report covering approximately thirty years of soil metals data collected in the City of Greater Sudbury. In these reports, the authors concluded that elevated levels of heavy metals (specifically nickel, copper, cobalt, and to a lesser extent, selenium) and arsenic were common in the area, and elevated near the three smelting centres of Copper Cliff, Coniston, and Falconbridge. They recommended that additional sampling be conducted to better determine the extent and magnitude of metal and arsenic concentrations in local areas, and its impact on human health and ecological risk management.[48]

As a follow-up to MOE's 2001 study, the Sudbury Soils Study was initiated in the same year. Covering an area of 40 000 square kilometres, it was one of the largest and most comprehensive of its kind in North America. The MOE, Inco, Falconbridge, the Sudbury and District Health Unit, and the City of Greater Sudbury joined forces with other major stakeholders under the umbrella of the Sudbury Soils Study Technical Committee to undertake a rigorous soil sampling and analysis program, and to oversee the study. The 8 400 soil samples already collected were expanded to 14 000 and analyzed for 20 elements. In addition to the sampling program, the study conducted an extensive survey of metal levels in local vegetable gardens and from 10 per cent of local residences.[49] The MOE also asked the mining companies to conduct both a Human Health Risk Assessment (HHRA) and an Ecological Risk Assessment (ERA) to determine if the elevated metal levels posed an unacceptable health risk to people and/or the environment. Other members of the team included a Technical Committee and a Public Advisory Committee to provide guidance for the process and to ensure public engagement. Early in 2003, a consortium of consulting firms working together as the Sudbury Area Risk Assessment (SARA) Group was retained to undertake the risk assessment portion of the study. To ensure scientific accountability, it was decided that all results and reports produced by SARA would be peer-reviewed by an expert international panel selected by Toxicology Excellence for Risk Assessment (TERA), a non-profit, independent corporation.

The three volumes of the Sudbury Study Reports were released in 2008 and 2009.[50] The eight-year study cost $15 million to complete, and concluded that while there were issues that needed to be dealt with, there were little to no health risks from exposure to metals mined in the area (according to the metal levels as defined in the study).[51] It also determined that terrestrial plant communities in the Greater Sudbury area continue to be affected by the six chemicals of concern (arsenic, cobalt, copper, lead, nickel, and selenium).[52] While a report commissioned by Mine Mill Local 598CAW and Local 6500 Steelworkers and prepared by Environmental Defence was critical of several aspects of the SARA study, it had limited impact.[53] In order to address some of the key ecological issues still unresolved, a Biodiversity Action Plan involving the City of Greater Sudbury, Vale Inco, and Xstrata Nickel was released late in 2009.

Nickel District Conservation Authority

While most attention has been given to the role of the re-greening process in the rehabilitation of the local landscape, it is important to note the existence of another body that has effectively served the environmental and natural resource needs of the Sudbury area since 1957, that is, the Nickel District Conservation Authority. Conservation authorities first appeared in Southern Ontario following the issuance of the Conservation Authorities Act in 1946, but it was not until the passage of Hurricane Hazel in 1954 that concrete steps were taken to meet the needs of municipalities in Northern Ontario.[54] By this time, the annual cycle of flood damage had been exacerbated by population growth and the construction of permanent structures in flood-prone areas.[55] Thus, the Junction Creek Conservation Authority was created on December 12, 1957. This was followed by the establishment of the Whitson Valley Conservation Authority on September 3, 1959. Together, these two watershed authorities encompassed seventeen municipalities. Unlike other planning bodies in the area at the time, the authorities featured strong leadership, and a greater sense of purpose.

These authorities existed as separate entities until June 6, 1973, when they were amalgamated to form the Nickel District Conservation Authority. At the same time, its area of jurisdiction was enlarged to 7 125 square kilometres, covering the Regional Municipality of Sudbury and parts of seventy-six unorganized townships in the remainder of the watershed to the north and south. In size, it ranked second among the thirty-nine conservation authorities in Ontario. Following the formation of the City of Greater Sudbury, the authority's area of jurisdiction expanded to 7 576 square kilometres, and it was placed under the administration of a board appointed by the new municipality.

Throughout its history, the authority has carried out a broad range of programs dedicated to flood and erosion control management, floodplain mapping, and prohibition of development in hazard areas. In 1964 and 1965, the authority undertook channel improvements involving the construction of a 3 840-metre-long concrete box culvert along Junction and Nolin Creeks in and around the downtown area. Since 1973, it has encouraged development on the basis of a master plan. Among the authority's major flood control measures can be included the construction of the Maley Dam in 1971 and the Nickeldale Dam in 1980. Other improvement projects were implemented along the

Whitson River in 1977 and elsewhere from 1987 to 2001. Protective works to correct erosion problems at various sites have been developed, and in the mid-1980s the authority installed a flood forecasting network. Another major project was the opening of the Lake Laurentian Conservation Area (soon to be known as Conservation Sudbury) in 1967. The authority acquired lands surrounding this area to maintain the man-made Lake Laurentian in 1958 and its watershed as a reservoir to augment Lake Ramsey, one of the main water sources of the City of Greater Sudbury. Now encompassing some 977 hectares, the park has been expanded to include a chalet and nature study site, a large green space, a self-guided nature trail, wetland areas, hiking trails, birdwatching areas, and snowshoeing and cross-country skiing trails in winter. Other features associated with the conservation area include the Lake Laurentian Environmental Education Program, and its annual summer nature camp Bitobig. The authority developed these activities in close cooperation with local school boards. It also owns Gaillard Island in the east bay of Ramsey Lake, which once served as the John Island summer camp,[56] and contains some of the oldest white pines in the city. Recently, new issues related to the quality of the water supply and climate change have come to the fore, which the authority plans to address.[57]

Junction Creek Waterway Park

Another example of the healing process going on in the Sudbury area was the creation of the Junction Creek Waterway Park. While the installation of covered culverts in the 1960s and the beginning of Sudbury's green revival in the 1970s and 1980s were steps in the right direction, there remained long stretches of polluted and unsightly waterways along Junction Creek. As stated in *The Sudbury Star*, the city had "simply turned its back on the creek."[58] To rectify this situation, the City of Sudbury's Secondary Plan, passed in 1987, recommended that an eighteen-kilometre stretch of Junction Creek be developed as an interconnected urban parkway. The plan's underlying philosophy was to turn this liability into a natural asset. This was followed in 1991 by the initiation of a Community Improvement Plan to develop a park across eighteen kilometres of walkways and parkettes from Maley Drive in New Sudbury to Kelly Lake Road in the west end. The trail's purpose was to develop a new community focus, and to provide an oasis of open space for recreational opportunities such as walking, bicycling, and jogging. By the end of the 1990s, more than nine kilometres of trail had been developed.[59]

The project took a major turn in 1999 when the Junction Creek Stewardship Committee (JCSC) was formed to deal with the entire length of the 52-kilometre creek and its 329-square kilometre upper watershed. The objectives of this citizen-based committee were laid out in clear terms: clean it up, rehabilitate it, and educate the public about it. In 2003, an ambitious five-year plan was approved calling for numerous rehabilitation projects, more removal of garbage from streams and banks, and the expansion of the trail to eighteen kilometres. By 2009, volunteers had hauled 50 000 kilograms of garbage from the creek, stocked it with over 9 600 brook trout, and planted 18 000 trees and shrubs.

Another initiative was the development of a master plan to remove creosote that had migrated from a former industrial site, and to address salt and pesticides in storm water runoff. Thanks to the committee, Junction Creek is now very much alive, attracting not only birds, plants, and twelve species of fish, but people as well. This has been achieved by the creation of resting spots, parks, and playgrounds along its route. An educational film was completed in 2009 aimed at inspiring groups and individuals to become involved with the project.[60] The committee has received many awards for its efforts. In strategic terms, this project is a classic example of what can happen when ordinary citizens seize the initiative to reclaim a creek as their own.

By 2011, eleven kilometres of the waterway trail had been completed.[61] In this year, the regional council agreed to complete the entire trail by 2015 and link the system to other trails so that it would be possible to traverse it from Garson to Whitefish. Proponents of this plan argued that it would provide a green backbone for the community, serve as a tourist destination, promote healthy lifestyles, offer an alternative transportation route through the core of the city, and be a precious resource for the citizenry. When completed, the rehabilitation of Junction Creek will represent another major step in the re-greening of Sudbury.[62]

Taming the Hilltops

The sight of blackened hilltops has been another traditional aspect associated with Sudbury. Denuded of trees and soils and stretching for kilometres along the skyline, these rugged-looking ridges did much to contribute to the city's negative environmental image. They affirmed the adage that Sudbury was simply "no place to build a city."[63] Until the Second World War, urban settlement and its accompanying road network used a "passive" form of planning, with houses, businesses, and roads following tenuous and snake-like patterns along the lower stretches of land strung between black hilltops. The Kingsway stretch is a classic example of this pattern.

Another feature associated with this phase was the myriad of bisected streets that were created, often with the same name, on both sides of a hill. This patchwork situation caused much confusion among drivers, especially first-time visitors to the city. Until the Second World War, rock was the enemy and it reigned supreme; however, there were anomalies to this spatial arrangement which could be found in places like the enclave of Little Italy. The community stubbornly clung around a great mound of rock, as did the higher reaches of the Kingsmount neighbourhood that developed south of the CPR tracks.

As the pace of urban development intensified in the 1950s and 1960s, many of the blackened surfaces disappeared as residences, lawns, and trees began to climb up the natural contours of the hillsides. This new trend appeared in different parts of the city, including the York Street and Grandview neighbourhoods. While more costly to build, these elevated neighbourhoods succeeded because they were attractive to higher-income residents. They were also among the first areas to introduce planning elements such as curvilinear roads, t-intersections, and the pre-installation of municipal services.

The tyranny of rock lasted until 1974, when the registration of the Moonglo Subdivision changed everything. A new planning process emerged that was more "active" rather than passive. This subdivision was the first to demonstrate a more aggressive approach toward construction and its linkage to the physical environment. Originally a landscape consisting of rugged black bedrock and stunted trees, the area was blasted, flattened, and transformed into a neighbourhood setting that included more than 400 single-family homes, a large apartment building, and tree-lined avenues.[64] It is now one of Sudbury's most sought-after neighbourhoods. Other areas that made a similar transformation followed. Maki Avenue and its peninsula reshaped a rocky wilderness zone into an elite residential area, where most homes look over water. More recently, the location of Sunrise Ridge Estates on the rocky ridge north of the Kingsway has altered the skyline. Other examples of hilltop flattening and/or rock elimination can be found associated with several commercial and transportation projects as well, including the big box retail complex, the realignment of the Kingsway, and the construction of the Highway 17 Bypass across Long Lake road. For these projects, the process of rock removal, rather than being a liability, constituted a financial asset as flattening the terrain provided the rubble as a base for highway construction.

While most people would agree that these newer developments reflect significant advances in terms of engineering technology and improved greenery and make Sudbury more attractive overall, the generic style of curved streets, cul-de-sacs, and mall-style buildings often reflect little architectural imagination in transforming these rockscapes into something that is truly Northern. Perhaps this is another appropriate area of study for Laurentian University's proposed School of Architecture.

CHAPTER 15

Beyond a Rock and a Hard Place

Sudbury serves as a good example of the geographical adage, "while all places are unique, some are more unique than others." Its trajectory, which began millions of years ago, has few parallels elsewhere on the planet. While other communities have distinctive backgrounds, few can lay claim to the complexity of Sudbury's setting. Where else can one find a place that has a physical environment shaped by meteorite impacts, the clashing of geological provinces, the rise and fall of mountain ranges, the implanting of great ore bodies, and submersion under a post-glacial lake? How many cities have a natural landscape that has undergone both devastation and rehabilitation? Where else can one find a human landscape encompassing a mix of anglophones, francophones, ethnic Europeans, and Aboriginals that, while metropolitan in population size, has been woven by numerous forces into a spatial shape known as a "Constellation City?" Here too, we have a community known throughout the world as the site of a titanic struggle between unions and giant multinational corporations. Few urban places have experienced the negative imagery and reputation that Sudbury has had to endure since mining began in 1886. The municipal restructuring that took place during the Harris regime produced an enormous city, which prompted former mayor John Rodriguez to devise his well-known "what fits into Sudbury" map in 2008 (fig. 15.1). The map depicts Sudbury's boundaries with fifteen Southern Ontario communities easily wedged inside, with room to spare. This map shows better than words can say just how big Sudbury really is. Given its scattered population base, some consider Sudbury an "inside-out city," one that is supersized, with no clear beginning or end. In contrast to other cities that have well-defined patterns and edges, Greater Sudbury's landscape reflects a shapeless diversity, with hills, wetlands, and lakes all serving to break the urban fabric into disparate settlements.[1] It is this constellation form of urban geography that sets Sudbury apart from other settlements. In other words, "Sudbury is no ordinary town."[2]

A Challenging Geography

As demonstrated in previous chapters, a high degree of intimacy has always existed between the physical and human environment in the area. The "Sudbury Event" has been the constant physical factor influencing the development of the region since permanent

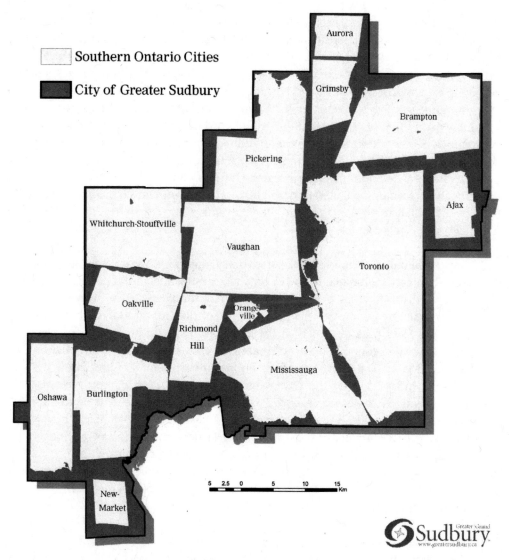

FIGURE 15.1 What Fits into Sudbury Map. From the Planning Services Division, City of Greater Sudbury.

settlement began in the 1800s. From a municipal perspective, this geologic setting has consistently given rise to major challenges with respect to expenditure and revenue streams. Since the 1880s, some ninety minesites and company towns have been established in the area, setting an early foundation for a dispersed spatial macro-form. This dispersal was exacerbated by company-owned mining lands that unwittingly created a major divide between the Valley and the Sudbury's urban core. When former agricultural settlements were added to mix, the result was a constellation population network featuring an urban core surrounded, not by a concentric pattern of suburbs as per the Canadian norm, but rather by smaller nodes at considerable distances from one another. The woeful lack of planning before the 1970s also left a legacy of low population density on many rural roads where residents expect to receive the same level of services available in the larger centres.[3] Per capita expenses thus became, and continue to remain, higher than they are for other Ontario municipalities. This combined "tyranny of distance" and "urban sprawl" makes the municipality one of the most difficult and costly to manage in Canada. Since this geographical fact is irrevocable, the long-standing tensions between Sudbury's core and its outlying centres and their ongoing fight for scarce infrastructure dollars will continue to dominate local decision-making processes. One consequence of this status quo will be the difficulty of fashioning a vibrant urban hub in the downtown such as those that exist in other Canadian metropolitan cities. One perceptive letter writer to *The Sudbury Star* succinctly makes this point:

> [Greater Sudbury] is now the largest city in Ontario by surface area, and the seventh largest in Canada, but with only a small fraction of the population of densely urbanized cities. This amalgamated city shouldered fiscal responsibilities offloaded by the province, and is now responsible for a wide, thinly spread municipal infrastructure ... which [is] harder to maintain on a thin tax base, yet vital to the economic and social fabric of each star in this "Constellation City."... It is evident that the rate of annual infrastructural depreciation significantly outpaces what the city budgets for normal capital improvements/replacements. We're losing ground. The spread-out infrastructure is crumbling faster than we can afford to rebuild, maintain and service it.... We need a mayor and council that understand, as responsible leaders, that they can no longer afford to ignore the elephant in the room.... We need a customized plan for a unique set of circumstances. If the stars in the constellation are not kept vital, the long term outlook for all of Greater Sudbury as a viable regional centre in Northeastern Ontario will grow dimmer.[4]

This letter indicates that the relationship problems that exist between downtown Sudbury and its satellites have not been resolved, and it raises a critical issue regarding the future political shape of the municipality. Should the political and planning focus be on revitalizing central Greater Sudbury into a metropolitan form and leave the "stars" in the

Constellation City to wither on the vine? Or should the emphasis be on making all of the stars in the constellation vital? While the latter is the path recommended by the Report of the Greater Sudbury Community Solutions Team, it will require innovative approaches to taxation and expenditure issues.[5]

To deal with the infrastructural setting in the future, the municipality will have to take into consideration the woeful tale of past mining taxation. Despite numerous attempts to gain an equitable portion from the province, the municipality of Sudbury has yet to succeed in getting a fair share of its industrial tax base like other municipalities in Southern Ontario, such as Hamilton or Oakville. According to *The Report of the Advisory Panel on Municipal Mining Revenues* issued in 2008, mining's tax contribution to Sudbury's municipal base declined from 25 per cent in the 1970s to a mere 6.5 per cent by 2006.[6] This figure hardly conforms with Sudbury's reputation of being a company town. Rather, it supports Michael Atkins' argument that the province's concern is still about "money, inequity, powerlessness and unfairness."[7] It is obvious that a municipality such as Greater Sudbury cannot be run when the underground facilities of mining companies are not taxable, surface facilities are under-assessed, and it receives little compensating revenue from the province. This untenable position has been affirmed by the Federation of Northern Ontario Municipalities and the Ontario Mining Association.[8] While the proposed Ontario Progressive Conservative Party growth plan titled *Changebook North* presented in 2011 to Sudbury gave some recognition to this inequity, their plan, like others before, is simply too little too late. The plan proudly touted the party's intention to hand over mineral royalties to host communities, but the fact that the proposed policy would apply only to "new" mines would do little to offset the negative effects of past taxation policies. The prestigious Conference Board of Canada came to the same conclusion, stating, "Greater Sudbury, whose land mass covers an area several times the size of Toronto with only about 160 000 people to support the infrastructure, needs *an extra $30 million annually* to achieve sustainability—a 'fraction' of what the provincial and federal governments reap from the city's mining companies."[9] It is a sad commentary that council in 2011 was so desperate for mining revenue, it felt it was necessary to plead for a small tax on mining trucks using municipal roads. This problem was partially solved in 2012, when local mining companies and Greater Sudbury initiated a process of cost-sharing for rebuilding roads to mine sites to free them from seasonal trucking restrictions.[10]

Overcoming Colonial Attitudes

While attempts were made in the past to rectify many of Sudbury's issues, they frequently met with limited success because of senior government neglect, and their "colonial" attitude to northern mining communities. This is partly a reflection of the traditional division between Canada's core (often referred to by geographers as the country's "heartland") and its peripheral areas known as the "hinterland."[11] Whatever little came to Sudbury often came late, and only after bitter fighting or intense political pressure. The litany of such examples is endless. While the devastation wrought on Sudbury's physical and human landscape was known from the time that smelting began in 1888, little was done to deal with the situation until the superstack was erected in 1972. It was only after this structure became a perceived threat to other Ontario residents and people around the globe that serious attempts were made to reduce emissions. This advance notwithstanding, concerns regarding the health and safety of workers in the mining industry remained a secondary issue. The belief that "the history of mining is written in premature death and debilitating disease" continued unabated until the publication of the Ham Report in 1976 and the creation of joint safety and health committees in the same decade.[12]

Other Ontarians also showed little sympathy for the workers who fought for union rights in order to improve their working conditions and increase wages. Efforts to establish workers' rights were continually opposed not only by the Province of Ontario and the Government of Canada but also by the RCMP. While it is true that unions were permitted after 1942, the mining companies have continued to oppose them to the present day. A sad result of the ongoing conflict has been a continuous series of strikes, and the development of a boom-and-bust economy. This situation notwithstanding, Sudbury's role in promoting the union movement in Canada against formidable opposition should not be forgotten.

Sudbury's "colonial" position in the minds of provincial and federal politicians continues to exist. Provincial bureaucracies are wedded to the status quo and a philosophy based on resource extraction rather than community development still prevails.[13] While it is true that there was considerable support for Sudbury's plight from the two senior levels of government during the 1970s and 1980s, such assistance was long overdue, and was only provided in reaction to the allegation that Canada could not allow one of its metropolitan-sized communities to wither away. It is worth remembering that it remained until the middle 1950s before a direct highway connection to Toronto was available, thereby preventing the costly and time-consuming need to travel by car to Sudbury via North Bay. Only after long delays and intense political pressure were other transportation improvements, such as bypasses around Sudbury and a four-lane highway to Parry Sound initiated. Efforts to improve the local infrastructure via major projects required heroic efforts. The role of the Lougheed family has been important to these efforts (biography 16).

As politicians and local boosters such as Judy Erola and Jim Gordon can attest, decisions to establish institutions such as Science North, the Northern Cancer Research Foundation, and the Northern Ontario School of Medicine met with strong opposition from forces located in Toronto and Ottawa. The federal attitude toward Sudbury and its current attempt to become a world-class mining and technology centre continues to reflect this colonial view, shown by Tony Clement's remark that the Sudbury area was on the verge of becoming a "Valley of Death." The existence of this attitude in other areas is exemplified by an article in *Saturday Night Magazine* in the late 1980s that slighted the existence of a university in Sudbury.[14] While some may claim that this neglect is due to Sudbury's tradition of voting for the wrong political parties at the wrong time, this in itself is not sufficient to justify the long-standing mistreatment of Sudburians by provincial and federal governments.

Changing the Status Quo between Company and Union

As recent events suggest, there needs to be a wake-up call regarding the existing status of company–union relationships, and the current approach to foreign investment and take-overs. While the five-year contract signed between Vale and the United Steelworkers in 2010 was seen by many as a defeat for the union, it also heralded the need to reassess the traditional approach to bargaining and the role that the two senior levels of government have played (or not played) in this strike.[15] The system of labour relations in Sudbury that has evolved since the First World War has been adversarial in nature, and the long-term result in most instances has been a lose–lose rather than a win–win situation for both parties, as well as the Sudbury community. The world is on the threshold of a new spatial order and, like it or not, globalization is reshuffling the economic pack, with new faces such as China, India, and Brazil at the top. It is inevitable that, in this new world geography, a clash of economic cultures will from time to time occur. Whatever problems these may create, however, they should not override the need for unions to take more interest in the economic viability of the industry that pays their wages. More cooperative approaches and new workplace strategies will be required to ensure corporate competitiveness, and the maximum number of job retentions. Such models exist elsewhere in the world.

The union movement needs a new public relations approach to make it a more integral part of the community, as was the case with the Mine Mill union after the Second World War. Changes are likewise needed at the provincial level, where the reactive anti-union legislation of the Harris period should be reformulated to protect innocent bystanders such as Sudbury and the revenue position of Ontario. At the federal level, there needs to be an acknowledgement that foreign-owned mining companies are different from those that are Canadian-owned in that they wish mainly to deplete our natural wealth and send profits elsewhere. While foreign ownership has always been an issue in Canada, there was nevertheless the sense that these companies were in tune with Canadian norms; the fact that they had major offices in Canada was positive. The welcoming mat needs to be kept out, but the federal government must at the same time establish and enforce clear rules of corporate behaviour that follow Canadian traditions, especially as they relate to union issues. The cult of secrecy surrounding the takeover of Canadian companies by foreign interests should be eliminated. The public surely has the right to know what net gain there will be for Canada from such takeovers.

Future Scenarios

Economic

What about the economic future? While the Sudbury Igneous Complex has often surprised prospectors in the past, it is unrealistic to think that mining per se will serve as the main generator of future employment. There are several reasons for this point of view. The presence of two foreign-owned mining giants in the Sudbury Basin is increasingly incongruous

Biography 15

THE LOUGHEED FAMILY
Pillars of the Community

The Lougheeds, one of Sudbury's most prominent business families, have distinguished themselves as dedicated supporters of community development and advocates for a better quality of life for all of its citizens. According to one Sudburian, they can be ranked among the most influential families in Ontario.[a] High praise indeed! The Lougheed story begins with Gerry Lougheed, Sr. and Marguerite Grace McIntosh (1929–2006), who were raised in Southern Ontario near Windsor. Gerry Sr. graduated from the Canadian School of Embalming in 1949, and served his apprenticeship with the Morris Funeral Home in Windsor. The funeral home had a flower shop where Marguerite, who had graduated as a teacher in 1948, would come to help from time to time. In 1950, Gerry and Marguerite were married. The couple moved to Sudbury in 1952 where they established the Lougheed Funeral Home, Flower, and Ambulance Service on the corner of Eyre and Spruce Streets. After a brief teaching career at College St. Public School, Marguerite joined the family firm to manage its flower shop division. The family ran an ambulance service for seventeen years, until the province took over this role. Lougheed's operated on Eyre Street until 1969, when they moved to larger quarters on Regent Street, with a separate building for a flower shop, fellowship room, and garage. In 1984, Lougheed's acquired the modern-looking Jackson and Barnard Funeral Home (designed by a local architect, J.B. Sutton) situated on the corner of Larch and Paris Streets. This new home, built in 1959, replaced the former Jackson and Barnard Funeral Home site on Larch Street that had been in existence since 1933. Later, Lougheed Funeral Homes and Chapels were established in Hanmer and Val Caron.[b]

Gerry Sr. and Marguerite soon established themselves as trailblazers in the funeral process. Following a suggestion from a Finnish minister, they started offering receptions after funerals long before the practice became the norm. They likewise instituted after-funeral visits, the audio recording of funeral services, and the use of video tributes.

The Lougheed family includes two children, Gerry Jr. and Geoffrey, both of whom have carried on the Lougheed legacy. Gerry Jr. graduated from Humber College in 1976, with the highest grades for the Board of Funeral Service exam in all of Ontario. Geoffrey later graduated from Humber College in 1979, with the highest standing in Ontario for Funeral Service Education. Colette Lougheed, Geoffrey's wife, graduated from Humber College in 1980, and received her funeral director's licence in 1980. In 1985, she joined the Lougheed and Jackson & Barnard Funeral Homes. Under the direction of the Lougheed family, the company operations have grown to employ some 100 people. The family has played a significant role in improving funeral service in Ontario through its participation on the Ontario Board of Funeral Services and the Northern Ontario Funeral Director's Association, and some family members have acted as examiners for the Funeral Board Licensing Committee.

Over the years, the Lougheed family initiated many positive changes in the Sudbury area, and their name has become synonymous with community service, generosity, and fundraising activities. Since 1952, Gerry Sr. served as a force in Sudbury's first Rotary Club; he then assumed the task of rebuilding St. John's Ambulance in the area. In 1990, he became the first Sudburian knighted by the Governor General of Canada. He has received two Paul Harris Fellowship Awards, the highest honour a Rotarian can receive. Gerry Jr. founded the second Rotary Club (Sunrisers) in Sudbury, and is an honourary Steward of Local 6500 of the United Steelworkers of America, the only non-trade unionist in North America to receive this distinction. He received an Honourary Doctorate of Laws Degree in 1989 from Laurentian University, and an Honourary Doctorate of Sacred Letters followed from Huntington University in 2010. His other initiatives include fundraising for the Regional Cancer Centre, Sudbury Regional Hospital, Maison Vale Inco Hospice, and St. Joseph Villa. He has also served as chair of the Bereavement Foundation of Sudbury, the Palliative Care Task Force, the Northern Cancer Research Foundation, and the Northern Ontario School of Medicine.

Gerry's son Geoffrey likewise is a strong Rotarian who has served as the Chair of the St. John Ambulance Board of Directors, the founding president of the Sudbury Food Bank, and Commander of the Most Venerable Order of the Hospital of St. John of Jerusalem. Geoffrey's wife, Colette (Blais) Lougheed, has been active in the French community, and in 1990 she was selected as the Woman of the Year in Sudbury. Both Gerry Jr. and Geoffrey have been involved with the Heart and Stroke Campaign, the United Way, and the Samaritan Centre.[c]

Among the honours bestowed upon the Lougheed family are the establishment of the Lougheed Teaching and Learning Centre of Excellence based at Huntington University, and having the south wing of the Sudbury Regional Hospital named the Lougheed Tower. In 2008, the Lougheed family was inducted into the Community Builders Hall of Fame.

Following a long battle with cancer, Marguerite passed away in 2006 at the age of seventy-six.[d] It was her experience of travelling to Toronto for treatment that inspired the family to become involved with the campaign to build a cancer treatment centre in Sudbury. In her memory, the family established the Marguerite Lougheed Community Centre in the former St. Clement's church located in the West End. The 2 800-square metre facility provides free space for non-profit and charitable meetings, as well as family celebrations. Gerry Lougheed, Sr. passed away in 2012.

---

[a]"The Lougheed Family," Community Builders Awards, accessed February 8, 2010, http://www .cbawards.ca/Winners/2008/
[b]For a review of the history of the Lougheed and Jackson and Barnard Funeral Homes, see "Featured Funeral Homes of Sudbury," *Canadian Funeral News*, November 1978, n.p. and "Lougheed Funeral Homes, Sudbury, Ontario," *Canadian Funeral News*, January 2000, n.p.
[c]"The Lougheed's History," Lougheeds.ca, accessed February 8, 2010, http://www.lougheeds.ca/ history.asp
[d]"Marguerite Grace Lougheed: 1929–2006," *Sudbury Star*, May 8, 2010, 9.

in a world dominated by the pressures of globalization. The likelihood of single ownership within the Sudbury Structure is strong, and it would bring with it corporate rationalizations and savings, resulting in fewer jobs. Technological changes that have prevailed in the industry since the 1970s will continue to minimize the mining labour force. The possibility of what has been termed the "minerless mine" is increasing with each passing year. That being said, it should not be forgotten that while Sudbury is emerging from the grip of mining, the industry remains as the soul of the community.[16] On a more optimistic note, the $2 billion AER project announced by Vale Canada in 2011 and its Clarabelle ($200 million) and Totten mine ($360 million) projects will provide a short-term boost for the local economy.

There is, however, a more dynamic side to the mining industry: regardless of the state of mining employment, the mineral presence will continue to be influential as this sector shifts away from extraction and toward the provision of mining services and technologies for international markets. In fact, by 2010 this trend had developed to such a degree that municipal revenues from the mining-related sector were equivalent to those from mining itself.[17] What is interesting about this phenomenon is that it is a result of vigorous private rather than public investment. This financing is in sharp contrast to the government assistance that came to the community during the latter decades of the previous century, when facilities such as the Taxation Data Centre and Science North were built, and improvements were made to the health care and education sector.

An even more impressive aspect of this new orientation relates to the distribution of its employment. Rather than a couple of large companies hiring hundreds of workers, there are now hundreds of supply and service firms hiring workers, many of them technologists, engineers, and skilled tradespeople.[18] According to the Greater Sudbury Development Corporation, mining-related employment is so important that it is being promoted as one of the five main economic engines driving Sudbury's future.[19] While this "bottom-up" approach is leading Greater Sudbury into a more sustainable economic position, one unfortunate consequence is that, unlike many European countries, this thrust has not been officially endorsed by senior levels of government. This situation indicates the lack of a federal or provincial resource policy for Sudbury and other mining communities. If the community was recognized by the two senior levels of government as a world-class "Centre of Mining Excellence," it would be a significant boost to Sudbury's global image and export/research potential.

The four other economic pillars of the future identified by the Greater Sudbury Development Corporation are (1) advancing education, research, and innovation; (2) serving as the regional centre for health services in Northeastern Ontario; (3) fostering a sustainable arts and culture community; and (4) promoting itself as a visiting destination.[20] These are worthwhile and achievable objectives, since many of the foundations for these initiatives exist in the city's university and two community colleges, Sudbury Regional Hospital (now Health Sciences North), the Regional Cancer Centre, the Northern Ontario School of Medicine, Science North, and Dynamic Earth. The presence of 330 lakes, millions of trees, and the city's unique geological history can likewise be used to attract tourists.

Greater Sudbury can exploit other Northern opportunities as well, and could become a centre for retirement living. This objective is feasible given the fact that the population of Northeastern Ontario is different from Southern Ontario in that it is aging more rapidly.[21] According to one provincial estimate, Northeastern Ontario, by 2036, will have from 25 to more than 35 per cent of its population consisting of seniors.[22] The 2011 Census reveals that the aging population is expanding, with the proportion of seniors in many municipalities in Northeastern Ontario exceeding that of Ontario and Canada.[23] Greater Sudbury's size, along with its bilingual character, thus has great potential as the retirement centre of choice for residents from other places in Northeastern Ontario, many of which are experiencing declining economic and population trends. Neither should it be forgotten that the city is becoming more attractive for Aboriginals, who are opting for an urban home rather than living in isolated reserves.

Sudbury too is well positioned to expand its role as the region's dominant shopping node. In 2011, it ended the year ranked fifteenth out of thirty-nine Canadian retail markets with a population of 100 000 or more.[24] This sector is so vibrant that *The Sudbury Star* claims Sudbury is "one of the best places to hang your hat in Ontario."[25] While some may assert that the growth of this sector with its lack of a strong union presence simply provides more part-time jobs at reduced salaries, it is equally true that these employment opportunities have helped to mitigate the out-migration of youth, and offer a stepping stone to higher-paying positions.[26]

There is, however, a caveat. It bears repeating that all of the future projections and scenarios have a commonality—they rely on the existence of extensive and integrated highway and road networks. While it was the mining industry that sustained Sudbury prior to the Second World War, it was the provincial highway network that allowed the city to expand its economic base in new directions, ranging from Kapuskasing to the north and Parry Sound to the south. The completion of four-laning Highway 69 to Southern Ontario must therefore continue to be a priority to expand Sudbury's spatial outreach and strengthen its regional centrality.

These economic potentials, however, need to be balanced by the locational constraints imposed by Sudbury's size, the peripheral location of the core with respect to the Sudbury Structure, the obstacles faced by those residing in distant places such as Wanup, Capreol, Whitefish, and Levack, the existing constellation network of settlements, the powerful dialectic that continues to exist between "the City" and "the Valley," and the lack of mining tax revenue (fig. 15.2).[27] Many political and economic matters continue to focus on attempts to mesh these settlements into a more efficient pattern. Issues such as expanded busing, the expansion of the existing 400 kilometres of fibre optic cable, the construction of new arenas, industrial parks, and corridors linking the Valley to Sudbury's core are always on the political agenda. Among the casualties of this reality has been Sudbury's downtown; it seems unable to promote itself as a vibrant regional core, one that could attract a larger cultural crowd and serve as an entertainment destination. While this possibility would be enhanced by the presence of an entertainment/convention-style complex capable of attracting major shows and large corporate gatherings, it is unlikely that much support for this costly initiative would be received from the outlying centres. The ongoing strength of the New Sudbury Shopping Centre, the development of the big box phenomenon nearby, and the growing significance of the South End also makes a stronger downtown difficult to achieve. Any downtown endeavour would likely have to rely heavily on funding from large corporations with interests in the area, foundations, or perhaps Sudbury expatriates. Local planners are hopeful that the downtown strategy completed in 2011 will go a long way toward solving this core problem.[28] Another interesting proposal for the downtown has been put forward by Nico Taus, a local designer, who suggests that a more visual form of signage, similar to what existed in the 1960s, would go a long way to reflect a more bustling downtown with vibrant Las Vegas–style streetscapes.[29]

Population

What about Sudbury's population future? Given the historical tendency for the Sudbury area to go through boom and bust cycles, making any projection is a hazardous exercise. Such predictions involve the interaction of many variables like birth and death rates,

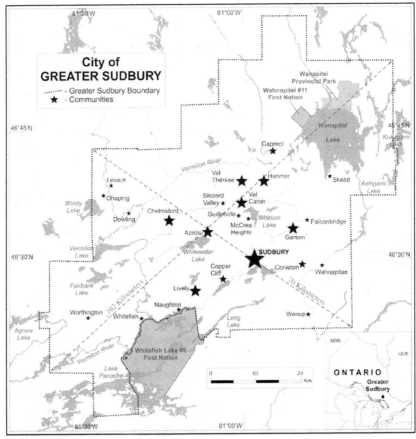

FIGURE 15.2   Sudbury as a Constellation City. From the Planning Services Division, City of Greater Sudbury.

provincial migration, and international immigration trends. Some of these demographic factors have shown some consistency since the turn of the new century. For instance, the rate of natural increase (the difference between births and deaths) has been in close balance in recent years. Between 2005 and 2010, the annual numerical difference between births and deaths in the Sudbury Metropolitan Area ranged from a high of 89 to a low of –23.[30] While the last few years have witnessed a small bump in the birthrate, it remains unlikely that a natural increase will play a major role for population change in the future. A similar scenario exists with respect to international immigration rates. In relation to this issue, one of Sudbury's major problems has been its inability to attract international immigrants. Greater Sudbury is simply not an attractive destination for people coming from foreign countries, a factor that planners need to deal with if the situation is to be rectified. While it is true that Greater Sudbury has benefited from small population growth in each of the last ten years which reversed a prolonged decline during much of the 1990s, this growth was still well below the Canadian and Ontario average.

The main variable that will determine Sudbury's population future will thus be the state of the local economy and its ability to reduce out-migration and increase in-migration from other parts of Canada. There is room for optimism here. As a result of the five-year contract signed between Vale Canada and the Steelworkers' Union in 2011, and the three-year contract between Xstrata Nickel and CAW/Local 598 in 2010, there will be a period of calm for the next few years that should bolster the stability of the local economy. Vale's new environmental and mining projects, the continued expansion of the retail and hotel sector, housing projects, and the strength of the world's metal markets all bode well for a sustainable pattern of development and a population increase not seen since the late 1980s.[31] This view is supported by the Conference Board of Canada, which forecasts small increases in Sudbury's mining-related and construction sectors, and by one commentator who claims that Sudbury is now the "luckiest city in North America."[32] These views of the future have been bolstered by a Bank of Montreal projection that nearly 4000 jobs could be added to the area by the end of 2016, and that "Sudbury's economy could outperform the province for the next ten years."[33] One local economist has gone even further and stated that he's "never been more optimistic about Sudbury's economy." According to him, "we're looking at at least another 20 years of being able to build a city and build a community."[34] This projection may be overly optimistic, but it sends another favourable message for the future. While the current economic situation in the United States and Europe may have some dampening effects on the future of the mining sector, Sudbury's strength as a regional centre and global mining supply and service hub should serve to mitigate any population slowdown.[35]

Re-greening, Urban Design, and Visual Enhancement

While the re-greening and rehabilitation measures implemented over the past thirty years are noteworthy and deserving of praise, more flexible re-greening strategies should be considered in the future. Specifically, there is a need for more "green" action focused on the beautification of the urban landscape itself. While much of the negative imagery related to Sudbury remains focused on the state of the natural environment, it must be acknowledged that parts of the city continue to be very drab. Examples of voluntary or municipally prescribed landscape improvements, such as the addition of trees, shrubs, or berms are rare. In some residential neighbourhoods, the lack of trees forms a sharp contrast to the setting that exists in other Ontario communities. What is required is a shift in re-greening plans from tourist-oriented highways to urban areas, where the majority of the local population resides. The pleasing effects of tree plantings in the Hawthorne-Gemmell-Westmount area of New Sudbury that took place after the Second World War can be used as a green model for other neighbourhoods. Similarly, the use of berms and grass-sowing implemented near or on the slag dumps by Vale Canada can be replicated elsewhere on a smaller scale to minimize the visual effects of unsightly industrial lands, such as the CPR railway lines in the urban core.

Along with these greening initiatives, the use of various art installations and outdoor wall murals, such as those found in Chemainus on Vancouver Island, could make the downtown a more pleasant environment for tourists and citizens alike. Gigantic murals on the silos in the Flour Mill, for instance, would add vitality to the downtown's northern entry from Notre Dame, thereby complementing the southern entry formed by The Bridge of Nations flag project completed in 2007. Unfortunately, the city has yet to formulate a well-defined public art policy, whether on streets, in buildings, or throughout parks and trails.[36] While there are examples of art installations, such as the mining monument in Bell Park, and the woman and child statue in front of the Sudbury and District Health Unit, public displays are otherwise rare.[37] The city is thus missing out on an opportunity to use iconic images to display its cultural and creative side. Herein lies an opportunity for Laurentian University's proposed School of Architecture to have an urban impact similar to that of Savannah's College of Arts & Design in Georgia.

Another critical visual issue is the lack of bylaw enforcement regarding property and transportation infrastructure. A cursory trip by automobile through Greater Sudbury reveals the existence of scarification and varying forms of unsightliness, notably along major arteries including Lorne Street, the Kingsway, parts of Paris Street, the edges of major industrial sites such as the CPR yards, the airport road connections, and along rural roads that continue to feature derelict buildings and junk-filled yards. The inability of the city's bylaw enforcement agencies to do their job properly is a cause for civic embarrassment and suggests that, in this connection, a "culture of mediocrity" exists at city hall. There needs to be a more determined use of the city's powers to enforce its rules governing properties in the areas of building construction, maintenance, fire codes, health inspections, and occupancy violations. The proper implementation of these bylaws has been reactive rather than proactive. The philosophy of city officials is to do nothing until it becomes a political issue. Derelict buildings and burned-out sites, many of them adjacent to major road networks or strategic intersections, have been allowed to remain unchanged for excessively long periods of time, often decades. A similar philosophy of "minimum standards" can be found with respect to road construction and maintenance. This was made clear in a recent Auditor General's report issued by Greater Sudbury that concluded many road construction projects have not met city specifications. The mayor even went further and indicated that the situation has existed for several decades.[38]

The Human Dynamic

While Sudbury's history has had many "ups and downs" and numerous challenges remain to be faced, the reality is that more than 160 000 people have firm roots in the city. In assessing the historical record, it needs to be remembered that while we all may look into the same mirror of the past, we have differing images. Thus, while some may plausibly argue that the historical record reflects a "half empty" argument (as propounded in the book *Mining Town Crisis*), there are others who suggest that the reverse is true.[39] For many Sudburians, the intimacy that exists between their everyday lives and the surrounding environment has become a dominant quality of life feature. The opportunity to own lakeside homes and cottages, the location of many beaches and campsites near residential areas, the myriad of blueberry-picking sites, and the simple pleasure of living in a contrasting environment all have a powerful appeal for those who prefer these attributes over more densely populated places like Toronto.

When all is said and done, however, the essential element of a place resides in its people. Narasim Katary, a former Sudbury planner and past member of the Ontario Municipal Board, has praised the special vitality associated with the people of Sudbury. He asserts that the decision of his family to make Sudbury their home "is a testament to the quality of life offered here," and that "the greatest indicator of quality of life is the endearing relationship with people who inhabit that environment."[40] Katary has used the accompanying poem to express his affection for this city:

Encomium to Sudbury

This crater city pulsates with vibrations of pneumatic drills in search of metal, while heavy water oscillates with the passage of invisible nutrinos in search of the origin of the universe.

Aeons ago a meteor carved out a basin. The same meteor continues to shape the configuration of neighbourhoods from Estaire to Cartier and from Nairn to Wahnapitae. 330 glacial lakes and rock outcroppings form the skeleton of the city while 43 neighbourhoods from Wanup to Levack and Naughton to Coniston form the body of the city.

A land survey mistake and a land suffused with a metal constituted the cradle of the city. Neither an infant nor a city can stay in the cradle for long. From rail town to mining town to mindfull town it has become. The mine-heads determined the first neighbourhoods. The mind-frames are filling in and completing the full expression of that built form.

For Desmarais, Muncaster and Campeau, Sudbury was a good place to come from. For thousands of others it has been and continues to be a good place to come to. Seventy-two nations and countries are places of origin for residents of Sudbury. For them, Sudbury has been a city of destination and destiny. People acknowledge their multiple roots but reinforce the common stem. This is neither a contrived mosaic nor an imagined melting pot but an authentic palimpsest of the past, present, and the future.

Heroes of sports, heroes of the imagination, and heroes of wars around the globe have adorned the cathedral as much as the shop class. Their collective endeavours have blossomed into the soulcraft of Sudbury. The greater city takes its name from an old town in England but is writing its story on the bedrock of knowledge. People here ask not where you came from but only care about where you are going and willingly lend a hand in your act of self-discovery and contribution to the community.

This is a city conceived as a frontier of possibilities, dedicated to the proposition that the shaping influence is neither geology nor fate but the toil of thousands rooted in reason. A community so conceived and so dedicated will endure so long as that spirit sustains it.[41]

Well said, Narasim!

Conclusion

For much of its history, Sudbury's trajectory could be described as existing "between a rock and a hard place." Rock defined the physical psyche of Sudbury's setting, whether in the form of rugged hills, hardrock mines, or forests and farms that were intimately linked with the Precambrian Shield. The human setting was equally hard. Since the 1880s, Sudburians have faced economic, political, and sociocultural settings that were frequently as rigid as the underlying rock. In forestry, agriculture, and mining, workers required great endurance and a tolerance for hazardous conditions. To improve their lot in life, many Sudburians turned to the only options available to them—the labour movement and left-wing political activism. Both of these options received little empathy from elsewhere in Ontario and Canada. Time and time again, efforts to improve the local setting were thwarted by outside forces with little interest in improving the lot of ordinary citizens who worked so hard to make Sudbury into one of Canada's major areas of wealth creation. Nonetheless, the local citizenry persisted in advancing their personal dreams and promoting the well-being of the community.

After the Second World War, the geographical setting of the area was transformed. Previously marginalized in Ontario, over the next few decades the area managed to evolve as the regional capital of Northeastern Ontario. Aided by a group of remarkable politicians, a unified community approach, and a firm belief in Sudbury's future, the difficulties of the 1970s and 1980s were met in remarkable fashion, culminating in a diversification of the economy and a greatly improved physical environment. While the transformations of the late twentieth century, the transitions that have emerged since the turn of the new century, and recent economic developments have opened the door for yet a more diversified and sustainable community, one that can be characterized as extending beyond "a rock and a hard place," the path forward will not be an easy one. This is indicated by polls conducted in 2011 and 2012 that ranked Sudbury at the bottom of a list of thirty municipalities when it came to satisfaction with municipal services.[42] Further, as one critic has observed, politicians in Sudbury appear to be "drifting along" instead of riding the wave of the new resource boom.[43] In another pessimistic assessment, a *Northern Life* writer suggested that Sudbury is in a "no-win" situation because the financial and infrastructure expectations of the 2001 amalgamation imposed by the province have yet to be achieved.[44]

It warrants mention, however, that pessimistic situations of this kind have often occurred in Sudbury's past, and that many of them were resolved by local politicians rising to the occasion and developing forward-looking strategies involving all levels of governance—municipal, provincial, and federal. It may again be time to revisit the progressive "golden" era of the early 1980s, and to relearn the lessons of what can be done with solid political leadership, community unity, and a firm vision of the future.

Appendix

Tables 5.1 and 5.2

TABLE 5.1 Population trends in the Sudbury area (1901–1971)

Year	1901	1911	1921	1931	1941	1951	1956	1961	1966	1971
Cities										
Sudbury	2 027	4 150	8 621	18 518	32 203	42 410	46 482	80 120	84 888	90 535
Towns										
Capreol			1 287	1 648	1 641	2 002	2 394	3 003	3 092	3 470
Chelmsford	493	550	561	725	905	1 210	2 142	2 559	2 752	
Coniston					2 245	2 292	2 478	2 692	2 692	2 907
Copper Cliff	2 500	3 082	2 597	3 173	3 732	3 974	3 801	3 600	3 505	4 089
Frood Mine				173	70	109	124			
Levack					895	1 883	2 929	3 178	3 025	2 948
Lively						2 840	3 211	3 169	3 000	
Improvement Districts										
Onaping							804	1 106	1 217	1 504
Organized Townships										
Balfour	437	557	534	758	747	724	1 440	1 907	2 341	9 101
Blezard	474	456	503	523	533	826	2 604	4 615	5 227	
Capreol		585	660	516	556	875	1 023	2 348	2 822	
Dowling	102	296	228	161	398	376	681	1 436	1 902	3 018
Drury, Denison & Graham	807	2 121	1 038	747	594	817	1 208	1 836	2 040	2 398
Falconbridge	20	15	28	445	905	1 066	1 369	1 349	1 239	1 269
Hanmer	189	643	760	695	710	855	1 512	4 007	5 687	
McKim	512	310	440	535	5 105	11 783	17 461			
Neelon & Garson	245	1 082	2 238	3 618	2 977	6 438	13 750	5 286	5 403	6 296
Rayside	820	865	952	1 067	962	1357	3 002	4 820	5 491	6 344
Valley East										17 937
Waters	129	227	455	524	772	991	1 469	2 064	2 485	2 936
Unorganized Townships										
Bowell					3					
Broder	20	238	448	528	573	836	1 609	2 857	3 367	4 486
Cleland		22	76	142	199	272	259	272	302	411
Creighton	94	57	46	50	9	12	109	114	14	
Dieppe					49	50	40	62	48	56
Dill	12	181	200	311	246	233	384	907	945	1 757
Dryden	186	177	146	231	574	564	874	1 201	1 245	1 891
Fairbank		82	66	76	37	96	14	3		
Hutton	553	297	149	75	14	13	6	7	2	

continued

TABLE 5.1   continued

Year	1901	1911	1921	1931	1941	1951	1956	1961	1966	1971
Levack	252	420	40	50	41					
Lorne	19	72	328	328	276	286	343	365	339	430
Louise	116	278	276	241	250	321	399	452	642	
Lumsden	7	85	37	49	59	14				
MacLennan	23	4	55	270	138	440	360	408	439	674
Morgan			62	81	85	76	97	145	149	
Norman	18	48	42	233	253	366	432	511	526	522
Snider	369	986	1 157	1 465	1 726	2 129	1 799	1 737	1 467	1 297
Trill	4	46	1	35	26	23	5	17	11	
Wisner					9	6	7			
TOTAL	9 507	17 524	24 409	38 469	60 604	85 597	116 304	138 152	148 291	169 961

Source: Statistics Canada, *Censuses of Canada* (1901–1971).
Note: This data refers to those settlements currently situated within the boundaries of the City of Greater Sudbury.

TABLE 5.2   Provincial and federal Members of Parliament from Sudbury (1905–2011)

Years	Electoral District	Member of Parliament	Party
PROVINCIAL MEMBERS			
Nipissing East Riding			
1905–1908	Nipissing East	Francis Cochrane	Conservative
Sudbury Riding (created in 1905)			
1908–1911	Sudbury	Francis Cochrane	Conservative
1911–1914	Sudbury	Charles McCrea	Conservative
1914–1919	Sudbury	Charles McCrea	Conservative
1919–1923	Sudbury	Charles McCrea	Conservative
1923–1926	Sudbury	Charles McCrea	Conservative
1926–1929	Sudbury	Charles McCrea	Conservative
1929–1934	Sudbury	Charles McCrea	Conservative
1934–1937	Sudbury	Edmond Lapierre	Liberal
1937–1943	Sudbury	James Maxwell Cooper	Liberal
1943–1945	Sudbury	Robert Carlin	CCF
1945–1948	Sudbury	Robert Carlin	CCF
1948–1951	Sudbury	Welland Gemmell	PC
1951–1954	Sudbury	Welland Gemmell	PC
1954–1955	Sudbury	Vacant (Death of Gemmell)	
1955–1959	Sudbury	Gerald Monaghan	PC
1959–1963	Sudbury	Elmer Sopha	Liberal
1963–1967	Sudbury	Elmer Sopha	Liberal
1967–1971	Sudbury	Elmer Sopha	Liberal

TABLE 5.2   continued

Years	Electoral District	Member of Parliament	Party
1971–1975	Sudbury	Bud Germa	New Democrat
1975–1977	Sudbury	Bud Germa	New Democrat
1977–1981	Sudbury	Bud Germa	New Democrat
1981–1985	Sudbury	Jim Gordon	PC
1985–1987	Sudbury	Jim Gordon	PC
1987–1990	Sudbury	Sterling Campbell	Liberal
1990–1995	Sudbury	Sharon Murdock	New Democrat
1995–1998	Sudbury	Rick Bartolucci	Liberal
1999–2003	Sudbury	Rick Bartolucci	Liberal
2003–2007	Sudbury	Rick Bartolucci	Liberal
2007–present	Sudbury	Rick Bartolucci	Liberal

Nickel Belt Riding (created in 1952)

Years	Electoral District	Member of Parliament	Party
1955–1959	Nickel Belt	Rhéale Bélisle	PC
1959–1963	Nickel Belt	Rhéale Bélisle	PC
1963–1967	Nickel Belt	Gaston Demers	PC
1967–1971	Nickel Belt	Gaston Demers	PC
1971–1975	Nickel Belt	Floyd Laughren	New Democrat
1975–1977	Nickel Belt	Floyd Laughren	New Democrat
1977–1981	Nickel Belt	Floyd Laughren	New Democrat
1981–1985	Nickel Belt	Floyd Laughren	New Democrat
1985–1987	Nickel Belt	Floyd Laughren	New Democrat
1987–1990	Nickel Belt	Floyd Laughren	New Democrat
1990–1995	Nickel Belt	Floyd Laughren	New Democrat
1995–1998	Nickel Belt	Floyd Laughren	New Democrat
1998 by–1999	Nickel Belt	Blair Morin	New Democrat
1999–2003	Nickel Belt	Shelley Martel	New Democrat
2003–2007	Nickel Belt	Shelley Martel	New Democrat
2007–present	Nickel Belt	France Gélinas	New Democrat

Sudbury East Riding (created in 1967 and merged with Nickel Belt in 1999)

Years	Electoral District	Member of Parliament	Party
1967–1971	Sudbury East	Elie Martel	New Democrat
1971–1975	Sudbury East	Elie Martel	New Democrat
1975–1977	Sudbury East	Elie Martel	New Democrat
1977–1981	Sudbury East	Elie Martel	New Democrat
1981–1985	Sudbury East	Elie Martel	New Democrat
1985–1987	Sudbury East	Elie Martel	New Democrat
1987–1990	Sudbury East	Shelley Martel	New Democrat
1990–1995	Sudbury East	Shelley Martel	New Democrat
1995–1999	Sudbury East	Shelley Martel	New Democrat

FEDERAL MEMBERS
Nipissing Riding (created in 1892)

Years	Electoral District	Member of Parliament	Party
1911–1917	Nipissing	Francis Cochrane	Conservative
1921–1925	Nipissing	Edmond A. Lapierre	Liberal
1925–1926	Nipissing	Edmond A. Lapierre	Liberal
1926–1930	Nipissing	Edmond A. Lapierre	Liberal
1930–1935	Nipissing	Dr. Joseph Hurtubise	Liberal

1935–1940	Nipissing	Dr. Joseph Hurtubise	Liberal
1940–1945	Nipissing	Dr. Joseph Hurtubise	Liberal
1945–1949	Nipissing	J. Leo Gauthier	Liberal

Sudbury Riding (created in 1947)

1949-1953	Sudbury	J. Leoda Gauthier	Liberal
1953–1957	Sudbury	Roger Mitchell	Liberal
1957–1958	Sudbury	Roger Mitchell	Liberal
1958–1962	Sudbury	Roger Mitchell	Liberal
1962–1963	Sudbury	Roger Mitchell	Liberal
1963–1965	Sudbury	Roger Mitchell	Liberal
1965–1967	Sudbury	Roger Mitchell	Liberal
1967–1968	Sudbury	Bud Germa	NDP
1968–1972	Sudbury	James Jerome	Liberal
1972–1974	Sudbury	James Jerome	Liberal
1974–1979	Sudbury	James Jerome	Liberal
1979–1980	Sudbury	James Jerome	Liberal
1980–1984	Sudbury	Doug Frith	Liberal
1984–1988	Sudbury	Doug Frith	Liberal
1988–1993	Sudbury	Diane Marleau	Liberal
1993–1997	Sudbury	Diane Marleau	Liberal
1997–2000	Sudbury	Diane Marleau	Liberal
2000–2004	Sudbury	Diane Marleau	Liberal
2004–2006	Sudbury	Diane Marleau	Liberal
2006–2008	Sudbury	Diane Marleau	Liberal
2008+	Sudbury	Glenn Thibeault	NPD

Nickel Belt (created in 1952)

1953–1957	Nickel Belt	J. Leo Gauthier	Liberal
1957–1958	Nickel Belt	J. Leo Gauthier	Liberal
1958–1962	Nickel Belt	O.J. Godin	Liberal
1962–1963	Nickel Belt	O.J. Godin	Liberal
1963–1965	Nickel Belt	O.J. Godin	Liberal
1965–1968	Nickel Belt	Norman Fawcett	NDP
1968–1972	Nickel Belt	Gaetan Serré	Liberal
1972–1974	Nickel Belt	John Rodriguez	NDP
1974–1979	Nickel Belt	John Rodriguez	NDP
1979–1980	Nickel Belt	John Rodriguez	NDP
1980–1984	Nickel Belt	Judy Erola	Liberal
1984–1988	Nickel Belt	John Rodriguez	NDP
1988–1993	Nickel Belt	John Rodriguez	NDP
1993–1997	Nickel Belt	Ray Bonin	Liberal
1997–2000	Nickel Belt	Ray Bonin	Liberal
2000–2004	Nickel Belt	Ray Bonin	Liberal
2004–2006	Nickel Belt	Ray Bonin	Liberal
2006–2008	Nickel Belt	Ray Bonin	Liberal
2008+	Nickel Belt	Claude Gravelle	NPD

Source: *Wikipedia*, s.v., "Sudbury (provincial electoral district)," accessed June 5, 2008, http://en.wikipedia.org/wiki/Sudbury_(provincial_electoral_district); "Nickel Belt (provincial electoral district)," accessed June 5, 2008, http://en.wikipedia.org/wiki/Nickel_Belt_%28provincial_electoral_district%29; "Sudbury East," accessed June 5, 2008, http://en,wikipedia.org/wikiSudbury_East; and Parliament of Canada, *History of Nipissing, Sudbury, and Nickel Belt Federal Ridings since 1867* (Ottawa, ON: Parliament of Canada, 2011).

Notes

NOTES TO PREFACE

1   "Forgotten Cemeteries," *Sudbury Star*, October 7, 2006, A1 and A6; and "Tales of Lives Lived," *Northern Life*, August 2, 2005, 3. Of these cemeteries, twenty-three are currently owned by the City of Greater Sudbury.

2   Floyd Laughren, *Constellation City: Building a Community of Communities in Greater Sudbury*, Report of the Greater Sudbury Community Solutions Team (Sudbury, ON: City of Greater Sudbury 2007), 7.

3   Ibid., 12.

NOTES TO CHAPTER 1

1   K.D. Card et al., "The Sudbury Structure: Its Regional Geological and Geophysical Setting," in *The Geology and Ore Deposits of the Sudbury Structure*, ed. E.G. Pye, A.J. Naldrett, and P.E. Giblin (Toronto, ON: Ontario Geological Survey, Special Volume 1, 1984), 26–43. The complexity is also apparent in Don H. Rousell and G. Heather Brown, eds., *A Field Guide to the Geology of Sudbury, Ontario,* Open File Report 6243 (Sudbury, ON: Ontario Geological Survey, 2009).

2   Environmental determinism was a doctrine whose adherents claimed that human activities are controlled by the natural environment. This point of view attracted many advocates, including Griffith Taylor, a distinguished geographer who taught at the University of Toronto and who used Sudbury as a case example of physical determinism. After the First World War, this doctrine was increasingly replaced by "possibilism," a principle that relegated the physical environment to a secondary role.

3   Guochun Zhao et al., "A Paleo-Mesoproterozoic Supercontinent: Assembly, Growth and Breakup," *Earth-Science Reviews* 67 (2004): 93.

4   Nick Eyles, *Ontario Rocks: Three Billion Years of Environmental Change* (Markham, ON: Fitzhenry & Whiteside, 2002), 88–89.

5   J.P. Golightly, "The Sudbury Igneous Complex as an Impact Melt: Evolution and Ore Genesis," in *Proceedings of the Sudbury-Noril'sk Symposium*, ed. P.C. Lightfoot and A.J. Naldrett (Toronto, ON: Ontario Geological Survey, Special Volume 5, 1994), 105–17.

6   Card, "The Sudbury Structure," 41.

7   An accounting of the geology and scenery of this part of Ontario is found in J.A. Robertson and K.D. Card, *Geology and Scenery: North Shore of Lake Huron Region*, Geological Guidebook No. 4 (Toronto, ON: Ontario Division of Mines, Ministry of Natural Resources, 1972), 1–224.

8   J.V. Guy-Bray, introduction to *New Developments in Sudbury Geology*, Special Paper Number 10, ed. J.V. Guy-Bray (Ottawa, ON: Geological Association of Canada 1972), 1–5.

9   E.C. Speers, "The Age Relation and Origin of the Common Sudbury Breccia," *Journal of Geology* 65 (1957): 513.

10   Bevan M. French, "Shock-Metamorphic Features in the Sudbury Structure Ontario: A Review," in Guy-Bray, *New Developments in Sudbury Geology*, 26.

11   K.O. Pope, S.W. Kieffer, and D.E. Ames, "Empirical and Theoretical Comparisons of the Chicxulub and Sudbury Impact Structures," *Meteoritics & Planetary Science* 39, no. 1 (2004): 97.

12   R.A.F. Grieve, "The Sudbury Structure: Additional Constraints on its Origin and Evolution," *Abstracts of the 25th Lunar Planetary Science Convention*, Houston, TX, March 14–18 (1994), 477. Others have estimated that there might be up to 35 000 cubic kilometres of melt.

13   William D. Addison et al., "Discovery of Distal Ejecta from the 1850 Ma Sudbury Impact Event," *Geology* 33, no. 3 (2005): 193–96; see also the Earth Impact Database (Impact Structures sorted by age and diameter; accessed August 28, 2012), http://www.passc.net/ Earth Impact Database.

14   Robert Bell, "Report on the Sudbury Mining District (1888–1890)," *Geological Survey of Canada, Annual Report*, n.s. 5 (1891), pt. 1, 5F–95F.

15   P.E. Giblin, "History of Exploration and Development of Geological Studies and Development of Geological Concepts," in *The Geology and Ore Deposits of the Sudbury Structure*, 1: 3–23.

16   See A.E. Barlow, "Report on the Origin, Geological Relations and Composition of the Nickel and Copper Deposits in the Sudbury Mining District, Ontario, Canada," *Geological Survey of Canada, Annual Report*, n. 873 (1904), 1–236; and Barlow, "On the Origin and Relations of the Nickel and Copper Deposits of Sudbury, Ontario, Canada," *Economic Geology* 1 (1906): 454–66.

17   Robert S. Dietz, "Sudbury Structure as an Astrobleme," *Journal of Geology* 72 (1964): 412–34; also see Dietz, "Sudbury Structure as an Astrobleme," *Transactions of the American Geophysical Union* 43 (1962): 445–46.

18   Guy-Bray, *New Developments in Sudbury Geology*, 1–5.

19   Don H. Rousell, Harold T. Gibson, and Ian R. Jonasson, "The Tectonic, Magmatic and Mineralization History of the Sudbury Structure," *Exploration and Mining Geology* 6, n. 1 (1997): 1.

20   M.R. Dence, "Meteorite Impact Craters and the Structure of the Sudbury Basin," in Guy-Bray, *New Developments in Sudbury Geology*, 15.

21   D.H. Rousell and K.D. Card, "Geologic Setting," in Rousell and Brown, *A Field Guide to the Geology of Sudbury, Ontario*, 1–200.

22   P.E. Giblin, "Glossary of Sudbury Geology Terms," in Pye, Naldrett, and Giblin, *The Geology and Ore Deposits of the Sudbury Structure*, 1: 571–74.

23   D.H. Rousell, W. Meyer, and S.A. Prevec, "Bedrock Geology and Mineral Deposits," in *The Physical Environment of the City of Greater Sudbury*, Ontario Geological Survey Special Vol. 6, ed. D.H. Rousell and K.J. Jansons (Sudbury, ON: Ontario Geological Survey, 2002), 21–55.

24   Ibid., 47.

25   Burkhard O. Dressler, "The Effects of the Sudbury Event and the Intrusion of the Sudbury Igneous Complex on the Footwall Rocks of the Sudbury Structure," in Pye, Naldrett, and Giblin, *The Geology and Ore Deposits of the Sudbury Structure*, 1: 99.

26   E.J. Cowan and W.M. Schwerdtner, "Fold Origin of the Sudbury Basin," in Lightfoot and Naldrett, *Proceedings of the Sudbury Noril'sk Symposium*, 45; and Anthony J. Naldrett, "From

Impact to Riches: Evolution of Geological Understanding as Seen at Sudbury, Canada," *GSA Today*, February 2003, 7.

27   A listing of these theories can be found in Robert Stephenson et al., *A Guide to the Golden Age: Mining in Sudbury, 1886–1977* (Sudbury, ON: Laurentian University, Department of History, 1979), 1–8. See also Giblin, "History of Exploration and Development, of Geological Studies and Development of Geological Concepts," 5–16.

28   J. E. Thompson, "Geology of the Sudbury Basin," *Sixty-Fifth Annual Report of the Ontario Department of Mines* 65, pt. 3 (1956): 6.

29   J. E. Thompson, and H. Williams, "The Myth of the Sudbury Lopolith," *Canadian Mining Journal* 80, no. 3 (1959): 57–62; K.D. Card, *Geology of the Sudbury-Manitoulin Area, Districts of Sudbury and Manitoulin, Ontario Geological Survey Report 166* (Toronto, ON: Ministry of Natural Resources, 1978), 151.

30   R.A.F. Grieve, "An Impact Model of the Sudbury Structure," in Lightfoot and Naldrett, *Proceedings of the Sudbury-Noril'sk Symposium*, 122.

31   Dietz, "Sudbury Structure as an Astrobleme," 412.

32   Golightly, "The Sudbury Igneous Complex as an Impact Melt," 105.

33   Grieve, "An Impact Model of the Sudbury Structure," 128.

34   Bevan M. French, "Sudbury Structure, Ontario: Some Petrographic Evidence for an Origin by Meteorite Impact," in *Shock Metamorphism of Natural Materials*, ed. Bevan M. French and N.M. Short (Baltimore, IL: Mono Book Corporation, 1968), 408.

35   Rousell, Gibson, and Jonasson, "The Tectonic, Magmatic and Mineralization History of the Sudbury Structure," 12.

36   Golightly, "The Sudbury Igneous Complex as an Impact Melt," 115.

37   W.V. Peredery and G.G. Morrison, "Discussion of the Origin of the Sudbury Structure," in *The Geology and Ore Deposits of the Sudbury Structure*, 504.

38   Grieve, "The Sudbury Structure: Additional Constraints on its Origin and Evolution," *Abstracts of the 25th Lunar & Planetary Science Conference*, 477.

39   Luann Becker et al., "Fullerenes in the 1.85-Billion-Year-Old Sudbury Impact Structure," *Science* n.s. 265, no. 5172 (July 29, 1994): 642–47.

40   Addison et al., "Discovery of Distal Ejecta from the 1850 Ma Sudbury Impact Event," 193–96.

41   B. Chadwick, P. Claeys, and B. Simonson, "New Evidence for a Large Paleoproterozoic Impact: Spherules in a Dolomite Layer in the Ketilidian Orogen, South Greenland," *Journal of the Geological Society* 158, no. 2 (March 2001): 331.

42   John F. Slack and William F. Cannon, "Extraterrestrial Demise of Banded Iron Formations 1.85 Billion Years Ago," *Geology* (November 2009): 1011–14. See also "Cradle of Life," *Sudbury Star*, November 21, 2009, A1 and A6.

43   Card, *Geology of the Sudbury-Manitoulin Area, Districts of Sudbury and Manitoulin,* 195–96.

44   Rousell, Gibson, and Johansson, "The Tectonic, Magmatic and Mineralization History of the Sudbury Structure," 1.

45   E.C. Speers cited in French, "Sudbury Structure, Ontario: Some Petrographic Evidence for an Origin by Meteorite Impact," in *Shock Metamorphism of Natural Materials*, ed. Bevan M. French and N.M. Short (Baltimore: Mono Book Corporation, 1968), 409.

46   M.R. Dence and J. Popelar, "Evidence for an Impact Origin for Lake Wanapitei Ontario," in Guy-Bray, *New Developments in Sudbury Geology*, 117–24.

47   For a review of the area's glacial history, refer to P.J. Barnett and A.F. Bajc, "Quaternary Geology," in Rousell and Jansons, *The Physical Environment of the City of Greater Sudbury*, 58–59; A.F. Bajc and P.J. Barnett, *Quaternary Geology and Geomorphology of the Sudbury Region*, Field Trip A5 Guidebook (Sudbury: Geological Association of Canada-Mineralogical Association of Canada, 1999), 1–68.

48   P.J. Barnett, "Quaternary Geology of Ontario," in *Geology of Ontario, Ontario Geological Survey*, Special vol. 4, part 2, ed. P.C. Thurston et al. (Toronto, ON: Ministry of Northern Development and Mines, 1992), 1015–17; G.J. Burwasser, *Quaternary Geology of the Sudbury Basin Area: District of Sudbury, Ontario Geological Survey Report 181* (Toronto, ON: Ministry of Natural Resources, 1979), 18–19.

49   Barnett, "Quaternary Geology of Ontario," 1029.

50   Robertson and Card, *Geology and Scenery North Shore of Lake Huron Region*, 87.

51   Burwasser, *Quaternary Geology of the Sudbury Basin Area*, 13.

52   Ibid., 16.

53   "Please Make It Rain," *Sudbury Star*, May 29, 2010, 6a.

54   D.A.B. Pearson, J.M. Gunn, and W. Keller, "The Past, Present and Future of Sudbury's Lakes, in Rousell and Jansons, *The Physical Environment of the City of Greater Sudbury*, 199.

55   Nickel District Conservation Authority, *Watershed Inventory* (Sudbury, ON: Nickel District Conservation Authority 1980), 4.5–4.26.

56   John. M. Gunn and W. Keller, "Urban Lakes: Integrators of Environmental Damage and Recovery," in *Restoration and Recovery of an Industrial Region*, ed. by John M. Gunn (New York: Springer-Verlag, 1995), 257.

57   City of Greater Sudbury, *Official Plan Natural Heritage Background Study* (Sudbury, ON: City of Greater Sudbury, February, 2005), 1–17.

58   Patrick J. Julig, "Archaoelogical Conclusions from the Sheguiandah Site Research," in *The Sheguiandah Site*, Mercury Series Archaeological Survey of Canada, Paper 161, ed. Patrick J. Julig (Ottawa, ON: Canadian Museum of Civilization, 2002), 308.

59   Kam-biu Liu, "Postglacial Vegetational History of Northern Ontario: A Palynological Study" (Ph. D. thesis, University of Toronto, 1982), 1–352. Liu's studies involved the study of Loon and Nina Lakes near Sudbury.

60   Nickel District Conservation Authority, *Watershed Inventory*, 1.4.

61   L.J. Chapman and M.K. Thomas, *The Climate of Northern Ontario*, Climatological Studies no. 6 (Toronto, ON: Department of Transport, Meteorological Branch, 1968), 2.

62   David Phillips, *The Climates of Canada* (Ottawa, ON: Minister of Supply and Services Canada, 1990), 97.

63   Chapman and Thomas, *The Climate of Northern Ontario*, 8 and 14.

64   Phillips, *The Climates of Canada*, 56. For a detailed examination of this tornado, refer to G.W. Gee and B.F. Findlay, *The Sudbury Tornado—August 20, 1970*, Technical Memoranda 764 (Ottawa, ON: Atmospheric Environment Service, Department of the Environment—Canada, 1972).

65 David Savageau with Ralph D'Agostino, *Places Rated Almanac* (New York: IDG Books Worldwide, 2000), 295.

66 David A.B. Pearson and J. Roger Pitblado, "Geological and Geographic Setting," in Gunn, *Restoration and Recovery of an Industrial Region*, 10–11; Savageau and D'Agostino, *Places Rated Almanac*, 286.

67 Liu, "Postglacial Vegetational History of Northern Ontario," 1–353.

68 Keith Winterhalder, "Early History of Human Activities in the Sudbury Area and Ecological Damage to the Landscape," in Gunn, *Restoration and Recovery of an Industrial Region*, 17.

69 Pearson and Pitblado, "Geological and Geographic Setting," 14. See also Geoffrey A.J Scott, *Canada's Vegetation: A World Perspective* (Montreal: McGill-Queen's University Press, 1995), 88–90.

70 Agricultural Economics Research Council, *Potentials for Agricultural Development* (Ottawa, ON: Agricultural Economics Research Council, 1979), 56–60; Nickel District Conservation Authority, *Watershed Inventory*, 6.5–6.10.

71 Nickel District Conservation Authority, *Watershed Inventory*, 8.1–8.38.

NOTES TO CHAPTER 2

1 See, for example, George Irving Quimby, *Indian Life in the Upper Great Lakes* (Chicago: University of Chicago Press, 1960), 1–8.

2 J.V. Wright, *Ontario Prehistory: An Eleven-Thousand-Year Archeological Outline* (Ottawa, ON: National Museum of Canada, National Museum of Man, 1972), 7.

3 P.J. Julig, *Laurentian University Field School at Lavase Site North Bay Ontario 1996*, Report No. 27 (Sudbury: Archeological Survey of Laurentian University 1998), 3–7. For cartographic information regarding the Aboriginal history of Northern Ontario, refer to R. Cole Harris, ed., *Historical Atlas of Canada*, vol. 1, *From the Beginning to 1800* (Toronto, ON: University of Toronto Press, 1987), plates 2–9; and R. Louis Gentilcore, ed., *Historical Atlas of Canada*, vol. 2, *The Land Transformed 1800–1891* (Toronto, ON: University of Toronto Press, 1987), plate 32.

4 K.C.A. Dawson, *Prehistory of Northern Ontario* (Thunder Bay, ON: Thunder Bay Historical Museum Society, 1983), 4.

5 For a detailed examination of the Sheguiandah Site, refer to Patrick J. Julig, ed., *The Sheguiandah Site: Archeological, Geological and Paleobotanical Studies at a Paleoindian Site on Manitoulin Island, Ontario*, Mercury Series Archaeological Survey of Canada Paper 161 (Ottawa, ON: Canadian Museum of Civilization, 2002), 1–314; especially chapter 11 by Julig, titled "Archaeological Conclusions from the Sheguiandah Site Research." See also Thor Conway, *Archaeology in Northeastern Ontario: Searching for our Past* (Toronto, ON: Ontario Ministry of Culture and Recreation, Historical Planning and Research Branch, 1981), 23.

6 Julig, *Laurentian University Field School at Lavase Site North Bay Ontario 1996*, 6.

7 Wright, *Ontario Prehistory*, 93–94.

8 Dawson, *Prehistory of Northern Ontario*, 21.

9 Robert P. Trott, *The Story of Onaping Falls* (Onaping Falls, ON: Town of Onaping Falls, ca. 1982), 106.

10 K.T. Buchanan, *An Archeological Survey of the Sudbury Area and a Site Near Lake of the Mountains*, Report No. 6 (Sudbury, ON: Laurentian University, 1979), 31.

11 Ibid.

12 P. Julig (professor and anthropologist), interviewed by author, July 26, 2000.

13 G.M. Asher, *Henry Hudson the Navigator* (New York: Burt Franklin, 1963), xlvii.

14 Tryggvi J. Oleson, *Early Voyages and Northern Approaches, 1000–1632* (Toronto, ON: McClelland and Stewart, 1963), 165.

15 W.P. Cumming, R.A. Skelton, and D.B. Quinn, *The Discovery of North America* (London: Elek, 1971). 228.

16 Each of these explorers produced a remarkable book. See Captaine Luke Fox, *North-West Fox* (London: B. Alsop and Tho. Fawcet, 1635), 79 and 100–269. This book also contains an account Button's voyage of 1612–1613. For a reprint of James' original account of 1633, see Captaine Thomas James, *The Strange and Dangerous Voyage* (New York: Da Capo Press, 1968), 1–120.

17 Morris Bishop, *White Men Came to the St. Lawrence* (Montreal: McGill University, 1961), 50.

18 J. Herbert Cranston, *Etienne Brûlé: Immortal Scoundrel* (Toronto, ON: Ryerson Press, 1949), 5.

19 Consul Willshire Butterfield, *History of Brûlé's Discoveries and Explorations 1610–1626* (Cleveland, OH: Helman-Taylor, 1898), 20.

20 Morris Bishop, *Champlain: The Life of Fortitude* (Toronto, ON: McClelland and Stewart, 1963), 192.

21 Charles Pomeroy Otis, *Voyages of Samuel de Champlain* (New York: Burt Franklin, 1882), v. For more information regarding Samuel de Champlain, see C.E. Heidenreich, *Explorations and Mapping of Samuel de Champlain, 1603–1632*, supplement no. 2, *Canadian Cartographer* 13 (1976).

22 Trudel, *Atlas de la Nouvelle France*, 86.

23 Reuben Gold Thwaites, ed., *The Jesuit Relations and Allied Documents*, 73 vols. (Cleveland: Burrows Bros., 1896–1901). These volumes consist of the annual reports sent from the Canadian mission of the Society of Jesus to its Paris office.

24 E.E. Rich, *The Fur Trade and the Northwest to 1857* (Toronto, ON: McClelland and Stewart, 1976), 17.

25 Nick and Helma Mika, *The Shaping of Ontario from Exploration to Confederation* (Belleville, ON: Mika Publishing, 1985), 248.

26 Edwin G. Higgins, *Whitefish Lake Ojibway Memories* (Cobalt, ON: Cobalt Highway Book Shop, 1982), 22.

27 David William Smyth, *A Map of the Province of Upper Canada*, 2nd ed. (London: W Faden, 1813).

28 Peter C. Newman, *Company of Adventurers*, vol. 1 (Markham, ON: Penguin Books of Canada, 1985), 1. See also George Bryce, *The Remarkable History of the Hudson's Bay Company* (Toronto, ON: William Briggs, 1910), 12.

29 J.P.G. de Lestard, *A History of the Sudbury Forest District* (Toronto, ON: Ontario Department of Lands and Forests, 1967), 13.

30 Bryce, *The Remarkable History of the Hudson's Bay Company*, 491.

31 Nick and Helma Mika, "Historic Sites of Ontario," in *Encyclopedia of Ontario*, vol. 1 (Belleville, ON: Mika Publishing Company, 1974), 180.

32 The local operations of the HBC can be traced through HBC inspection reports. See, for instance, Hudson Bay Company Archives, *Post Reports for Whitefish Lake and Sudbury*, Reel No. 1M1259, B.364/e/1, 1888 and B.364/e/2, 1897; and Reel No. 1M1260, B.355/e/1, 1888 and B. 355/e/6, 1891.

33 Edward M. Burwash, "Geology of the Nipissing-Algoma Line," *Sixth Report of the Bureau of Mines 1896* (Toronto, ON: Warwick Bros. & Rutter, 1897), following page 176.

34 John S. Galbraith, *The Hudson's Bay Company as an Imperial Factor 1821–1869* (Berkeley: University of California Press, 1957), 27.

35 Higgins, *Whitefish Lake Ojibway Memories*, 77–82. See also J.L. Morris, *Indians of Ontario* (Toronto, ON: Ontario Department of Lands and Forests, 1943), 32–34.

NOTES TO CHAPTER 3

1 George W. Spragge, "Colonization Roads in Canada West, 1859–1867," *Ontario History* 44 (1957): 3.

2 Commissioner of Crown Lands of Canada, *Annual Report for the Year 1866* (Toronto, ON: Hunter, Rose & Company, 1867), x and 208.

3 Commissioner of Crown Lands of Canada, *Annual Report for the Year 1856* (Toronto, ON: Stewart Derbishire and George Desbarats, 1857), Appendix 25.

4 Geological Survey of Canada, *Report of Progress for the Year 1848–49* (Toronto, ON: Lovell and Gibson, 1849), 63.

5 Commissioner of Crown Lands of Canada, *Annual Report for the year 1866*, 211. In an accompanying map, this sawmill is referred to as Waddle's Mill.

6 Cited in Don W. Thomson, *Men and Meridians*, vol. 1 (Ottawa, ON: Queen's Printer, 1966), 165.

7 Peter V.K. Krats, *The Sudbury Area to the Great Depression: Regional Development on the Northern Resource Frontier* (Ph.D. thesis, University of Western Ontario, 1988), 47.

8 A. Murray, "Report for the Year 1856," in *Geological Survey of Canada, Report of Progress for the Years 1853–54–55–56*, Appendix No. 52 (Toronto, ON: John Lovell, 1857), 80–106.

9 A brief sketch of Salter's life can be found in: Association of Ontario Land Surveyors, *Annual Report* (Toronto, ON: Association of Ontario Land Surveyors, 1915), 60–62.

10 Commissioner of Crown Lands of Canada, *Annual Report for the Year 1856*, Appendix 25.

11 Ladell, *They Left Their Mark*, 10.

12 H.V. Nelles, The *Politics of Development: Forests, Mines & Hydro-electric Power in Ontario, 1849–1941* (Toronto, ON: Macmillan of Canada, 1973), 44.

13 Krats, *The Sudbury Area to the Great Depression*, 11.

14 A history of early lumbering on the North Shore is found in Gwenda Hallsworth's, "'A Good Paying Business': Lumbering on the North Shore of Lake Huron, 1850–1910 with Particular Reference to the Sudbury District" (M.A. thesis, Laurentian University, 1983).

15   J. Howard Richards, "Lands and Policies: Attitudes and Controls in the Alienation of Lands in Ontario during the First Century of Settlement," *Ontario History* 1 (1958): 207.

16   L.M. Sebert, "The Land Surveys of Ontario 1750–1980," *Cartographica* 17, no. 3 (1980): 89–90 and 92–93.

17   Cited in Commissioner of Crown Lands of the Province of Ontario, *Annual Report for the Year 1883* (Toronto, ON: "Grip" Printing and Publishing Co., 1884), 46.

18   See, for example, D.W. Smith's Map of the Province of Upper Canada published in 1800, Archives of Ontario, Cartographic Records, Collection A-9, Repro. no. 492.

19   George W. Spragge, "The Districts of Upper Canada," *Ontario History* 34 (1947): 94.

20   *Statutes of Canada*, 1852–53, 16 Vict., c. 176, 720–24.

21   *Statutes of Canada*, 1857, 20 Vict., c. 60, 283–303.

22   *The Canada Gazette*, 17 (April 17, 1858), 676–677.

23   Ashley Thomson, "The 1890s," in *Sudbury: Rail Town to Regional Capital*, ed. C.M Wallace and Ashley Thomson (Toronto, ON: Dundurn Press, 1993), 38–39 and 49.

24   *Statutes of Ontario*, 1907, 7 Edw.VII, c. 25, 200–6.

25   A history of the Geological Survey of Canada is found in Morris Zaslow, *Reading the Rocks: The Story of the Geological Survey of Canada 1842–1972* (Toronto, ON: Macmillan Company of Canada, 1975).

26   R. Bell, "Report on the Sudbury Mining District (1888–1890)," *Geological Survey of Canada, Annual Report*, n.s. 5, pt. 1 (1891), 5F–95F; and P.E. Giblin, "History of Exploration and Development of Geological Studies and Development of Geological Concepts," in *The Geology and Ore Deposits of the Sudbury Structure, Ontario Geological Survey*, Special vol. 1, ed. E.G. Pye, A.J. Naldrett, and P.E. Giblin (Toronto, ON: Ontario Ministry of Natural Resources, 1984), 6.

27   Barlow, "Report on Origin, Geological Relations and Composition of the Nickel and Copper Deposits in the Sudbury Mining District, Ontario, Canada," *Geological Survey of Canada, Annual Report*, no. 873 (1904), 1–236.

28   E.G. Pye, "The First Hundred Years: A Brief History of the Ontario Geological Survey," in *Geology of Ontario, Ontario Geological Survey*, Special vol. 4, pt. 1, ed. P.C. Thurston et al. (Toronto, ON: Ministry of Northern Development and Mines, 1992), 30.

29   A.P. Coleman, "The Sudbury Nickel Region," in *Report of the Ontario Bureau of Mines*, vol. 14, pt. 3 (1905), 1–183.

NOTES TO CHAPTER 4

1   Royal Ontario Nickel Commission, *Report of the Royal Ontario Nickel Commission* (Toronto, ON: King's Printer, 1917), 29–30.

2   Aeneas McCharles, *Bemocked of Destiny* (Toronto, ON: William Briggs, 1908), 102.

3   "First House in Cliff Built 60 Years Ago," *Inco Triangle*, January 1945, 4; and "The Copper Cliff Mine, Now Market Street," *Inco Triangle*, March 1951, 4.

4   O.W. Main, *The Canadian Nickel Industry: A Study in Market Control and Public Policy*, Canadian Studies in Economics no. 4 (Toronto, ON: University of Toronto Press, 1955), 13.

5   Matt Bray (professor and historian), interviewed by author, October 16, 2010.

6   Main, *The Canadian Nickel Industry*, 13–17.

7   Ritchie, cited in Royal Ontario Nickel Commission, *Report of the Royal Ontario Nickel Commission*, 61.

8   "Rare Historical Pictures Taken in 1892 Show First Smelting Operations at Copper Cliff," *Inco Triangle*, July 1960, 8–10.

9   For a review of the early years of the Canadian Copper Company, refer to Main, *The Canadian Nickel Industry*, 7–17.

10 Dianne Newell, *Technology on the Frontier* (Vancouver: University of British Columbia Press, 1986), 87.

11 Capt. W.W. Folger, and Lieut. B.W. Buckingham, *Report to the Secretary of the United States Navy upon Nickel and Copper Deposits of Sudbury* (Ottawa: n.p., 1898). Reprint.

12 Mineral Resources Commission, *Report of the Royal Commission*, 405.

13 Eileen Alice Goltz, "Genesis and Growth of a Company Town: Copper Cliff: 1886–1920" (M.A. thesis, Laurentian University, 1983), 42.

14 Wallace Clement, *Hardrock Mining: Industrial Relations and Technological Changes at Inco* (Toronto, ON: McClelland and Stewart, 1981), 44.

15 John F. Thompson, and Norman Beasley, *For the Years to Come: A Story of International Nickel of Canada* (Toronto, ON: Longmans, Green & Company, 1960), 121.

NOTES TO CHAPTER 5

1   Cited in Morris Zaslow, *The Opening of the Canadian North 1870–1914* (Toronto, ON: McClelland and Stewart, 1971), 155.

2   Canada, *Sessional Papers*, 1882, No. 48, Appendix 9, 35–36.

3   C.M. Wallace, "The 1880s," in *Sudbury: Rail Town to Regional Capital*, ed. C.M Wallace and Ashley Thomson (Toronto, ON: Dundurn Press, 1993), 30n8.

4   Florence R. Howey, *Pioneering on the C.P.R.* (Ottawa: Mutual Press, 1938), 11. See also "When Sudbury Was Village of Shacks, Tents," *Sudbury Star*, March 4, l922, 5. Her book was reprinted in its entirety in the 1953 Progress Edition of *The Sudbury Daily Star*; "Mrs. W.H. Howey—Sudbury's First Historian," *Sudbury Daily Star*, August 4, 1953, 33; and "Pioneer Woman, Mrs. W.H. Howey, Called by Death," *Sudbury Star*, November 11, 1936, 1 and 5.

5   Pierre Berton, *The Impossible Railway: The Building of the Canadian Pacific* (New York: Alfred A. Knopf, 1972), 429.

6   Canada, *Sessional Papers*, 1883, No. 27, 15.

7   See, for example, D.M. LeBourdais, *Sudbury Basin: The Story of Nickel* (Toronto, ON: Ryerson Press, 1953), 23; and Charles Dorian, *The First 75 Years: A Headline History of Sudbury Canada* (Devon: Arthur H. Stockwell, 1958), 1.

8   Statement made by Rev. Lorenzo Cadieux in Robert Evans, *An Eye on Everything* (Sudbury: Laurentian University Press, 1966), 27.

9   Peter V.K. Krats, "The Sudbury Area to the Great Depression: Regional Development on the Northern Resource Frontier" (Ph.D. thesis, University of Western Ontario, 1988), 81.

10 Ibid., 82–83. See also Gilbert A. Stelter, "Origins of a Company Town: Sudbury in the Nineteenth Century," *Laurentian University Review* 3, no. 3 (1971): 3–37.

11 Patrick Boyle, *Saint Stanislaus Parish, Copper Cliff, The Early Years: 1886–1914* (Toronto, ON: Jesuit Archives, 1998), 3–7.

12 This geographical depiction of Sudbury's first plan of subdivision has been described in Wallace, "The 1880s," 23.

13 Eileen Alice Goltz, *Genesis and Growth of a Company Town: Copper Cliff, 1886–1920* (M.A. thesis, Laurentian University, 1983), 19.

14 "Reminiscences of Sudbury, 1886," *Inco Triangle*, February 1951, 4.

15 "How Sudbury Looked to Scribe Back in 1890," *Inco Triangle*, January 1948, 14.

16 Wallace, "The 1880s," 25. The Nickel Range Hotel was demolished in 1975.

17 "Reminiscences of Sudbury, 1886," 3.

18 LeBourdais, *Sudbury Basin*, 77.

19 *Statutes of Ontario*, 1892, 55 Vic., c. 88.

20 This collusion is a dominant theme in H.V. Nelles, *The Politics of Development: Forests, Mines & Hydro-electric Power in Ontario, 1849–1941* (Toronto, ON: Macmillan of Canada, 1974).

21 "Adam was a Happy-Go-Lucky Kid," *Sudbury Star*, May 7, 2009, A1 and A4.

22 The early history of high school education in Sudbury is found in: Sudbury Secondary School Reunion Committee, *Sudbury Secondary School: 100 Years Alumni* (Sudbury: Alumni Association, 2008). In Capreol, a Continuation School was started in 1923 and replaced by a high school in 1926. Capreol High School was closed in 1997.

23 Donald Dennie, "Sudbury 1883–1946: A Social Historical Study of Property and Class" (Ph.D thesis, Carleton University, 1989), 52.

24 Ray Thoms, "The N.W. Corner of Elm and Elgin Streets," *Snap Sudbury*, vol. 1, no. 9, 2008, 8.

25 Dorian, *The First 75 Years*, 263.

26 Ontario Motor League-Nickel Belt Club, *Sixty Golden Years 1915–1975* (Sudbury: Nickel Belt Club, 1975), n.p.

27 Dale Wilson, *Algoma Eastern Railway* (Sudbury: Nickel Belt Rails, 1979), 12–14.

28 Sudbury's streetcar history is covered in John D. Knowles, *The Sudbury Streetcars* (Sudbury: Nickel Belt Rails, 1983). See also "The Fate of Sudbury's Streetcars," *Sudbury Sun*, April 12, 2000, 2 and 5.

29 "Has Grown with Sudbury," *Sudbury Star*, June 23, 1923, 9.

30 "Baseball Found Fertile Ground in Sudbury's Early Years," *Sudbury Star*, May 31, 1983, C29.

31 "Cornetist Was a Pioneer," *Globe and Mail*, May 4, 1978, 15.

32 Krats, *The Sudbury Area to the Great Depression*, 200 and 228.

33 "Built in 1908, Has Featured the Famous," *Sudbury Star*, November 25, 1948, 11.

34 Cited in Thomson, "The 1890s," in Wallace and Thomson, *Sudbury: Rail Town to Regional Capital*, 50.

35 Thomas Henry Nicholson, "A Sordid Boon: The Business of State and the State of Labour at the Canadian Copper Company, 1890 to 1918" (M.A. thesis, Queen's University, 1991), 19.

36 Dennie, "Sudbury 1883–1946," 41.

37   "Francophone Businesses Flourished in Early Days," *Sudbury Star*, November 3, 2008, A2; and "J.A. Laberge," in *Greater Sudbury 1883–2008: The Story of Our Times,* ed. Vicki Gilhula (Sudbury, ON: Laurentian Media, 2008), 134.

38   See, for instance, "Brains vs. Bins—Blast, Batter, Bury Bulky Building?" *Sudbury Star*, January 10, 1956, 1–3.

39   L.P. Héroux, "Aperçu Sur Les Origines de Sudbury." *La Société Historique du Nouvel-Ontario*, no. 2 (Sudbury, ON : n.p., 1943), 1–20. This is a reprint of a book originally completed in 1905.

40   Stelter, "The Origins of a Company Town," 12.

41   Gaétan Gervais, "Sudbury, 1883–1914," in *To Our City/Á Notre Ville*, ed. Sudbury Centennial Foundation (Sudbury, ON: Sudbury Centennial Foundation, 1983), 28.

42   Gail Cuthbert Brandt, "The Development of French-Canadian Social Institutions in Sudbury, Ontario, 1883–1920," *Laurentian University Review* 11, no. 2 (1979): 12–13. Bishop Scollard also faced controversy because he appointed Irish priests to the churches at Capreol, Coniston, and Copper Cliff during the First World War.

43   The early history of St. Anne's church is covered in F.A. Peake and R.P. Horne, *The Religious Tradition in Sudbury 1883–1983* (Sudbury: Downtown Churches Association, 1983), 10–20.

44   "Sisters Played Crucial Role at Hospital," *Sudbury Star*, July 18, 2008, A2.

45   Alphonse S.J. Raymonde, *Paroisse Sainte-Anne de Sudbury 1883–1953*. Sudbury, ON : La Société Historique du Nouvel-Ontario, no. 26 (1953), 12. See also "Dollar Day Is Tag Day for St. Joseph's Hospital," *Sudbury Star*, May 16, 1923, 3.

46   St. Joseph's Hospital, *Golden Jubilee 1896–1946* (Sudbury, ON: St. Joseph's Hospital, 1946), 46.

47   Jeannette Bouchard, *Seven Decades of Caring* (Sudbury, ON: Laurentian University Press, 1984), 17–24.

48   St. Joseph's Hospital, *Golden Jubilee*, 24, 26, and 28; and "Grey Nuns Built Hospital in 1895," *Sudbury Star*, May 31, 1983, C32.

49   "Sisters of St. Joseph Open Convent on Saturday," *Sudbury Star*, December 5, 1923, 5.

50   'Flour Mill Has Deep Francophone Roots," *Sudbury Star*, August 23, 2008, A2.

51   Guy Gaudreau, "The Origins of Laurentian University," in *Laurentian University: A History*, ed. Matt Bray (Montreal: McGill-Queen's University Press, 2010), 4.

52   Gwenda Hallsworth, *A Venture into the Realm of Higher Education* (Sudbury: Laurentian University, 1985), 5.

53   "College du Sacre-Coeur," *Sudbury Star*, March 8, 1978, 1.

54   "Jesuits Brought First University to Area," *Sudbury Star*, May 31, 1983, C25.

55   "New Ukrainian Church Bound to Christianity," *Sudbury Star*, December 19, 1928, 3.

56   Claire Pilon, *Le Moulin à Fleur* (Sudbury, ON: Pilon, 1983), 8–10.

57   "Flour Mill Has Deep Francophone Roots," A2.

58   Dennie, "Sudbury 1883–1946," 40–42, and 228.

59   Brandt, "The Development of French-Canadian Social Institutions," 5.

60   A.D. Gilbert, "The 1920s," in Wallace and Thomson, *Sudbury: Rail Town to Regional Capital*, 116.

61   For an appreciation of the role played by the count in Sudbury's history, refer to *Frédéric Romanet du Caillaud «Comte» de Sudbury (1847–1919)* (Montréal: Les Éditions Bellarmin,

1971); "The Count of Sudbury," *South Side Story*, March, 1995, 1 and 3; and "Legend Surrounds Motives of 'Count' Building Grotto," *Sudbury Star*, July 5, 1958, 14; Robert Evans, "Half of Sudbury Once Owned by French Count," in *An Eye on Everything,* 38–42. An indication of the lack of historical appreciation of the count's role in Sudbury was the subsequent renaming of many of his original streets. See Pundit Joe, "The 'Old Count' Gave Sudbury Some Novel Names," *Sudbury Star*, Monday, October 22, 1956, 4.

62   The history of the Church of the Epiphany is recounted in F.A. Peake, *The Church of the Epiphany, Sudbury, Ontario: A Century of Anglican Witness* (Sudbury: Church of the Epiphany, 1982), 1–9.

63   Cited in Peake and Horne, *The Religious Tradition in Sudbury: 1883–1983*, 32.

64   "Anniversary of Church Recalls Early struggle," *Sudbury Star*, June 20, 1923, 5.

65   Linda M. Ambrose, *Glad Tidings Tabernacle: 70 Years of Pentecostal Ministry in Sudbury, 1937–2007* (Sudbury, ON: Glad Tidings Tabernacle, 2007), 2–4.

66   Dennie, "Sudbury 1883–1946," 40, 205 and 215; and Krats, *The Sudbury Area to the Great Depression*, 285.

67   Matt Bray, "1910–1920," in Wallace and Thomson, *Sudbury: Rail Town to Regional Capital*, 87.

68   Lionel Bonin and Gwenda Hallsworth, *Street Names of Downtown Sudbury* (Sudbury, ON: Your Scrivener Press, 1997), x and xiii.

69   Gilbert, "The 1920s," in Wallace and Thomson, *Sudbury: Rail Town to Regional Capital*, 115–16; see also Frederick Simpich, "Ontario, Next Door," *National Geographic* August 1932, 158.

70   "To Make Early Start on Memorial Park," *Sudbury Star*, February 17, 1926, 1.

71   "CPR Releases Last Group of Lots for Sale," *Sudbury Star*, June 4, 1927, 3. All of the properties are listed by street on page 2.

72   *Statutes of Ontario*, 1930, c. 102. The city was later permitted by the Ontario Municipal Board to annex a small part of McKim Township lying south and east of the Algoma Branch of the CPR. See Ontario Municipal Board, *Order P.F.A.—3258*, February 11, 1931.

73   Leslie Roberts, "Sudbury Looks to the Future," *Maclean's*, March 15, 1931, 13, 48 and 50.

74   Wallace, "The 1930s," Wallace and Thomson, *Sudbury: Rail Town to Regional Capital*, 155.

75   Advisory Committee on Reconstruction, *Final Report of the Housing and Community Planning Subcommittee* (Ottawa: King's Printer, 1946), 94, 96, 102 and 105.

76   Cited in Wallace, "The 1930s," Wallace and Thomson, *Sudbury: Rail Town to Regional Capital*, 157.

77   Judith E. Harris, "Well-Being in Sudbury 1931–1971: A Social Indicator Analysis" (M.Sc. thesis, University of Guelph, 1977), iii.

78   "Sudbury Worst City in North America Records Disclose," *Sudbury Star*, July 17, 1935, 1 and 12.

79   Don Delaplante, "Sudbury: Melting Pot for Men and Ore," *Maclean's*, April 15, 1951, 12–13, 5–54.

80   "Local YMCA Celebrating 75 Years," *Sudbury Star*, December 24, 2011, C1.

81   "Fine Range of Facilities at Inco Employees' Club," *Inco Triangle*, February 1938, 1.

82   Gatchell History Committee, *Memories of Gatchell* (Sudbury, ON: Gatchell History Committee, 1997), 7.

83 "Business Establishments Give Superior Service to Residents," *Sudbury Daily Star*, August 5, 1950, 8.

84 "Waives Claims on Land, Gets 198 Lots," *Sudbury Daily Star*, March 17, 1951, 1.

85 "Gatchells Did Live in Sudbury Subdivision," *Sudbury Star*, August 14, 1993, A7. A rivalry developed between the Gatchell and Holditch families over who was more important to the development of the area.

86 "New Civic Thinking behind Plans for Donovan Heights," *Sudbury Star*, November 7, 1956, 4.

87 "Gatchell Subdivision Shows Faith in Nickel District," *Sudbury Star*, June 13, 1928, 28.

88 Ibid., 2–11.

89 "Co-operation Sets Development Pace," *Sudbury Daily Star*, August 5, 1950, 8.

90 "Gatchell Then and Now," *Sudbury Star*, January 21, 1989, 17.

91 Maurizio A. Visentin, "The Italians of Sudbury," *Polyphony* 5, no. 1 (1983): 33.

92 "Settlement First Started Around Old Brick Yard," *Sudbury Star*, June 23, 1950, 8.

93 "The Best Investment in the Land is Land Itself: Donovan Sub-Division," *Sudbury Journal*, March 19, 1908, 3.

94 "Donovan Residents take Great Pride in District," *Sudbury Daily Star*, June 22, 1950, 8.

95 G. Vlahovich and S. Moutsatsos, "Serbians in Sudbury," *Polyphony* 5, no. 1 (1983): 119. See also "Favourite Watering Hole Closes in the Donovan," *Northern Life*, February 9, 2000, 11.

96 Mary Stefura, "The Ukrainian Co-operative Movement in Sudbury," *Polyphony*, v. 2, no. 1 (1979), 45–46.

97 Vlahovich and Moutsatsos, "Serbians in Sudbury," 120.

98 Henry Radecki, "Polish Immigrants in Sudbury, 1883–1980," *Polyphony* 5, no. 1 (1983): 51.

99 Cited in Stacey R. Zembrzycki, "Memory, Identity, and the Challenge of Community among Ukrainians in the Sudbury Region, 1901–1919" (Ph.D. thesis, Carleton University 2007), 152.

100 "Good Services Provided," *Sudbury Daily Star*, June 23, l950, 9.

101 "Nickel Park," *Sudbury Star*, April 28, 1923, 3.

102 "West End Section of City Opened When Homes Built Near Brewery," *Sudbury Daily Star*, July 7, 1950, 7.

103 "Forgotten Cemeteries," *Sudbury Star*, October 7, 2006, A6.

104 "Many Interests Plan Erection of Residences," *Sudbury Star*, April 23, 1930, 13.

105 "Many Fine Homes in Sudbury District, *Sudbury Star*, June 13, 1938, 27 and "Kingsmount Residents Win First Round over Building," *Sudbury Star*, May 30, 1938, 1.

106 "10 Dwellings May Be Erected in Kingsmount," *Sudbury Star*, February 26, 1930, 3.

107 "May Prohibit Apartments in Kingsmount," *Sudbury Star*, May 13, 1938, 17. For information on street name changes, see Bonin and Hallsworth, *Street Names of Downtown Sudbury: A Historical Directory*, 76 and 79.

108 "Minnow Lake during the 1930s," *Sudbury Sun*, July 5, 2000, 3.

109 Ibid.

110 "Faith in Nickel District Founded Minnow Lake," *Sudbury Star*, June 13, 1938, 30.

111 Arnold Michel (owner of A & J Home Hardware) interviewed by author, September 10, 2003.

112 The material for this part of the book has been derived largely from O.W. Saarinen, *Between a Rock and a Hard Place: A Historical Geography of the Finns in the Sudbury Area* (Waterloo, ON: Wilfrid Laurier University Press, 1999), 65–67.

113 Dorian, *The First 75 Years*, 123.

NOTES TO CHAPTER 6

1 For a review of the history of Copper Cliff, refer to Margaret Bertulli and Rae Swan, eds., *A Bit of the Cliff: A Brief History of the Town of Copper Cliff, Ontario 1901–1972* (Copper Cliff, ON: Copper Cliff Museum, 1982).

2 "Sudbury Almost Changed Name One Day in 1891," *Inco Triangle*, January 1949, 15.

3 "Copper Cliff Fifty Years Ago," *Inco Triangle*, August 1949, 13.

4 "Reminiscences of the Early Days," *Inco Triangle*, January 1953, 4.

5 The various stages in the early history of Copper Cliff has been derived from Eileen Alice Goltz, "Genesis and Growth of a Company Town: Copper Cliff: 1886–1920" (M.A. thesis, Laurentian University, 1983), i. See also Eileen Goltz, *The Exercise of Power in a Company Town: Copper Cliff, 1886–1980* (Ph.D thesis, University of Guelph, 1989), 44.

6 "Reminiscences of the Early Days," 12.

7 Patrick Boyle, *Saint Stanislaus Parish, Copper Cliff, The Early Years: 1886–1914* (Toronto, ON: Jesuit Archives), 1998, 25.

8 Diana Iuele-Colilli, *Italian Faces: Images of the Italian Community of Sudbury* (Welland, ON: Soleil Publishing, 2000), 25.

9 Karey Reilly, *Les Italiens de Copper Cliff, 1886–1914* (Honours B.A. thesis, Laurentian University, 1994), 17.

10 Maurizio A. Visentin, "The Italians of Sudbury," *Polyphony* 5, no. 1 (1983): 30.

11 An account of the history of the Italian section of Copper Cliff can be found in Copper Cliff Heritage Group, *Up the Hill: The Italians of Copper Cliff* (Sudbury, ON: Journal Printing, 1997).

12 "Eddie Santi a Dedicated Man," *Sudbury Star*, December 13, 2008, C9.

13 Goltz, ""The Exercise of Power in a Company Town," 44

14 Thomas Henry Nicholson, "A Sordid Boon: The Business of State and the State of Labour at the Canadian Copper Company, 1890 to 1918" (M.A. thesis, Queen's University, 1991), 54, 55, and 72–73.

15 *Statutes of Ontario*, 1901, 1 Edw. VII, c. 51, 250–52.

16 Nicholson, "A Sordid Boon," 57.

17 Goltz, "Genesis and Growth of a Company Town," 147.

18 Ibid., 94.

19 "Presided over Transformation of Swamp into Inco Park at Copper Cliff," *Inco Triangle*, January 1957, 7; and "Green Thumb," *Inco Triangle*, August/September 1972, 8.

20 The early history of the club is recorded in: "Copper Cliff Club: for 30 Years a Hub of Good Entertainment," *Inco Triangle*, July 1946, 10–11; and "Recall Highlights at Popular Now in 40th Year," *Inco Triangle*, June 1955, 5 and 9.

21 Cited in O.W. Saarinen, *Between a Rock and a Hard Place: A Historical Geography of the Finns in the Sudbury Area* (Waterloo, ON: Wilfrid Laurier University Press), 127.

22   Copper Cliff Italian Heritage Group, *Up the Hill*, 9. See also John Zucchi, "Società Italiana di Copper Cliff," *Polyphony* 2, no.1 (1979): 29–30.

23   Bertulli and Swan, *A Bit of the Cliff*, 43.

24   "Stanley Stadium," *Inco Triangle*, February 1967, 7 and 16.

25   "New Curling Rink Proves Very Popular," *Inco Triangle*, February, 1950, 5.

NOTES TO CHAPTER 7

1   H.V. Nelles, *The Politics of Development: Forests, Mines & Hydro-electric Power in Ontario, 1849–1941* (Toronto, ON: Macmillan of Canada, 1973), 107.

2   Jamie Swift and The Development Education Centre, *The Big Nickel: Inco at Home and Abroad* (Kitchener, ON: Between the Lines, 1977), 23.

3   O.W. Main, *The Canadian Nickel Industry: A Study in Market Control and Public Policy*, Canadian Studies in Economics no. 4 (Toronto, ON: University of Toronto Press, 1955), 67–68.

4   "Inside Inco's Power Plants," *Inco Triangle*, February 1972, 4–5.

5   Matt Bray and Angus Gilbert, "The Mond-International Nickel Merger of 1929: A Case Study in Entrepreneurial Failure," *Canadian Historical Review* 76, no. 1 (1995): 19–42.

6   Ibid., 37.

7   Much of this information on the Mond Company was derived from The Research Library, The International Nickel Company of Canada Limited, *History of Nickel Extraction from Sudbury Ores Part I: 1846–1920* and *Part II: 1921–1956* (Copper Cliff, ON: International Nickel Company of Canada Limited, 1956.)

8   James Colussi, "The Rise and Fall of the British America Nickel Corporation 1913–1924," (M.A. thesis, Laurentian University, 1989), 198–99, 221–22, and 229.

9   "Moldering Ruins Recall Early Days of the Nickel Industry," *Inco Triangle*, June 1959, 8.

10   Stephenson et al., *A Guide to the Golden Age: Mining in Sudbury, 1886–1977* (Sudbury, ON: Laurentian University, Department of History, 1979), 56.

11   There was some support for this position. In 1914, for example, more than 37 per cent of total nickel product exports from the United States went to Germany. In 1913 the proportion was only around 8 per cent. Royal Ontario Nickel Commission, *Report of the Ontario Nickel Commission* (Toronto, ON: King's Printer, 1917), 505.

12   Nelles, *The Politics of Development*, 354; and Main, *The Canadian Nickel Industry*, 84.

13   The Research Library, *History of Nickel Extraction from Sudbury Ores*, 40.

14   John F. Thompson and Norman Beasley, *For the Years to Come: A Story of International Nickel of Canada* (Toronto, ON: Longmans, Green & Company, 1960), 199.

15   Bray and Gilbert, "The Mond-International Nickel Merger of 1929," 34.

16   The War Plan Red document was declassified in 1974 and first made public in Robert Preston, *The Defence of the Undefended Border: Planning for War In North America 1867–1939* (Montreal: McGill-Queen's University Press, 1977). See also Floyd Rudmin, *Bordering on Aggression: Evidence of US Military Preparations Against Canada* (Hull, QC: Voyageur Publications, 1993); and Peter Carlson, "Raiding the Icebox: Behind Its Warm Front, the

United States Made Cold Calculations to Subdue Canada," *Washington Post*, December 30, 2005, CO1.

17   "Inco Water Storage around Bisco Has System of 11 Dams," *Inco Triangle*, December 1946, 11.

18   "Sequestered High Falls Is Nerve-Centre of Industry," *Inco Triangle*, August 1937, 4. See also "High Falls School and Scholars," *Inco Triangle*, January 1951, 15.

19   "The 1175-Foot Dam at Big Eddy," *Inco Triangle*, December 1946, 11.

20   "Five Plants in Inco's Huronian Hydro-Electric Power System," *Inco Triangle*, August 1952, 9.

NOTES TO CHAPTER 8

1   A review of the early settlement of the Sudbury area can be found in Peter V.K. Krats, "The Sudbury Area to the Great Depression: Regional Development on the Northern Resource Frontier" (Ph.D. thesis, University of Western Ontario, 1988), 104–5 and 206–11.

2   The history of the Algoma Eastern Railway (AER) is reviewed in Dale Wilson, *Algoma Eastern Railway* (Sudbury, ON: Nickel Belt Rails, 1979).

3   "Frank Dennie 1880–1991," *South Side Story*, January 2005, 30.

4   Frank Dennie's recollections of this period are recorded in "Railway Shops Must Stay in Capreol by agreement," *Sudbury Star*, August 4, 1951, 9.

5   Sudbury Land Registry Office, *Directory of M-Plans*, M65, M69 and M71.

6   Diane Iuele-Collili, *Italian Faces: Images of the Italian Community of Sudbury* (Welland, ON: Soleil, 2000), 16–17. This ethnic connection was enhanced by the fact that Capreol was named after one of the first Italian settlers in Ontario, F.C. Capreol.

7   For a more extensive history of the Town of Capreol, refer to: Capreol Public Library Board, *Capreol: The First 75 Years* (Capreol, ON: Capreol Library Board, n.d.).

8   "Bartolucci Calls Cliffs Agreement 'Historic,'" *Northern Life*, May 10, 2012, 4.

9   "With a Future, Appropriate to Shed Light on Worthington's Past," *Northern Life*, May 9, 2007, 8.

10   W.H. Makinen, 'The Mond Nickel Company and the Communities of Victoria Mines and Mond," in *Industrial Communities of the Sudbury Basin: Copper Cliff, Victoria Mines, Mond and Coniston*, ed. F.A. Peake (Sudbury, ON: Sudbury and District Historical Society, 1986), 23–44.

11   G.O. Tapper, ed., *Wa-Shai-Ma-Gog: Memories of Fairbank Lake and Surrounding Area* (Sudbury, ON: Fairbank Lake Camp Owners Association, 2000), 220.

12   Krats, "The Sudbury Area to the Great Depression," 209.

13   Ibid., 193.

14   Information regarding the Victoria Mines townsite has been largely obtained from Makinen, *Industrial Communities of the Sudbury Basin*. See also Ron Brown, *Ghost towns of Ontario*, vol. 2 (Toronto, ON: Cannon Books, 1983), 78–80.

15   "Victoria Mines Soon to Be Lost," *Sudbury Star*, November 10, 1952, 3.

16   Royal Ontario Nickel Commission, *Report of the Royal Ontario Nickel Commission* (Toronto, ON: King's Printer, 1917), 80.

17   Krats, "The Sudbury Area to the Great Depression," 349.

18 Creighton Mine and the O'Donnell roast yards are remembered in Jim Fortin, ed., *There Were No Strangers: A History of the Village of Creighton Mine* (Walden, ON: Anderson Farm Museum, 1989).

19 Royal Ontario Nickel Commission, *Report of the Royal Ontario Nickel Commission*, 37.

20 Ibid., 5.

21 Tom Davies, "Spirit of Creighton Mine Lives On," *Sudbury Star*, November 26, 1991, CM-2. This article is part of a special ninetieth-anniversary edition commemorating Creighton Mine.

22 "Town of Creighton Mine Rejuvenated by Reviving Industrial Activity," *Sudbury Star*, September 29, 1934, 10.

23 The International Nickel Company of Canada, Limited, Research Library, *History of Nickel Extraction from Sudbury Ores, Part II*, 61.

24 "Optimism High as Plant Close to Completion," *Sudbury Star*, September 26, 1960, 18.

25 Cited from Ray Kaattari, *Voices from the Past: Garson Remembers* (Garson, ON: Garson Historical Group, 1992), 3–4.

26 Trent Black et al., *Nickel Centre Yesterdays* (Sudbury, ON: Northern Heritage Nickel Centre, 1974), 20.

27 "Garson Proud of Handsome Employee Club," *Inco Triangle*, March 1950, 9–10.

28 Krats, "The Sudbury Area to the Great Depression," 208.

29 Information regarding Coniston's history has been largely derived from Coniston Historical Group, *The Coniston Story* (Coniston, ON: Coniston Historical Group, 1983) and Trent Black et al., *Nickel Centre Yesterdays*.

30 Robert P. Trott, *The Story of Onaping Falls* (Onaping Falls, ON: n.p., ca. 1982), 32.

31 Department of History, *Index to The Sudbury Star*, Vols. 1 and 2 (Sudbury, ON: Department of History, Laurentian University, 1980).

32 Ibid., 45.

33 "Six Different Dwelling Styles in '49 Program," *Inco Triangle*, July 1949, 5–6; "$1 200 000 for Levack Housing," *Inco Triangle*, July 1952, 12; "Full Speed Ahead on New Housing at Levack," *Inco Triangle*, October 1952, 15; and "Levack Is Proud of Smart New Curling Centre," *Inco Triangle*, February 1953, 15.

34 "Levack District Takes Pride in Its Beautiful New High School," *Inco Triangle*, February 1959, 13.

35 Much of this information concerning Falconbridge has been derived from Black et al., *Nickel Centre Yesterdays*, 56–57.

36 "Happy Valley Residents Not Happy," *Sudbury Star*, June 3, 1970, 3.

37 Cited in John Deverell and the Latin American Working Group, *Falconbridge: Portrait of the Canadian Mining Multinational* (Toronto, ON: James Lorimer, 1975), 97.

38 E. Lindeman, *Moose Mountain Iron-Bearing District Ont.* (Ottawa, ON: Department of Mines, 1914), 3.

39 "Nickel Belt Was to Be Iron Belt but Sellwood Bubble Burst in 1919," *Sudbury Star*, June 12, 1951, 9.

40 "Sellwood," Ghosttownpix.com, accessed February 5, 2011, http://www.ghosttownpix.com/ontario/towns/sellwood.html; and "Sellwood," Ghosttownpix.com, accessed February 5, 2011, http://www.ontarioabandonedplaces.com/sellwood/sellwood.asp.

41 "Will Sellwood See Its Second Boom?" *Sudbury Star*, November 26, 2011, A1 and A6.

42 "Cliffs Picks Sudbury," *Sudbury Star*, May 9, 2012, A1 and A4.

43 *Wikipedia*, s.v. "CFS Falconbridge," accessed January 31, 2011, http://en.wikipedia.org/wiki/CFS Falconbridge; and "CFS Falconbridge," Ghosttownpix.com, accessed January 31, 2011, http://www.ghosttownpix.com/ontario/towns/falconbridge.html.

44 "CFB Falconbridge," Ontario abandoned places.com, accessed February 2, 2011, http://www.ontarioabandonedplaces.com/articles/cfbfalconbridge.html.

NOTES TO CHAPTER 9

1 A.R.M. Lower, *The North American Assault on the Canadian Forest: A History of the Lumber Trade between Canada and the United States* (Toronto, ON: Ryerson Press, 1938), 28, 171, and 178.

2 Cited in Gwenda Hallsworth, "A Good Paying Business: Lumbering on the North Shore of Lake Huron, 1850–1910 with Particular Reference to the Sudbury District" (M.A. thesis, Laurentian University, 1983), 34–35.

3 J.P.G. de Lestard, *A History of the Sudbury Forest District* (Toronto, ON: Ontario Department of Lands and Forests, 1967), 23.

4 Peter V.K. Krats, "The Sudbury Area to the Great Depression, Regional Development on the Northern Resource Frontier" (Ph.D. thesis, University of Western Ontario, 1988), 92 and 250.

5 "Some Camps Still in Preliminary Preparation," *Sudbury Star*, October 23, 1936, 11.

6 Ibid., 248.

7 de Lestard, *A History of the Sudbury Forest District*, 23–33.

8 Ibid., 24.

9 The names of these licences were obtained from T. Thorpe, *A Review of the Logging and Pulp and Paper Operations in the Sudbury District during the Years 1901–1950* (Sudbury, ON: Ontario Department of Lands and Forests, 1950).

10 Ibid., 10–11.

11 Morris Zaslow, *The Opening of the Canadian North 1870–1914* (Toronto, ON: McClelland and Stewart, 1971), 151.

12 "T-i-m-b-e-r!!!" *Inco Triangle*, August 1944, 3.

13 James H. Gray, "Big Nickel," *Maclean's*, October 1, 1947, 49.

14 "More Action Needed to Improve Lake," *Sudbury Star*, July 16, 2007, 1–2.

15 Brief histories of Skead and Boland's Bay can be found in Trent Black et al., *Nickel Centre Yesterdays* (Sudbury, ON: Northern Heritage Nickel Centre, 1974), 2–17.

16 Jack Hambleton, *Fire in the Valley* (Toronto, ON: Longmans, Green & Company, 1960).

17 "Black Smoke Hid the Sun," *Sudbury Star*, May 5, 2007, A1 and A6.

18 Ray Kaattari, *Voices from the Past: Garson Remembers* (Garson, ON: Garson Historical Group, 1992), 4–6.

19 Gary Peck, "Logging at Wahnapitae Before the Century Dawned," *Sudbury Star*, January 7, 1978, 3.

20 Gary Peck, "Wahnapitae in the Late 19th Century Termed 'a Hustling and Bustling Village,'" *Sudbury Star*, December 3, 1977, 3.

21   Wayne F. LeBelle, *Valley East 1850–2002* (Field, ON: WFL Communications, 2002), 21.

22   Donald Dennie, *Á L'Ombre de L'Inco: Étude de la transition d'une communauté canadienne-française de la région de Sudbury 1890–1972* (Ottawa, ON: Les Presses de l'Université d'Ottawa, 2001), 104–10. This is a study of French-Canadian settlement in the Townships of Rayside and Balfour.

23   de Lestard, *A History of the Sudbury Forest District*, 25.

24   Guy Gaudreau, "Les activités forestières dan deux communautés agricoles du Nouvel-Ontario, 1900–1920," *Revue d'Histoire de L'Amérique Française* 54, no. 4 (2001): 528–29. This study focuses on the two Townships of Blezard and Hanmer.

25   O.W. Saarinen and G.O. Tapper, "The Beaver Lake Finnish-Canadian Community: A Case Study of Ethnic Transition as Influenced by the Variables of Time and Spatial Networks, ca. 1907–1983," in *Finns in North America*, ed. Michael Karni, Olavi Koivukangas, and Edward W. Laine (Turku, Finland: Institute of Migration, 1988), 175. For the significance of farming in the Long Lake and Wanup areas, see O.W. Saarinen, *Between a Rock and a Hard Place: A Historical Geography of the Finns in the Sudbury Area* (Waterloo, ON: Wilfrid Laurier University Press), 89 and 92.

26   The role of lumbering in the lives of the early Finnish settlers in Lorne Township is evident in the family histories found in G.O. Tapper and O.W. Saarinen, eds., *Better Known as Beaver Lake: An History of Lorne Township and Surrounding Area* (Walden, ON: Walden Public Library, 1998).

27   The importance of the lumber camps as economic opportunity sites for Finnish women is discussed in Varpu Lindström, "Finnish Women's Experience in Northern Ontario Lumber Camps, 1920–1939," in *Changing Lives*, ed. Margaret Kechnie and Marge Reitsma-Street (Toronto: Dundurn Press, 1996), 107–22.

28   Royal Ontario Nickel Commission, *Report of the Royal Ontario Nickel Commission* (Toronto: King's Printer, 1917) 1.

29   Gatchell History Committee, *Memories of Gatchell* (Sudbury, ON: Gatchell History Committee, 1997), 141.

30   H.V. Nelles, *The Politics of Development: Forests, Mines & Hydro-Electric Power in Ontario, 1849–1941* (Toronto, ON: Macmillan of Canada, 1973), 44.

31   Commissioner of Crown Lands of the Province of Ontario, *Annual Report for the Year 1901* (Toronto, ON: L.K. Cameron, 1902), iv.

32   Krats, "The Sudbury Area to the Great Depression," 88. Much of the ensuing information regarding the history of agriculture in the Sudbury area has been derived from this work.

33   Ibid., 88–89. See also Richard S. Lambert and Paul Pross, *Renewing Nature's Wealth: A Centennial History of the Public Management of Lands, Forests & Wildlife in Ontario 1763–1967* (Toronto, ON: Department of Lands and Forests, 1967), 136.

34   Saarinen, *Between a Rock and a Hard Place*, 69.

35   Cited in Krats, "The Sudbury Area to the Great Depression," 89.

36   Cited in Saarinen, *Between a Rock and a Hard Place*, 93.

37   Ibid.

38   Krats, "The Sudbury Area to the Great Depression," 161.

39   Statistics Canada, *Census of Canada (1891–2011)*.

40   Krats, "The Sudbury Area to the Great Depression," 163–64.

41   Statistics Canada, *Census of Canada (1911)* and Krats, "The Sudbury Area to the Great Depression," 252.

42   During the Great Depression, the DND was merged into the Department of Highways, where it remained until the 1970s.

43   Dennie, *À L'Ombre de L'Inco*, 101. See also "Farming Fame of Nickel District Spreads," *Sudbury Star*, December 30, 1955, 2.

44   The use of this term has been derived from Dennie, *À L'Ombre de L'Inco*, 171.

45   A history of Azilda can be found in Denis Landry, *Azilda, Comme Je L'ai Connu: Document Historique 1890–1972* (Azilda, ON: Denis Landry, 2001).

46   "Former Convent Gets Reprieve," *Sudbury Star*, August 20, 2008, A3. In 2008 the convent was closed and the site sold to the local French Catholic School Board.

47   Cited in Dennie, *À L'Ombre de L'Inco*, 32.

48   Le Club 50 de Chelmsford, *Chelmsford 1883–1983* (Chelmsford, ON : Le Club 50 de Chelmsford, 1983), 273.

49   Cited in Dennie, *À L'Ombre de L'Inco*, 61.

50   "District Due for Vigorous Expansion," *Sudbury Star,* June 1, 1927, 1.

51   Le Club 50 de Chelmsford, *Chelmsford 1883–1983*, 141–43.

52   Landry, *Azilda, Comme Je L'ai Connu*, 366–67.

53   "Horsemen Blast Cash Grab," *Sudbury Star*, March 20, 2012, A1 and A4; and "City's Slot Share Up in the Air," *Northern Life*, March 20, 2012, 1 and 4.

54   "Uncertainty Looming," *Sudbury Star*, April 28, 2012, A1 and A6.

55   Huguette Parent, "Le Township de Hanmer 1904–1969," *La Société Historique du Nouvel-Ontario*, Documents Historiques No. 70 (Sudbury, ON : Université de Sudbury, 1979), 12–13.

56   Lebelle, *Valley East 1850–2002*, 17.

57   Cited in Huguette Parent, "Le Township de Hanmer 1904–1969," 30.

58   Lebelle, *Valley East 1850–2002*, 21.

59   Ibid., 153.

60   For a discussion concerning the role of the church and left-wing organizations in the Sudbury area, see O.W. Saarinen and G.O Tapper, "Sudbury in the Great Depression: The Tumultuous Years," in *Karelian Exodus: Finnish Communities in North America and Soviet Karelia During the Depression Era*, ed. Ronald Harpelle, Varpu Lindstrom, and Alexis Pogorelskin (Beaverton, ON: Aspasia Press, 2004), 48–66. See also O.W. Saarinen, "Finnish Adaptation and Cultural Maintenance: The Sudbury Experience," in *Entering Multiculturalism: Finnish Experience Abroad*, ed. Olavi Koivukangas (Turku, Finland: Institute of Migration, 2002), 199–215.

61   Parent, "Le Township de Hanmer 1904–1969," 28.

62   Eino Nissilä, *Pioneers of Long Lake* (Sudbury, ON: Eino Nissilä, 1987), 2 and 11–12.

63   For a more detailed history of Beaver Lake, see G.O. Tapper and O.W. Saarinen, eds., *Beaver Lake II: Sisu, Stumps, and Sugar Lumps–A Way of Life* (Sudbury, ON: Laurentian University, 2003).

64   Higgins, *Whitefish Lake Ojibway Memories*, 82. See also J.L.Morris, *Indians of Ontario* (Toronto: Ontario Department of Lands and Forests, 1943), 32–34.

65   The following section has been adapted from Higgins, *Whitefish Lake Ojibway Memories*, 1–208.

66   Department of Indian Affairs and Northern Development, *Registered Indian Population by Sex and Residence 2000* (Ottawa, ON: Department of Indian Affairs and Northern Development, Information Management Branch, March, 2001), 6.

67   "First Nation Aims for Excellence," *Sudbury Star*, May 19, 2012, A3.

68   Much of this early history of the Burwash Industrial Farm has been culled from Province of Ontario, *Annual Reports of the Inspector of Prisons and Reformatories (1917–1946)*.

69   An extensive listing of articles pertaining to the Burwash Industrial Farm can be found in Sudbury Public Library, *Burwash: A History 1911–1981* (Sudbury, ON: Sudbury Public Library, 1982).

70   "The Death of a Prison Brings Cries of Protest," *Toronto Star*, November 20, 1974, B3.

NOTES TO CHAPTER 10

1   See Sawchuk & Peach Associates, *Nickel Basin Planning Study* (Toronto, ON: Department of Municipal Affairs, 1967), map following page 46 and appendix table 4.

2   Ibid., 112–13; Charles Dorian, *The First 75 Years: A Headline History of Sudbury Canada* (Devon, ON: Arthur H. Stockwell, 1958), 170; and D.M. LeBourdais, *The Sudbury Basin: The Story of Nickel* (Toronto, ON: Ryerson Press, 1953), 138.

3   Sawchuk & Peach Associates, *Nickel Basin Planning Study*, 159.

4   "The Falconbridge Story," *Canadian Mining Journal* 80, no. 6 (1959): 111.

5   John Deverell and the Latin American Working Group, *Falconbridge: Portrait of a Canadian Mining Multinational* (Toronto, ON: James Lorimer, 1975), 38; and O.W. Main, *The Canadian Nickel Industry: A Study in Market Control and Public Policy,* Canadian Studies in Economics no. 4 (Toronto: University of Toronto Press, 1955), 113.

6   Falconbridge Nickel Company, *Celebrating 75 Years: 1928–2003* (Falconbridge, ON: Falconbridge Limited, 2003), 10.

7   The Research Library, The International Nickel Company of Canada Limited, *History of Nickel Extraction from Sudbury Ores Part I: 1846–1920 and Part II: 1921–1956* (Copper Cliff, ON: International Nickel Company of Canada Limited, 1956), 36 and 38.

8   O.W. Saarinen, "The 1950s," *Sudbury: Rail Town to Regional Capital*, ed. C.M. Wallace and Ashley Thomson (Toronto, ON: Dundurn Press, 1993), 192.

9   Deverell and the Latin American Working Group, *Falconbridge: Portrait of a Canadian Mining Multinational*, 43.

10   Ibid., 10.

11   Main, *The Canadian Nickel Industry*, 120.

12   "Three Stacks of Copper Cliff World Famous," *Inco Triangle*, September 1951, 7.

13   The Research Library, The International Nickel Company of Canada Limited, *History of Nickel Extraction*, 19–44.

14 Ibid., 44.

15 A review of the history of women working at Inco can be found in Jennifer Keck and Mary Powell, "Working at Inco: Women in a Downsizing Male Industry," in *Changing Lives,* ed. Margaret Kechnie and Marge Reitsma-Street (Toronto: Dundurn Press, 1996), 147–61.

16 Sandra P. Battaglini, "Don't Go Down the Mine Mamma!" Women in Production Jobs at Inco During World War II, 1942–1945" (M.A. thesis, Laurentian University, 1996), 148–49.

17 "Mother Nature's Dark Secrets Probed by the 'Flying X-Ray,'" *Inco Triangle,* January 1953, 4–5 and 13.

18 Mount, "The 1940s," in Thomson and Wallace, *Sudbury: Rail Town to Regional Capital,* 176.

19 "US Govt. to Buy 120 Million Pounds of Inco Metallic Nickel," *Inco Triangle,* July 1953, 14–15.

20 "Inco Iron Ore Goes to Mills by Rail, Boat," *Inco Triangle,* October 1970, 13.

21 "Process of New Iron Ore Plant Makes History," *Inco Triangle,* October 1953, 2.

22 "How Iron Became the Fourteenth Element on the Inco Product Team," *Inco Triangle,* January 1965, 4–5.

23 "Plant's Capacity Will Be Tripled," *Inco Triangle,* November 1960, 13.

24 Val Ross, "The Arrogance of Inco," *Canadian Business,* May 1979, 131.

25 Heidi Ulrichsen, "Remembering a Community Builder," Northern Life.ca, January 10, 2011, accessed January 23, 2011, http://www.northernlife.ca/news/localNews/2011/01grassby11011.aspx.

26 For a review of these activities in the 1960s, refer to Jamie Swift and The Development Education Centre, *The Big Nickel: Inco at Home and Abroad* (Kitchener, ON: Between the Lines,1977).

27 John F. Thompson and Norman Beasley, *For the Years to Come: A Story of International Nickel of Canada* (Toronto, ON: Longmans, Green & Company, 1960), 274–77.

28 Ross, "The Arrogance of Inco," *Canadian Business,* 125 and 126.

29 Ibid., 128.

30 "Bitter Standoff Grinds Down Company Town," *Sudbury Star,* June 6, 2010. A8.

31 David Humphreys, *Nickel: An Industry in Transition* (paper presented at the World Stainless Steel Conference, Dusseldorf, September 17–19, 2006).

32 Energy, Mines and Resources, Canada, *Canada's Nonferrous Metals Industry: Nickel and Copper—A Special Report* (Ottawa, ON: Energy, Mines and Resources, Canada, 1984), 8.

33 Anthony J. Naldrett, "From Impact to Riches: Evolution of Geological Understanding as Seen at Sudbury, Canada," *GSA Today,* February 2003, 4.

34 For a detailed review of these technological changes, see Wallace Clement, *Hardrock Mining: Industrial Relations and Technological Changes at Inco* (Toronto, ON: McClelland and Stewart, 1981), 94–218.

35 "Inco Building the World's Leading Nickel Company," Vale, accessed October 31, 2005, http://www.inco.com/about/history/default.aspx.

36 Humphreys, *Nickel: An Industry in Transition,* 3.

37 SARA Group, *Sudbury Area Risk Assessment,* Final Report (Guelph, ON: SARAH Group, 2008), 2–55.

38   Gord Slade, "One Man, 27 Different Jobs," *Falconbridge 75 Years* (Falconbridge, ON: Falconbridge Limited, 2003), 23–24.

39   Ross, "The Arrogance of Inco," 135.

40   Wallace, "The 1980s," in Thomson and Wallace, *Sudbury: Rail Town to Regional Capital*, 276.

41   "Bots Go Down the Mine Shaft," *Toronto Star*, August 6, 2001, D2.

42   See, for instance, "Happily Ever After," *Northern Life*, October 12, 2005, 1 and 12; and "Unions Happy with Takeover," *Northern Life*, October 12, 2005, 1 and 12.

43   "Sun Sets on Old Inco," *Sudbury Star*, October 12, 2005, A1–A2. Refer also to "This Town's Only Big Enough for One of Us," *Toronto Star*, October 15, 2005, D1 and D4.

44   "Goodbye Inco, 'bem-vindos' to Sudbury CVRD," *Northern Life*, October 25, 2006, 8; and "Iconic Inco Rides off into the Sunset," *Toronto Star*, October 23, 2006, C1–C2.

45   "Union Leaders Upbeat about Changes at Inco," *Sudbury Star*, October 26, 2006, A4; and "CVRD Boss Makes Nice with Union Officials," *Northern Life*, October 27, 2006, 1 and 8.

46   "Vale Inco Mulls Deep Cuts at Sudbury Mine," *Toronto Star,* accessed December 8, 2010, http://www.thestar.com/business/article/7522989.

47   "Mining Industry Begins Slow Recovery," *Sudbury Star*, March 26, 2010, A10.

48   "Bitter Standoff Grinds Down Company Town," *Sudbury Star*, June 6, 2010, A8.

49   "Government to Blame for Vale's Intrusion," *Sudbury Star*, June 3, 2010, 11; and "Vale and Inco Don't Go Together," *Sudbury Star*, June 7, 2010, A11.

50   "Wallbridge Mining Company Limited," Wallbridge Mining Co. Ltd., accessed December 1, 2010, http://www.wallbridgemining.com/s/Sudbury.asp; and "Wallbridge Mining Company Limited," Wallbridge Mining Co. Ltd., accessed December 1, 2010, http://www.wallbridgemining.com/s/North Range.asp.

51   For information regarding this company, see Wallbridge Mining Company, *Annual Report 2009* (Sudbury, ON: Wallbridge Mining Company, 2010).

52   "Quadra, FNX Both Thriving," Sudbury Star.com, November 11, 2010, http://www .thesudburystar.com/ArticleDisplay.aspx?e=2841299

53   "Sagamok, Quadra Ink Deal," *Sudbury Star*, January 18, 2012, A3.

54   "Joint Ventures," Sudbury Star.com, accessed February 5, 2011, http://www.thesudburystar .com/ArticleDisplay.aspx?archive=true&e=261. See also "Xstrata Working on Innovative Mining Projects," *Sudbury Star*, Mining Supplement, May 10, 2012, 4.

55   "Lockerby Up and Running," *Sudbury Star*, May 1, 2012, A1.

56   "Sudbury Business Moving into the Big Leagues," TheSudburyStar.com, accessed January 23, 2011, http://www.thesudburystar.com/ArticleDisplay.aspx?e=2942129. See also "Full Production Planned for Lockerby Mine," *Sudbury Star*, January 30, 2012, A1.

57   "Ursa Major Minerals Set to Resume Production," TheSudburyStar.com, accessed February 5, 2011, http://www.thesudburystar.com/ArticleDisplay.aspx?archive=true&e=218.

58   "First Nations Fuming over Smelter Decision," *Northern Life*, May 10, 2012, 4.

59   "Our Management of Resources Second Rate," *Northern Life*, May 2, 2007, 10.

60   Judy Erola (former Member of Parliament and Director, International Nickel Company of Canada) interviewed by author, August, 2008.

61   Cited in Ian Ross, "Xstrata PLC Sparking Merger Mania," *Northern Ontario Business*, July 18, 2008.

62   Val Ross, "The Arrogance of Inco," 44.

63   Nelles, *The Politics of Development*," 327.

64   "Why the Leafs Stink," *Maclean's*, April 14, 2008, 49.

65   "Recent Wave of Takeovers Raises Foreign Eyebrows," *Toronto Star*, October 24, 2006, F1 and F8.

66   "Iconic Inco Rides Off into the Sunset," *Toronto Star*, C2.

NOTES TO CHAPTER 11

1   The importance of the royal visit is made clear in "Royal Visit Edition," *Sudbury Star*, June 2, 1939.

2   Graeme S. Mount, "The 1940s," in *Sudbury: Rail Town to Regional Capital*, ed. C.M. Wallace and Ashley Thomson (Toronto, ON: Dundurn Press, 1993), 172.

3   Information obtained from the Ontario Municipal Board. An important reference is Ontario Municipal Board, *Order P.F.M.–787*, Toronto, July 18, 1957. This is a quieting order establishing the legal existence, corporate status, area, and boundaries of the Township of McKim. It also makes reference to boundary changes affecting the Towns of Copper Cliff and Frood Mine.

4   John Bland and Harold Spence-Sales, *A Report on the City of Sudbury and Its Extensions* (Sudbury, ON: City of Sudbury, 1950), 5.

5   Mount, "The 1940s," in Wallace and Thomson, *Sudbury: Rail Town to Regional Capital*, 170.

6   Ray Thoms and Kathy Pearsall, *Sudbury* (Toronto, ON: Stoddart Publishing, 1994), 80.

7   Mary Stefura, "Ukrainians in the Sudbury Region," *Polyphony* 5, no. 1 (Spring/Summer 1983): 71–81 and Varpu Lindström-Best, "Central Organization of the Loyal Finns in Canada," *Polyphony* 3, no. 2 (Fall 1981): 97–103.

8   Mount, 'The 1940s," in Wallace and Thomson, *Sudbury: Rail Town to Regional Capital*, 170–71.

9   For more details regarding Sudbury in the 1950s, refer to O.W. Saarinen, "Sudbury: A Historical Case Study of Multiple Urban-Economic Transformation," *Ontario History*, 82, no. 1 (1990).

10   See Honourable William S. Paley, Chairman, *President's Materials Policy Commission* (Washington, DC: Government Printing Office, 1952).

11   L. Carson Brown, "Elliot Lake: The World's Uranium Capital," *Canadian Geographical Journal* 75, no. 4 (1967): 125.

12   O.W. Saarinen, "The 1950s," in Wallace and Thomson, *Sudbury: Rail Town to Regional City*, 20.

13   These townsite developments are recorded in Inco's publication, *Inco Triangle*, and Falconbridge Nickel Mines' publication, *Falcon*.

14   Maurizio A. Visentin, "The Italians of Sudbury," *Polyphony* 5, no. 1 (1983): 36.

15   "Polish Hall Celebrates 60th Birthday," *Sudbury Star*, May 28, 2012, A1 and A4.

16   Gertrud Jaron Lewis, "German-Speaking Immigrants in the Sudbury Region," *Polyphony* 5, no. 1 (1983): 82.

17   Ibid.

18   The strong work ethic of the German population is portrayed in Gudrun Jahns Closs, *Canadian Mosaic: Life Stories of Post-War German-Speaking Immigrants to Sudbury* (Sudbury, ON: German-Canadian Association, 2003).

19   Pierre Berton, "Dionne Quintuplets," in *The Canadian Encyclopedia*, vol. 1, 2nd ed. (Edmonton, AB: Hurtig Publishers, 1988), 599.

20   For a review of the events that occurred during this period, see Matt Bray, "The Founding of Laurentian University, 1958–1960," in *Laurentian University: A History*, ed. Matt Bray (Montreal, QC: McGill-Queen's University Press, 2010), 17–30. Early classes were held in the Empire Block on Elgin Street and the former Jackson and Barnard funeral parlour on Larch Street.

21   The early history of Laurentian University is traced in Gwenda Hallsworth, *A Brief History of Laurentian University* (Sudbury, ON: Laurentian University, 1985).

22   "Historical Highlights," *Sudbury Star*, February 7, 1993, CC6. Cambrian moved out of the Sacred Heart College campus in 1977.

23   Sister Bonnie MacLelland, "A Message to the Citizens of Sudbury," press release, June 17, 2010. Sisters of St. Joseph of Sault Ste. Marie.

24   Much of the information on nursing was compiled from Jeannette L. Bouchard, *Seven Decades of Caring* (Sudbury, ON: Laurentian University Press, 1984).

25   Gwenda Hallsworth and Peter Hallsworth, "The 1960s," in Wallace and Thomson, *Sudbury: Rail Town to Capital*, 230.

26   "Pine Street Water Tower Doomed," *Sudbury Star*, July 16, 2011, A3; and "Sudbury Water Tower Redevelopment Project," accessed August 12, 2011, http://sudburywatertower.com/. Both towers were decommissioned in 1998 when a new reservoir was built.

27   "Mall Led Retail Revolution," *Sudbury Star*, October 20, 2007, D1.

28   "Celebrates 50 Years," *Sudbury Star*, May 29, 2010, 3.

29   For a more detailed review of the early planning history of the Sudbury area, see O.W. Saarinen, "Planning and Other Developmental Influences on the Spatial Organization of Urban Settlement in the Sudbury Area," *Laurentian University Review* 3, no. 3 (1971): 38–70.

30   "Ottawa Post for Faintuck; Was First Sudbury Planner," *Sudbury Star*, June 9, 1957, 13.

31   "Sudbury Council Approves Law Restricting Building Outside Commercial Area," *Sudbury Star*, August 3, 1955, 13.

32   "Pick Polish-Born Planner for Sudbury Area Post," *Sudbury Star*, August 27, 1957, 13.

33   A summary of this urban renewal project is found in Stephen Michael Sajatovic, "The Borgia Area Redevelopment Project: A Case Study in Urban Renewal" (Honours B.A. thesis, Laurentian University, 1973), 1–98.

34   "Main Downtown Mall Gets Multi-Million-Dollar Facelift," *Sudbury Star*, April 26, 1988, Review 35.

35   Gwenda and Peter Hallsworth, "The 1960s," in Wallace and Thomson, *Sudbury: Rail Town to Regional Capital*, 226.

36 See, for example, H.V. Nelles, *The Politics of Development: Forests, Mines & Hydro-electric Power in Ontario, 1849–1941* (Toronto, ON: Macmillan of Canada, 1973), 434–35.

37 Cited from the Royal Ontario Nickel Commission, *Report of the Royal Ontario Nickel Commission* (Toronto, ON: King's Printer, 1917), 511.

38 "Tax Revenue Available for Nickel Belt Needs without Amalgamation," *Sudbury Star*, June 16, 1958, 1 and 3.

39 Ibid., 1.

40 "Township Can't Tax Inco Iron Plant," *Sudbury Star*, February 13, 1957, 1 and 5.

41 The Mayor's Committee on Sudbury's Financial Problems, *1964: The Year of Dilemma* (Sudbury, ON: City of Sudbury, 1964), 1 and 3; and International Union of Mine, Mill and Smelter Workers (Canada) in the City of Sudbury, *Submission before the Select Committee of the Ontario Legislature on Mining* (Sudbury, ON: International Union of Mine, Mill and Smelter Workers [Canada], 1964), 26–27.

42 The International Nickel Company of Canada, Limited, *A Submission by The International Nickel Company of Canada, Limited on The Report of the Ontario Committee on Taxation* (Copper Cliff, ON: International Nickel Company of Canada, Limited, 1968), 2–2.

43 Ibid., 2–6 and 2–7.

44 Ontario Municipal Board, *P.C.F. – 4756*, January 16, 1951. Refer also to "Annexation Adds 171 McKim Acres to Size of Sudbury," *Sudbury Star*, January 17, 1951, 1 and 5.

45 For a listing of the various amalgamation attempts made by the city between 1951 and 1958, see "Once Upon a Time, Back in 1951 …," *Sudbury Star*, February 8, 1958, 15; and "Nearly Nine Years since Secret Meeting of City Council Started the Whole Thing," *Sudbury Star*, November 16, 1959, 9.

46 Ontario Municipal Board, *P.F.M.—5847*, August 12, 1957.

47 Ontario Municipal Board *P.F.M.—5143–6*, November 12, 1959 and "Sudbury Becomes City of 75 000 as McKim, Frood, Half of Neelon Absorbed in Amalgamation Move," *Sudbury Star*, November 16, 1959, 1, 3 and 9.

48 C.M. Wallace, "Sudbury: The Northern Experiment with Regional Government," *Laurentian University Review* 17, no. 2 (1985): 92.

49 City of Sudbury, *1964: The Year of Dilemma*, 1–15.

50 "Mines Less of a Tax Motherlode for Sudbury," *Sudbury Star*, October 15, 1988, 17.

51 J.R. Winter, *Sudbury: An Economic Survey* (Sudbury, ON: Sudbury & District Industrial Commission, 1967). For an earlier and similar view on company towns, see "Rap 'Company Towns,'" *Sudbury Star*, November 9, 1958, 1.

52 Sawchuk & Peach Associates, *Nickel Basin Planning Study* (Sudbury, ON: Sawchuk & Peach, 1967), 92–110.

53 Province of Ontario, Committee on Council, *Order-in-Council Appointing Commission* (Toronto, ON: Lieutenant-Governor, February 13, 1969).

54 J.A. Kennedy, *Sudbury Area Study* (Toronto: Department of Municipal Affairs, 1970).

55 Wallace, "Sudbury: The Northern Experiment," 94.

56 Ontario Department of Municipal Affairs, *Sudbury: Local Government Reform Proposals* (Toronto: Ontario Department of Municipal Affairs, 1971).

57   Ontario, Legislative Assembly, *Bill 164: An Act to Establish the Regional Municipality of Sudbury*, 29th Legislature, 2nd session (June 29, 1972). Toronto, ON: Queen's Printer, 1972.

58   "Mines Assessment Irks Mayor," *Sudbury Star*, March 12, 1970, 1.

NOTES TO CHAPTER 12

1   "Saga of Lily Creek Plaza Finally Ended by Cabinet," *Sudbury Star*, May 4, 1976, 1.

2   Much of this chapter has been derived from O.W. Saarinen, "Sudbury: A Historical Case Study of Multiple Urban-Economic Transformation," *Ontario History* 82, no. 1 (1990): 54–81.

3   James Simmons and Brian Speck, *Spatial Patterns of Social Change: The Return of the Great Factor Analysis*, Research Paper No. 160 (Toronto, ON: Centre for Urban and Community Studies, University of Toronto, 1986), 35 and 37.

4   Dieter K. Buse, "The 1970s," in *Sudbury: Rail Town to Regional Capital*, ed. C.M. Wallace and Ashley Thomson (Toronto, ON: Dundurn Press, 1993), 252.

5   Peter C. Newman, "Sudbury's Sunny Renaissance," *Maclean's*, April 1, 1991, 40.

6   Robert Dickson, *Sudbury Iron Bridge* (Sudbury, ON: Canned Collective Works, 1978).

7   C.M. Wallace, "Sudbury: The Northern Experiment with Regional Government," *Laurentian University Review* 17, no. 2 (1985): 98.

8   "New Church of the Epiphany Rising from the Ashes," *Sudbury Star*, April 26, 1988, Review, 10; "Creighton Reunion Being Planned," *Sudbury Star*, April 26, 1988, Review, 13; "Last Chance for the Grand Theatre," *Northern Ontario Business*, February, 1989, 65; and "Our 100th Year," *Sudbury Star*, October 29, 2008, 51.

9   Wieslaw Michalak, "Economic Changes of the Canadian Urban System 1971–1981," *Alberta Geographer* 22 (1986): 65.

10   Community Planning Programs Division, Ontario Ministry of Municipal affairs and Housing, *Project Group on Urban Economic Development—Case Study Report: Sudbury, Ontario, Canada* (Paris: OECD, 1985).

11   The history of this planning initiative can be traced in Sudbury 2001, *Sudbury 2001 Foundations* (Sudbury, ON: Sudbury 2001, 1983); *Proceedings Sudbury 2001* (Sudbury, ON: Sudbury 2001, 1983); *A Conference on Economic Development: April 6, 7 and 8, 1978* (Sudbury, ON: Sudbury 2001, 1978) and *Retrospect: Prospect* (Sudbury, ON: Sudbury 2001, 1981). See also "Goats Are Gone, Understanding of 2001 'Better,'" *Sudbury Star*, April 28, 1981, Progress 3.

12   An assessment of Sudbury 2001 can be found in "Sudbury 2001: Legend of Job Creation Program Lives on, but Spirit That Fuelled It Seems Dead," *Northern Life*, May 29, 1996, 1A and 3A. See also "Scaled-Down 2001 Carries on with Job Mandate," *Northern Life*, March 20, 1985, 1 and 16.

13   Pierre Filion, "Potentials and Weaknesses of Strategic Community Development Planning: A Sudbury Case Study," *Canadian Journal of Regional Science* 11, no. 3 (1988): 402.

14   See O.W. Saarinen, *Between a Rock and a Hard Place: A Historical Geography of the Finns in the Sudbury Area* (Waterloo, ON: Wilfrid Laurier University Press, 1999), 275; and "Judy Erola: The One-Woman Energy Resource Who's Minister of State for Mines," *Chatelaine*, April 1981, 52 and 146–53.

15 See, for example, "Private Sector Urged to Play Leading Role," *Sudbury Star*, April 26, 1988, Review, 5.

16 This response is outlined in Regional Municipality of Sudbury, *Toward Economic Diversification: The Key to Long-Term Job Creation in the Regional Municipality of Sudbury* (presentation, Ontario Provincial Legislature, November 26, 1982).

17 "The Sudbury Community Adjustment Project: It Works," *Matrix* 1, no. 1 (1988): 1.

18 Filion, "Potentials and Weaknesses of Strategic Community Planning," 401.

19 Matt Bray, *Science North: Evolution of a Northern Dream* (Sudbury, ON: Science North, 2011), 20 and 38–39.

20 Peter C. Newman, "Sudbury's Sunny Renaissance," 40.

21 Narasim Katary, personal communication, October 12, 2012.

22 "Ukrainian Seniors' Centre Will Be Ready by Fall," *Sudbury Star*, April 26, 1988, Review 43.

23 "OGS Expects to Be Swamped with Applications for Positions," *Northern Life*, April 1, 1992, Progress '92, 50.

24 See "Sudbury to Play Leading Role in Treatment of Cancer," *Sudbury Star*, April 27, 1988, Review, 17; and "Against all Odds: Northern Cancer Centre Flourishes," *Sudbury Star*, March 31, 1992, Br-A7.

25 See Sudbury Regional Development Corporation, *The Next Ten Years: A Conference Discussion Paper* (Sudbury, ON: Sudbury Regional Development Corporation, 1992); and *Patterns for Prosperity: Final Report of the Next Ten Years Project* (Sudbury, ON: Sudbury Regional Development Corporation, 1993).

26 Regional Municipality of Sudbury, *Sudbury 20/20 Focus on the Future*, supplement, *Northern Life*, April 14, 1999, 3.

27 Regional Municipality of Sudbury, *Sudbury Tomorrow: The New Way* (Sudbury, ON: Regional Municipality of Sudbury, 2000).

28 "It Doesn't Pay to Take Little Things for Granted," *Northern Life*, June 6, 1993, 4; and "SRDC Fears Project Will Lose Momentum," *Northern Life*, April 21, 1993, 3.

29 "Call Centres Major Force in City," *Sudbury Star*, March 28, 2008, C3.

30 "Boréal is on to Promising Projects," *Sudbury Star*, March 28, 2008, A9.

31 Greater Sudbury Development Corporation, *Coming of Age in the 21st Century: An Economic Development Strategic Plan for Greater Sudbury 2015* (Sudbury, ON: Greater Sudbury Development Corporation, 2003), 13.

32 Maureen Woodrow, *Challenges to Sustainability in Northern Ontario* (Toronto: Environmental Commission of Ontario, 2002), 13.

33 "Searching for the Universe's Secrets," *Sudbury Star*, May 19, 2012, A1 and A6.

34 "City's Retail Sector Booming," *Sudbury Star*, March 28, 2008, A6.

35 "Downtown Needs Rethought," *Sudbury Star*, January 23, 2012, A11.

36 "Evolutionary Blockbuster," *Sudbury Star*, January 17, 2012, A1 and A4.

37 Hugh J. Thomas, *Report to the Minister of Municipal Affairs and Housing on Local Government Reform for Sudbury* (Sudbury, ON: Office of the Special Advisor on Local Government Reform in the Sudbury Region, 1999).

38 Floyd Laughren, *Constellation City: Building a Community of Communities in Greater Sudbury*, Report of the Greater Sudbury Community Solutions Team (Sudbury, ON: City of Greater Sudbury, 2007).

39 For a history of this saga, refer to "Questions Arise from Sale," *Sudbury Star*, June 26, 2010, A1 and A6.

40 Association of Municipalities of Ontario and City of Toronto, *Provincial-Municipal Fiscal and Service Delivery Review* (Toronto: Association of Municipalities of Ontario and City of Toronto, 2008).

41 Ontario North Economic Development Corporation, *Pan Northern Mining Supply and Services Strategy* (Sudbury, ON: Ontario North Economic Development Corporation, 2009), 4; and "Sudbury Luckiest City," *Sudbury Star*, January 23, 2012, A1 and A4.

42 Dick DeStefano (Executive Director of Sudbury Area Mining Supply and Services Association), interviewed by author, January 23, 2012.

43 Greater Sudbury Development Corporation, *Coming of Age in the 21st Century: Digging Deeper* (Sudbury, ON: Greater Sudbury Development Corporation, 2009), 3–6, 6.

44 "Real Success Is Measured by the Quality of Succession," *Northern Life*, May 3, 2012, 10.

45 Ibid.

46 See "Greater Sudbury Has Done Well under Bartolucci's Tenure," *Sudbury Star*, February 9, 2013, A10, and "Sudbury Has Lost a Powerful Ally," *Northern Life*, February 12, 2013, 8.

47 Greater Sudbury Development Corporation, *Coming of Age in the 21st Century: Digging Deeper*, 3 and 17.

48 "We Want to Be a City of Learning," *Sudbury Star*, October 20, 2011, A1 and A4.

49 City of Greater Sudbury, *The City of Greater Sudbury Official Plan* (Sudbury, ON: City of Greater Sudbury, 2008).

50 Dick DeStefano, interviewed by author, September 3, 2011.

51 "Steel to City Businesses: 'You're either on Our Side, or Vale's,'" *Sudbury Star*, September 21, 2009, A1 and A4.

52 Shawn Van Sluys et al., eds., *Sudbury: Life in a Northern Town* (Sudbury, ON: Laurentian Architecture, 2011), 68–69.

53 "Historic Flag Celebrates 35th Anniversary," *Sudbury Star*, September 28, 2010, A2.

54 "Games Leave Legacy," *Sudbury Star*, July 25, 2011, A1 and A4.

55 "Building a Performing Arts Centre," *Sudbury Star*, May 25, 2012, Downtown Sudbury, 7.

56 "City Celebrates Aboriginal Culture," *Sudbury Star*, September 30, 2010, A3.

57 Urban Aboriginal Task Force, *Sudbury Final Report* (Sudbury, ON: Ontario Federation of Indian Friendship Centres, 2007), 27.

58 Statistics Canada, *Single and Multiple Ethnic Origins for the City of Greater Sudbury*, Catalogue no. 97-562-XCB2006006 (Ottawa, ON: Statistics Canada, 2011).

59 Urban Aboriginal Task Force, *Sudbury Final Report*, 38.

60 See, for example, "Caruso Club Welcomes First Female President," *Northern Life*, December 30, 2010. 11.

NOTES TO CHAPTER 13

1   John Thompson and Norman Beasley, *For the Years to Come: A Story of International Nickel of Canada* (Toronto, ON: Longmans, Green, 1960), 371.

2   Unless otherwise indicated, the information from this section has been gleaned from Mike Solski and John Smaller, *Mine Mill: The History of the International Union of Mine, Mill and Smelter Workers in Canada since 1895* (Ottawa, ON: Steel Rail Publishing, 1984); John B. Lang, "A Lion in a Den of Daniels: A History of the International Union of Mine Mill and Smelter Workers in Sudbury Ontario, 1942–1962" (M.A. thesis, University of Guelph, 1970); and Jason A. Miller, "Divided We Stand: A Study of the Development of the Conflict between the International Union of Mine, Mill and Smelter Workers and the United Steelworkers of America in Sudbury, Ontario, 1942–1969" (M.A. thesis, McMaster University, 2003).

3   O.W. Saarinen, *Between a Rock and a Hard Place: A Historical Geography of the Finns in the Sudbury Area* (Waterloo, ON: Wilfrid Laurier University Press), 188.

4   Royal Commission on Industrial Relations, *Report of Commission* (Ottawa, ON: Committee of the Privy Council, 1919), 47.

5   Jim Tester, *The Shaping of Sudbury: A Labour View* (Sudbury, ON: Sudbury & District Historical Society, 1979), 29. "Whisky seniority" meant bringing a "bottle" to the boss on regular occasions to gain favours.

6   Ontario Bureau of Mines, *Annual Report for 1913* (Toronto: Legislative Assembly of Ontario, 1914), 67.

7   This is a major conclusion made by Laurel Sefton MacDowell in her book, *"Remember Kirkland Lake": The History and Effects of the Kirkland Lake Gold Miners' Strike, 1941–1942* (Toronto: University of Toronto Press, 1983).

8   Solski and Smaller, *Mine Mill*, 108.

9   Tester, *The Shaping of Sudbury*, 27.

10 Hans Brasch, *A Miner's Chronicle: Inco Ltd. and the Unions 1944–1997* (Sudbury, ON: Hans Brasch, 1997), 19.

11 For a listing of Local 902's contracts in 1959, see Donald Dennie, "Le local 902 du Mine Mill: Les dix premières années (1949–1959) du syndicate des travailleurs de la ville et du district de Sudbury," in *Hard Lessons: The Mine Mill Union in the Canadian Labour Movement*, ed. Mercedes Steedman, Peter Suschnigg, and Dieter K. Buse (Toronto: Dundurn Press, 1995), 64–65.

12 Weir Reid's contribution to enriching Sudbury's cultural life has been extensively covered by Dieter K. Buse, "Weir Reid and Mine Mill: An Alternative Union's Cultural Endeavours," in Steedman, Suschnigg, and Buse, *Hard Lessons*, 284–85.

13 Ibid. For other assessments of Reid's life, see, "Milestones," *Time*, August 16, 1971, 49; refer also to "Robeson's 'beloved friends in Canada,'" *Toronto Star*, April 9, 1998, G10.

14 For a more detailed assessment of the work of the auxiliaries, see Ruth Reid, "We're Still Here," in Steedman, Suschnigg, and Buse, *Hard Lessons*, 151–57. Ruth Reid was a leading figure in the Ladies' Auxiliary Local 117. See also Mercedes Steedman, "The Red Petticoat Brigade: Mine Mill Women's Auxiliaries and the Threat from Within, 1940s–70s," in *Whose*

National Security? Canadian State Surveillance and the Creation of Enemies, ed. Gary Kinsman, Dieter K. Buse, and Mercedes Steedman (Toronto, ON: Between the Lines, 2000), 55–72.

15   Miller, "Divided We Stand," iii.

16   Frank Southern, *The Sudbury Incident* (Toronto: York Publishing, 1978), 32.

17   Ibid., 48.

18   The idea that the inter-union rivalries that took place during the 1950s and 1960s were more about empire-building rather than the Communist factor has been put forward by John B. Lang. See Lang, "One Hundred Years of Mine Mill," in Steedman, Suschnigg, and Buse, *Hard Lessons*, 17.

19   Miller, "Divided We Stand," 40.

20   "As the Community Sees It," *Sudbury Star*, December 5, 1958, 1.

21   Cited in Robert A. Stephenson, "To Strike—or Not to Strike?" (Honours B.A. thesis, Laurentian University, 1978), 65

22   Miller, "Divided We Stand," 36.

23   Gilbert H. Gilchrist, *As Strong as Steel* (Sudbury, ON: USWA, 1999), 17.

24   Ibid., 50–51.

25   Jim Tester, "We're Still Here," in Steedman, Suschnigg, and Buse, *Hard Lessons*, 165.

26   Southern, *The Sudbury Incident*, 54–55.

27   Stephenson, "To Strike—or Not to Strike?" 33.

28   Ibid., 55; and Brasch, *A Miner's Chronicle*, 48.

29   Miller, *Divided We Stand*, 46.

30   Ibid., 49.

31   Alexandre Boudreau to Claude Jodoin, June 5, 1959. Jim Tester Papers, PO59 File 25, Laurentian University Extension Dept., Workers Educational Association Canadian Labour Congress 50/67, Laurentian University Archives Centre, Sudbury, ON.

32   Lang, "A Lion in a Den of Daniels," 222

33   Miller, "Divided We Stand," 54.

34   Cited by David Lewis Stein in Stephenson, "To Strike—or Not to Strike?" 80.

35   Ibid., 67.

36   Lang, "A Lion in a Den of Daniels," 255.

37   Gilchrist, *As Strong as Steel*, 22.

38   United Steelworkers of America, "Trade Unionism in the Nickel Industry: The Sudbury Story," *Information* 12, no. 3 (1964): 5.

39   "Mine-Mill Declared a Communist Organization," *Sudbury Star*, February 13, 1962, 12.

40   "Mine Mill to Build New Union Hall," *Sudbury Star*, November 4, 2009, A3. In 2009, Mine Mill decided to construct a new union hall on the 34 acres of land that it owned at Richard Lake. In the 1950s the land was held in trust for union members, but in 1984 it was purchased outright.

41   Derived from Brasch, *A Miner's Chronicle*, 54.

42   Sheila McLeod Arnopoulos, *Voices from French Ontario* (Montreal, QC: McGill-Queen's University Press, 1982), 99.

43 Donald Dennie makes the same point in "French Ontario: Two Realities," in *Mining Town Crisis; Globalization, Labour and Resistance in Sudbury*, ed. David Leadbeater (Halifax, NS: Fernwood, 2008), 200–201.

44 Ibid., 105–21.

45 Prior to the 1960s, French-speaking members referred to themselves as French Canadians. Since then, this term has been replaced by the term Franco-Ontarians. See Dennie, "French Ontario: Two Realities," 196.

46 Gagnon's achievements have been recorded in Adelle Larmour's, *Until the End: Memoirs of Sinter-plant Activist Jean L. Gagnon* (North Bay, ON: Wynterblue Publishing Canada, 2010).

47 Elie Martel, "The Name of the Game Is Power: Labour's Struggle for Health and Safety Legislation," in Steedman, Suschnigg, and Buse, *Hard Lessons*, 195–209.

48 Ibid., 204.

49 James M. Ham, Peter Riggin, and John Beaudry, *Report of the Royal Commission on the Health and Safety of Workers in Mines* (Toronto, ON: Ministry of the Attorney-General, 1976).

50 Val Ross, "The Arrogance of Inco," *Canadian Business*, May 1979, 44–55 and 116–42; and Henry Radecki, *The 1978–79 Strike at Inco: The Effects on Families*, Strike Impact Study (SIS) Analysis Report (Sudbury, ON: Laurentian University, 1979).

51 "Value of a Strike," *Sudbury Star*, July 14, 2010, A10.

52 Brasch, *A Miner's Chronicle*, 112.

53 Peter Suschnigg, "Talking Peace but Preparing for War," in *Mine Mill Fights Back: Mine Mill/CAW Local 598 Strike 2000–2001 Sudbury*, ed. Kaili Beck et al. (Sudbury, ON: Mine Mill/AW Local 598, 2005), 75–76.

54 See, for example, "Labour Council's Legacy Runs Deep in Sudbury," *Sudbury Star*, May 19, 2007, A6.

55 Suschnigg, "Talking Peace but Preparing for War," in Beck et al., *Mine Mill Fights Back*, 88.

56 "A Company with Global Ambitions," *Sudbury Star*, September 19, 2009, A6.

57 "Calls Erupt for Clement's Resignation," *Sudbury Star*, July 21, 2009, A1 and A3.

58 "Nobody Voted for That Contract," *Sudbury Star*, July 10, 2010, 6A.

59 "Steel to City Businesses: You're Either on Our Side or Vale's," *Sudbury Star*, September 21, 2009, A1.

60 "Labour Minister Urges Deadlocked Parties to Continue Talks, *Northern Life*, July 2, 2010, 1.

61 "A New Mayor, A New Era: Old Rules," *Northern Life*, October 28, 2010, 12.

62 "A Brand New Home," *Sudbury Star*, January 28, 2012, A1 and A6.

63 "A Helping Hand," *Sudbury Star*, November 28, 2012, A1.

NOTES TO CHAPTER 14

1 Nicola Ross, *Healing the Landscape: Celebrating Sudbury's Reclamation Story* (Sudbury, ON: Vegetation Enhancement Technical Advisory Committee, 2001), 8.

2 Keith Winterhalder, "Environmental Degradation and Rehabilitation in the Sudbury Area," *Laurentian University Review* 16, no. 2 (1984): 17.

3   Claude Laroche, Glen Sirois, and W.D. McIlveen, *Early Roasting and Smelting Operations in the Sudbury Area—An Historical Outline* (Sudbury, ON: Ontario Ministry of the Environment, Experience '79 Program, 1979), n.p.

4   Sudbury Area Risk Assessment, *Summary of Volume III: Ecological Risk Assessment* (Sudbury, ON: SARA, March 2009), 14.

5   For a description of the early roasting process, see Royal Ontario Nickel Commission, *Report of the Ontario Royal Ontario Nickel Commission* (Toronto, ON: King's Printer, 1917), 430–32.

6   William E. Lautenbach, *Land Reclamation Program 1978–1984* (Regional Municipality of Sudbury, ON: Vegetation Enhancement Technical Advisory Committee, 1985), 3.

7   Laroche, Sirois, and McIlveen, *Early Roasting and Smelting Operations in the Sudbury Area*, n.p.

8   For a review of the sulphur fume problem in the Sudbury area, see Matthew Bray, "The Province of Ontario and the Problem of Sulphur Fumes Emissions in the Sudbury District: An Historical Perspective," *Laurentian University Review* 16, no. 2 (1984): 81–90.

9   Royal Commission on the Mineral Resources of Ontario and Measures for their Development, *Report* (Toronto, ON: Warwick & Sons, 1890), 379.

10   A.E. Barlow, "Report of the Nickel and Copper Deposits of the Sudbury Mining District," *Geological Survey of Canada*, 14, Part H (1904), 191.

11   Cited in Laroche, Sirois and McIlveen, *Early Roasting and Smelting operations in the Sudbury Area*, n.p.

12   T.H. Peters, "Rehabilitation of Mine Tailings: A Case of Complete Ecosystem Reconstruction and Revegetation of Industrially Stressed Lands in the Sudbury Area, Ontario, Canada," in *Effects of Pollutants at the Ecosystem Level*, ed. P.J. Sheehan et al. (New York: John Wiley & Sons, 1984), 408.

13   A.P. Gagnebin, *Shaping the Natural Environment* (address, Canadian Manufacturers' Association, June 2, 1969), 13.

14   T.H. Peters, *The Use of Vegetation to Stabilize Mine Tailings at Copper Cliff* (Copper Cliff, ON: International Nickel, 1969), 7.

15   Winterhalder, "Environmental Degradation and Rehabilitation in the Sudbury Area," 32.

16   T.H. Peters, "Revegetation of the Copper Cliff Tailings Area," in *Restoration and Recovery of an Industrial Region*, ed. John M. Gunn (New York: Springer-Verlag, 1995), 128.

17   B. Michelutti and M. Weiseman, "Engineered Wetlands as a Tailings Rehabilitation Strategy," in Gunn, *Restoration and Recovery of an Industrial Region*, 135–41.

18   M.J. Puro et al., "Inco's Copper Cliff Tailings Area," in *Proceedings of Sudbury '95—Mining and the Environment*, vol. 1, ed. Thomas P. Hynes and Marcia C. Blanchette (Ottawa, ON: CANMET, 1995).

19   "Slag Makes Sudbury Unique," *Sudbury Star*, July 14, 2002, A1.

20   "Inside-Out City," *Sudbury Star*, July 30, 2011, A6.

21   "Black Slag Pile Will Soon Be Green," *Northern Life*, August 7, 2007, 7.

22   R.H. Murray and W.R. Haddow, *First Report of the Sub-Committee on the Investigation of Sulphur Smoke Conditions and Alleged Forest Damage in the Sudbury Region*, unpublished

report (Toronto: Ontario Department of Mines and Department of Lands and Forests, 1945).

23 "An Alternatives Interview with Elie W. Martel," *Alternatives* 2, no. 3 (1979): 10. See also B.R. Dreisinger, *Sulphur Dioxide Levels in the Sudbury Area and the Effects of the Gas on Vegetation* (Sudbury, ON: Ontario Department of Mines Yearly Reports, 1952–71). Later reports were issued by the Ministry of the Environment created in 1970.

24 J.P.G. de Lestard, *A History of the Sudbury Forest District* (Toronto: Ontario Department of Lands and Forests, 1967), 7.

25 Lautenbach, *Land Reclamation Program 1978–1984*, 7.

26 "An Alternatives Interview with Elie W. Martel," 10.

27 Potvin Air Management Consulting, *Air Quality Trends: City of Greater Sudbury, Ontario 1953–2002* (Sudbury, ON: Potvin Air Management Consulting, 2004), 28.

28 Sudbury Area Risk Assessment, *Background, Study Organization and 2001 Soils Survey*, Vol. 1 (Sudbury, ON: SARA), 2–51.

29 Raymond R. Potvin, and John J. Negusanti, "Declining Industrial Emissions, Improving Air Quality, and Reduced Damage to Vegetation," in Gunn, *Restoration and Recovery of an Industrial Region*, 54–56.

30 Potvin Air Management Consulting, *Air Quality Trends*, 32–33.

31 Ibid., 21.

32 "Vale Pumps $2B into Environmental Project," *Northern Life.ca*, accessed November 25, 2011, http://www.northernlife.ca/news/localNews/2011/11/25-Vale-AER-project.

33 "It's the Right Thing to Do," *Sudbury Mining Week 2012* (Sudbury, ON: Northern Life, 2012), 9.

34 Lautenbach, *Land Reclamation Program 1978–1984*, 44–45.

35 Vegetation Enhancement Technical Advisory Committee, *Annual Report 2010 Regreening Program* (Sudbury, ON: VETAC, 2011), 1. Other sources suggest that more than 12 million trees have been planted. See City of Greater Sudbury, *Living Landscape: A Biodiversity Action Plan for Greater Sudbury* (Sudbury, ON: City of Greater Sudbury, December 23, 2009), 9.

36 "Boosting Biodiversity in Greater Sudbury," *Green Living*, Summer 2011, 23.

37 "Re-greening and Beautification," *Greater Sudbury.ca*, accessed May 26, 2012, http://www.greatersudbury.ca/cms/index.cfm?app=div_earthcare&lang=en&currID=6855.

38 Vegetation Enhancement Technical Advisory Committee (VETAC), *Annual Report 2010 Regreening Program* (Sudbury, ON: VETAC, 2011), 1.

39 Ibid., 26.

40 For a review of Sudbury's water history, see D.A.B. Pearson, J.M. Gunn, and W. Keller, "The Past, Present and Future of Sudbury's Lakes," in *The Physical Environment of the City of Greater Sudbury, Ontario Geological Survey*, Special Vol. 6, ed. D.H. Rousell and K.J. Jansons (Sudbury, ON: Ontario Geological Survey, 2002),195–215. Much of this section has been derived from this work.

41 Ibid., 210–11.

42 "Fishing Hole Shows Promise," *Sudbury Star*, April 21, 2012, A1 and A6.

43   "Cooperative Freshwater Ecology Unit," *Laurentian University,* accessed August 30, 2012, http://www3.laurentian.ca/livingwithlakes/about/cooperative-freshwater-ecology-unit.

44   "A Building to Call Their Own," *Sudbury Star,* accessed July 20, 2011, http://www .thesudburystar.com/ArticleDisplay,aspx?e=2865797.

45   "Ribbon Cut for Vale Living with Lakes Centre," *Sudbury Star*, August 26, 2011, A1 and A4.

46   "Living with Lakes Recognized," *Sudbury Star,* May 8, 2012, A3.

47   Ibid.

48   Ontario Ministry of the Environment, *Metals in Soil and Vegetation in the Sudbury Area: Survey 2000 and Additional Historic Data* (Toronto: Ontario Ministry of the Environment, September 2001), iii.

49   Ontario Ministry of the Environment, *City of Greater Sudbury 2001 Urban Soil Survey* (Toronto: Ontario Ministry of the Environment, July 2004).

50   Sudbury Area Risk Assessment, vol. 1, *Background, Study Organization and 2001 Soils Study;* vol. 2, *Human Health Risk Assessment;* and vol. 3, *Ecological Risk Assessment* (Sudbury, ON: SARA, 2008–09).

51   Sudbury Area Risk Assessment, *Summary of Volume II: Human Health Risk Assessment* (Sudbury, ON: SARA, May 2008), 3.

52   Sudbury Area Risk Assessment, *Summary of Volume III: Ecological Risk Assessment* (Sudbury, ON: SARA, March 2009), 3.

53   Kapil Khatter, *Sudbury Human Health Risk Assessment Briefing* (Toronto, ON: Environmental Defence, September 22, 2008).

54   Nickel District Conservation Authority, *A Tour through 25 Years of Watershed Management* (Sudbury, ON: Nickel District Conservation Authority, 1982), 7.

55   For a historical assessment of flood damage in the Sudbury area, refer to Nickel District Conservation Authority, *Watershed Inventory* (Sudbury, ON: Nickel District Conservation Authority, 1980), 4–6 and 4–24.

56   "Let's Be Friends," *Sudbury Star*, November 12, 2011, C1.

57   For a review of the history of the Nickel District Conservation Authority, see "Caring Beyond Tomorrow," *Sudbury Star*, December 12, 2007, A6–A7.

58   "Junction Creek Key to City," *Sudbury Star*, June 25, 2011, A6.

59   O.W. Saarinen, and W.A. Tanos, "The Physical Environment of the Sudbury Area and Its Influence on Urban Development," in Rousell and Jansons, *The Physical Environment of the City of Greater Sudbury*, 14–15.

60   The film, titled *Junction Creek Clearwater Revival*, premiered at the Cinéfest Sudbury International Film Festival in 2009.

61   "From Creosote to Critters," *Green Living*, 8.

62   "Junction Creek Key to City," A1 and A6–A7.

63   This phrase was taken from Shawn Van Sluys, Kenneth Hayes, and Jocelyn Laurence, eds., *Sudbury: Life in a Northern Town* (Sudbury, ON: Laurentian Architecture, 2011), 67.

64   Saarinen and Tanos, "The Physical Environment of the Sudbury Area and Its Influence on Urban Development," in Rousell and Jansons, *The Physical Environment of the City of Greater Sudbury*, 13.

NOTES TO CHAPTER 15

1   "Inside-Out City," *Sudbury Star*, July 30, 2011, A1 and A6.

2   Kenneth Hayes, "Be Not Afraid of Greatness or Sudbury: A Cosmic Accident," in *Sudbury: Life in a Northern Town*, edited by Shawn Van Stuys, Kennth Hayes, and Jocelyn Laurence. (Sudbury, ON: Musagetes and Laurentian Architecture, 2011), 17.

3   "Changes Would Cause More Urban Sprawl," *Sudbury Star*, February 6, 2012, A10.

4   "Let's Hear a Viable Plan for Sudbury," *Sudbury Star*, October 11, 2010, A11.

5   See Floyd Laughren, *Constellation City: Building a Community of Communities in Greater Sudbury*, Report of the Greater Sudbury Community Solutions Team (Sudbury, ON: City of Greater Sudbury, 2007), 12.

6   Advisory Panel on Municipal Mining Revenues, *Report of the Advisory Panel on Municipal Mining Revenues* (Sudbury, ON: City of Greater Sudbury, 2008), 29.

7   "Here's What's Wrong with Municipal Mining Revenue in Ontario," *Republic of Mining*, last modified June 4, 2008, http://www.republicofmining.com/2008/06/04.

8   "Groups Want Cut of Mining Tax," *Sudbury Star*, July 6, 2011, A3.

9   Emphasis mine. "Northern Plans Fail to Hit the Mark," *Sudbury Star*, July 21, 2011, A10; and "Growth Plan Short on Details," TheSudburyStar.com, accessed July 21, 2011, http://www.thesudburystar.com/ArticleDisplay.aspx?archive=true&e=225.

10 "Agreement Awaits Approval," *Sudbury Star*, January 23, 2012, A3.

11 This historical theme is explored by Maurice Yeates in his book, *Main Street: Windsor to Quebec City* (Toronto, ON: Macmillan Company of Canada, 1975) and D. Michael Ray, "The Spatial Structure of Economic and Cultural Differences," *Papers in Regional Science* 23, no. 1 (1969): 7–24.

12 Gilbert Gilchrist, *As Strong as Steel* (Sudbury, ON: USWA, 1999), 54. See also Hans Brasch, *A Miner's Chronicle: Inco Ltd. and the Unions 1944–1997* (Sudbury, ON: Hans Brasch, 1997), 138–52; and "Sudbury Remembers," *Sudbury Star*, June 18, 2009, B9.

13 "The Way Ahead Lies North," *Sudbury Star*, November 14, 2009, A1 and A6.

14 Cited in "Laurentian Is Leaner and Fitter: Daniel," *Northern Ontario Business*, February 1989, 47.

15 This is the theme presented in "Vale Strike Marks Threshold of Brave New World for Company, Union: Horn," *Sudbury Star*, December 21, 2009, A1 and A3.

16 "Tragedy at Stobie Mine a Reminder of Our Spirit," *Sudbury Star*, June 10, 2011, A10.

17 Dick DeStefano (Executive Director, Sudbury Area Mining Supply and Services Association), interviewed by author, August 4, 2010.

18 "Boom Goes Sudbury," *Sudbury Star*, May 31, 2012, A1 and A4.

19 Greater Sudbury Development Corporation, *Coming of Age in the 21st Century: An Economic Development Strategic Plan for Greater Sudbury 2015*, and *Coming of Age in the 21st Century: Digging Deeper* (Sudbury, ON: Greater Sudbury Development Corporation, 2003 and 2009).

20 Ibid.

21 Northern Ontario Local Training and Adjustment Boards, *An Aging Population in Northern Ontario* (Thunder Bay, ON: Northern Ontario Local Training and Adjustment Boards, October 31, 2002).

22   Ontario Ministry of Finance, *Ontario Population Projections Update 2010–2036* (Toronto: Ontario Ministry of Finance, 2011), 14.

23   Statistics Canada, *2011 Census* (Ottawa, ON: Statistics Canada, 2012).

24   "The Buck Stops Here," *Sudbury Star*, November 25, 2011, A3.

25   "Sudbury's Business, Retail Community Is Thriving," *Sudbury Star*, November 25, 2011, A2.

26   This union perspective is voiced by David Leadbeater in *Mining Town Crisis: Globalization, Labour and Resistance in Sudbury* (Halifax, NS: Fernwood, 2008), 25–26.

27   Several of these themes can be found in Hayes, "Be Not Afraid of Greatness or Sudbury," in Van Stuys et al., *Sudbury: Life in a Northern Town*, 16–25.

28   Bill Lautenbach (General Manager of Growth and Development, City of Greater Sudbury) interviewed by author, August 26, 2011.

29   "Sign of the Times Makes for Boring Core," *Sudbury Star*, February 6, 2012, A1 and A4.

30   The demographic data in this section has been obtained from Statistics Canada, *Annual Demographic Estimates: Subprovincial Areas: 2005 to 2010*, Catalogue no. 91-214-X (Ottawa, ON: Statistics Canada, 2011).

31   See, for example, "Metal Market Boom Coming: City Poised to Reap Rewards," *Northern Life*, September 1, 2011, 1.

32   "Think Tank Bullish on Sudbury in 2012," *Sudbury Star*, January 14, 2012, A3; and "Sudbury Luckiest City," *Sudbury Star*, January 23, 2012, A1.

33   BMO Financial Group, *Sudbury: Mining for Economic Growth* (Toronto, ON: BMO Capital Markets Economic Research, May 31, 2012), 1–2; and "Boom Goes Sudbury," A1.

34   "Sudbury Economy on the Move," *Northern Life*, May 15, 2012, 1 and 5.

35   This position has been put forward by the Central 1 Credit Union's Chief economist. See "Sudbury in Good Position," *Sudbury Star*, November 28, 2011, A1.

36   This is a point forcefully made in "Art and the City," *Sudbury Star*, August 30, 2008, A1 and A7.

37   "Public Art Spaces Help Define Cities," *Sudbury Star*, August 8, 2008, B7.

38   "'Alarming' Findings in Road Audit." *Northern Life*, August 16, 2012, 1 and 5.

39   Leadbeater, *Mining Town Crisis*, 7 and 10.

40   Narasim Katary to O.W. Saarinen, July 20, 2010. Private collection.

41   Ibid.

42   "Not Happy Campers," *Sudbury Star*, June 8, 2012, A1 and A4. The polls were conducted by Forum Research.

43   "Sudbury Needs to Be Aggressive," *Sudbury Star*, June 9, 2012, A1 and A4.

44   "Sudbury in No-Win Situation," *Northern Life*, June 12, 2012, 1 and 4.

Bibliography

ARCHIVES

Alexandre Boudreau to Claude Jodoin, June 5, 1959. Jim Tester Papers. Laurentian University Extension Dept., Workers Educational Association, Sudbury, ON.

Boyle, Patrick. *Saint Stanislaus Parish, Copper Cliff, The Early Years: 1886–1914.* Jesuit Archives. Toronto, ON, 1998.

Canadian Labour Congress 50/67. Sudbury, ON: Laurentian University Archives Centre, 1959.

Ontario Ministry of Natural Resources. *Survey Records.* OSG and Information Resource Management Branch.

Post Reports for Whitefish Lake and Sudbury. Reel No. 1M1259, B.364/e/1, 1888 and B.364/e/2, 1897; and Reel No. 1M1260, B.355/e/1, 1888 and B.355/e/6, 1891. Hudson Bay Company Archives, Provincial Archives of Manitoba, Winnipeg.

Smith, D.W. Map of the Province of Upper Canada, published in 1800. Cartographic Records, Collection A-9, Repro. no. 492. Archives of Ontario, Toronto.

PUBLIC DOCUMENTS

Municipal

Advisory Panel on Municipal Mining Revenues. *Report of the Advisory Panel on Municipal Mining Revenues.* Sudbury, ON: City of Greater Sudbury, 2008.

Association of Municipalities of Ontario and City of Toronto. *Provincial–Municipal Fiscal and Service Delivery Review.* Toronto: Association of Municipalities of Ontario and City of Toronto, Fall 2008.

Bland, John, and Harold Spence-Sales. *A Report on the City of Sudbury and Its Extensions.* Sudbury, ON: City of Sudbury, 1950.

"Boosting Biodiversity in Greater Sudbury." *Green Living*, Summer 2011, 23.

Capreol Public Library Board. *Capreol: The First 75 Years.* Capreol, ON: Capreol Library Board, n.d.

City of Greater Sudbury. *The City of Greater Sudbury Official Plan.* City of Greater Sudbury, ON, 2008.

City of Greater Sudbury. *Living Landscape: A Biodiversity Action Plan for Greater Sudbury.* City of Greater Sudbury, ON, December 23, 2009.

City of Greater Sudbury. *Official Plan Natural Heritage Background Study.* City of Greater Sudbury, ON, February 2005.

City of Sudbury. *1964: The Year of Dilemma.* Sudbury, ON: City of Sudbury, 1964.

Greater Sudbury Development Corporation. *Coming of Age in the 21st Century: An Economic Development Strategic Plan for Greater Sudbury 2015.* Sudbury, ON: Greater Sudbury Development Corporation, 2003.

———. *Coming of Age in the 21st Century: Digging Deeper*. Sudbury, ON: Greater Sudbury Development Corporation, 2009.

Greater Sudbury Public Library. *Biographical Materials for Max Silverman (August 25, 1899–October 5, 1966)*.

Laughren, Floyd. *Constellation City: Building a Community of Communities in Greater Sudbury*. Report of the Greater Sudbury Community Solutions Team. Sudbury, ON: City of Greater Sudbury, 2007.

Lautenbach, William E. *Land Reclamation Program 1978–1984*. Regional Municipality of Sudbury, ON: Vegetation Enhancement Technical Advisory Committee, 1985.

Nickel District Conservation Authority. *A Tour through 25 Years of Watershed Management*. Sudbury, ON: Nickel District Conservation Authority, 1982.

———. *Watershed Inventory*. Sudbury, ON: Nickel District Conservation Authority, 1980.

Regional Municipality of Sudbury. *Toward Economic Diversification: The Key to Long-Term Job Creation in the Regional Municipality of Sudbury*. A presentation to the Government of Ontario, Toronto, November 26, 1982.

———. *Sudbury 20/20 Focus on the Future*. A *Northern Life* supplement, April 14, 1999, 3.

———. *Sudbury Tomorrow: The New Way*. Sudbury, ON: Regional Municipality of Sudbury, 2000.

Sudbury Public Library. *Burwash: A History 1911–1981*. Sudbury, ON: M.C. Shantz Collection, Sudbury Public Library, 1982.

Sudbury Regional Development Corporation. *The Next Ten Years: A Conference Discussion Paper*. Sudbury, ON: Sudbury Regional Development Corporation, November 1992.

———. *Patterns for Prosperity: Final Report of the Next Ten Years Project*. Sudbury, ON: Sudbury Regional Development Corporation, April 1993.

Thomas, Hugh, J. *Report to the Minister of Municipal Affairs and Housing on Local Government Reform for Sudbury*. Sudbury, ON: Office of the Special Advisor on Local Government Reform Sudbury Region, 1999.

Trott, Robert P. *The Story of Onaping Falls*. Onaping Falls, ON: Town of Onaping Falls, ca. 1982.

Vegetation Enhancement Technical Advisory Committee (VETAC). *Annual Report 2010 Regreening Program*. Sudbury, ON: VETAC, 2011.

Ontario

Barnett, P.J. "Quaternary Geology of Ontario." In P.C.Thurston, H.R. Williams, R.H. Sutcliffe, and G.M.Stott, eds. *Geology of Ontario*. Special vol. 4, pt. 2 of *Ontario Geological Survey*. 1015–17.

Barnett, P.J., and A.F. Bajc. "Quaternary Geology." In *The Physical Environment of the City of Greater Sudbury*. Special vol. 6 of *Ontario Geological Survey*. D.H. Rousell and K.J. Jansons, eds. Sudbury, ON: Ontario Geological Survey, 2002, 58–59.

Boissonneau, A.N. *Algoma, Sudbury, Timiskaming and Nipissing Surficial Geology*. Toronto: Ontario Department of Lands and Forests, 1965. Map No. S465.

Burwash, Edward M. "Geology of the Nipissing-Algoma Line." *Sixth Report of the Bureau of Mines 1896*. Toronto, ON: Warwick Bros. & Rutter, 1897.

Burwasser, G.J. *Quaternary Geology of the Sudbury Basin Area: District of Sudbury*. Report 181 of *Ontario Geological Survey*. Toronto, ON: Ministry of Natural Resources, 1979.

Card, K.D. *Geology of the Sudbury-Manitoulin Area, Districts of Sudbury and Manitoulin.* Report 166 of *Ontario Geological Survey.* Toronto, ON: Ministry of Natural Resources, 1978.

Card, K.D., V.K. Gupta, P.H. McGrath, and F.S. Grant. "The Sudbury Structure: Its Regional Geological and Geophysical Setting." In *The Geology and Ore Deposits of the Sudbury Structure.* Special vol. 1 of *Ontario Geological Survey.* Edited by E.G. Pye, A.J. Naldrett, and P.E. Giblin, 25–43. Toronto: Ontario Ministry of Natural Resources, 1984.

Chapman, L.J., and M.K. Thomas. *The Climate of Northern Ontario.* Climatological Studies No. 6. Toronto, ON: Department of Transport, Meteorological Branch, 1968.

Coleman, A.P. "The Sudbury Nickel Region." In *Report of the Ontario Bureau of Mines.* Vol. 14, pt. 3, 1–183, 1905.

Commissioner of Crown Lands of the Province of Ontario. *Annual Report for the Year 1883.* Toronto, ON: "Grip" Printing and Publishing, 1884.

———. *Annual Report for the Year 1901.* Toronto: L.K. Cameron, 1904.

Community Planning Programs Division, Ontario Ministry of Municipal affairs and Housing. *Project Group on Urban Economic Development—Case Study Report: Sudbury, Ontario, Canada.* Paris: OECD, 1985.

Conway, Thor. *Archaeology in Northeastern Ontario: Searching for our Past.* Toronto: Ontario Ministry of Culture and Recreation, Historical Planning and Research Branch, 1981.

Cowan, E.J., and W.M. Schwerdtner. "Fold Origin of the Sudbury Basin." In Lightfoot and Naldrett, *Proceedings of the Sudbury Noril'sk Symposium.* Special vol. 5 of *Ontario Geological Survey.* Toronto: Ontario Geological Survey, 1994: 45–56.

Crown Lands Department. *Reduction of Mr. Salter's Plan of the North Shore of Lake Huron.* Toronto: Crown Lands Department, 1857.

de Lestard, J.P.G. *A History of the Sudbury Forest District.* Toronto: Ontario Department of Lands and Forests, 1967.

Dreisinger, B.R. *Sulphur Dioxide Levels in the Sudbury Area and the Effects of the Gas on Vegetation.* Sudbury, ON: Ontario Department of Mines. Yearly Reports, 1952–1971.

Dressler, Burkhard, O. "The Effects of the Sudbury Event and the Intrusion of the Sudbury Igneous Complex on the Footwall Rocks of the Sudbury Structure." In *The Geology and Ore Deposits of the Sudbury Structure.* Special vol. 1 of *Ontario Geological Survey,* 97–136. Toronto: Ontario Ministry of Natural Resources, 1984.

Giblin, P.E. "History of Exploration and Development of Geological Studies and Development of Geological Concepts." In *The Geology and Ore Deposits of the Sudbury Structure.* Special vol. 1 of *Ontario Geological Survey,* 3–23.

———. "Glossary of Sudbury Geology Terms." In *The Geology and Ore Deposits of the Sudbury Structure.* Special vol. 1 of *Ontario Geological Survey,* 571–74.

Golightly, J.P. "The Sudbury Igneous Complex as an Impact Melt: Evolution and Ore Genesis." In Lightfoot and Naldrett, *Proceedings of the Sudbury-Noril'sk Symposium.* Special vol. 5 of *Ontario Geological Survey,* 105–17.

Grieve, R.A.F. "An Impact Model of the Sudbury Structure." In Lightfoot and Naldrett, *Proceedings of the Sudbury-Noril'sk Symposium.* Special vol. 5 of *Ontario Geological Survey,* 119–132.

Ham, James M., Peter Riggin, and John Beaudry. *Report of the Royal Commission on the Health and Safety of Workers in Mines.* Toronto, ON: Ministry of the Attorney-General, 1976.

Kennedy, J.A. *Sudbury Area Study*. Toronto, ON: Department of Municipal Affairs, 1970.

Laroche, Claude, Glen Sirois, and W.D. McIlveen. *Early Roasting and Smelting Operations in the Sudbury Area—An Historical Outline*. Sudbury, ON: Ontario Ministry of the Environment, Experience '79 Program, 1979.

Lambert, Richard S., and Paul Pross. *Renewing Nature's Wealth: A Centennial History of the Public Management of Lands, Forests & Wildlife in Ontario 1763–1967*. Toronto: Department of Lands and Forests, 1967.

Mineral Resources Commission. *Report of the Royal Commission on the Mineral Resources of Ontario and Measures for Their Development*. Toronto: Legislative Assembly, 1890.

Morris, J.L. *Indians of Ontario*. Toronto: Ontario Department of Lands and Forests, 1943.

Murray, R.H., and W.R. Haddow. *First Report of the Sub-Committee on the Investigation of Sulphur Smoke Conditions and Alleged Forest Damage in the Sudbury Region*. Unpublished report. Toronto: Ontario Department of Mines and Department of Lands and Forests, 1945.

Northern Ontario Local Training and Adjustment Boards. *An Aging Population in Northern Ontario*. Thunder Bay, ON: Northern Ontario Local Training and Adjustment Boards, October 31, 2002.

Ontario Bureau of Mines. *Annual Report for 1913*. Toronto: Legislative Assembly of Ontario, 1914.

Ontario Department of Municipal Affairs. *Sudbury: Local Government Reform Proposals*. Toronto: Ontario Department of Municipal Affairs, 1971.

Ontario. Legislative Assembly. *Bill 164: An Act to Establish the Regional Municipality of Sudbury*. 29th Legislature, 2nd session (June 29, 1972). Toronto, ON: Queen's Printer, 1972.

Ontario Ministry of Finance. *Ontario Population Projections Update 2010–2036*. Toronto: Ontario Ministry of Finance, 2011.

Ontario Ministry of the Environment. *Metals in Soil and Vegetation in the Sudbury Area: Survey 2000 and Additional Historic Data*. Toronto: Ontario Ministry of the Environment, September 2001.

———. *City of Greater Sudbury 2001 Urban Soil Survey*. Toronto: Ontario Ministry of the Environment, July 2004.

Ontario Municipal Board. *Order P.F.A.—3258*. February 11, 1931.

———. *Order P.C.F.—4756*. January 16, 1951.

———. *Order P.F.M.—787*. July 18, 1957.

———. *Order P.F.M.—5847*. August 12, 1957.

———. Order *P.F.M.—5143-6*. November 12, 1959.

Pearson, D.A.B., J.M. Gunn, and W. Keller. "The Past, Present and Future of Sudbury's Lakes." In Rousell and Jansons, *The Physical Environment of the City of Greater Sudbury*. Special vol. 6 of *Ontario Geological Survey*, 195–215.

Peredery, W.V., and G.G. Morrison. "Discussion of the Origin of the Sudbury Structure." In *The Geology and Ore Deposits of the Sudbury Structure*. Special vol. 1 of *Ontario Geological Survey*, 491–511.

Province of Ontario. *Annual Reports of the Inspector of Prisons and Reformatories (1917–1946)*. Toronto: Ontario Reformatories and Prisons Branch, 1918–1947.

Province of Ontario Committee on Council. *Order-in-Council Appointing Commission*. Toronto, ON: Lieutenant-Governor, February 13, 1969.

Pye, E.G. "The First Hundred Years: A Brief History of the Ontario Geological Survey." In

P.C.Thurston, H.R. Williams, R.H. Sutcliffe, and G.M. Stott, eds. *Geology of Ontario*. Special vol. 4, pt. 1 of *Ontario Geological Survey*, 27–57.

Pye E.G., A.J. Naldrett, and P.E. Giblin, eds. *The Geology and Ore Deposits of the Sudbury Structure*. Special vol. 1 of *Ontario Geological Survey*. Toronto: Ontario Ministry of Natural Resources, 1984.

Robertson, J.A., and Card, K.D. *Geology and Scenery: North Shore of Lake Huron Region*, Geological Guidebook No. 4. Toronto: Ontario Division of Mines, Ministry of Natural Resources, 1972.

Rousell, D.H., and K.J. Jansons, eds. *The Physical Environment of the City of Greater Sudbury*. Special vol. 6 of *Ontario Geological Survey*. Sudbury, ON: Ontario Geological Survey, 2002.

Rousell, D.H., W. Meyer, and S.A. Prevec. "Bedrock Geology and Mineral Deposits." In Rousell and Jansons, *The Physical Environment of the City of Greater Sudbury*. Special vol. 6 of *Ontario Geological Survey*, 21–55.

Rousell, D.H., and K.D. Card. "Geologic Setting." In Rousell and Brown, *A Field Guide to the Geology of Sudbury, Ontario*, 1–200.

Rousell, D.H., and G. Heather Brown. *A Field Guide to the Geology of Sudbury, Ontario*. Open File Report 6243. Sudbury, ON: Ontario Geological Survey, 2009.

Royal Ontario Nickel Commission. *Report of the Royal Ontario Nickel Commission*. Toronto, ON: King's Printer, 1917.

Saarinen, O.W., and W.A. Tanos. "The Physical Environment of the Sudbury Area and Its Influence on Urban Development." In Rousell and Jansons, *The Physical Environment of the City of Greater Sudbury*. Special vol. 6 of *Ontario Geological Survey*, 3–18.

Sawchuk & Peach Associates. *Nickel Basin Planning Study*. Toronto, ON: Department of Municipal Affairs, 1967.

Statutes of Ontario, 1892, 55 Vic., c. 88.

Statutes of Ontario, 1901, 1 Edw. VII, c. 51, 250–52.

Statutes of Ontario, 1907, 7 Edw. VII, c. 25, 200–6.

Statutes of Ontario, 1930, c. 102.

Sudbury Land Registry Office. *First Registration Books*.

Sudbury Land Registry Office. *Directory of M-Plans*, M65, M69 and M71.

Thompson, J.E. "Geology of the Sudbury Basin." *Sixty-Fifth Annual Report of the Ontario Department of Mines*, vol. 65, pt. 3, 1956. 1–56.

Thorpe, T. *A Review of the Logging and Pulp and Paper Operations in the Sudbury District during the Years 1901–1950*, Sudbury, ON: Ontario Department of Lands and Forests, 1950.

Thurston, P.C, H.R. Williams, R.H. Sutcliffe, and G.M. Stott, eds. *Geology of Ontario*. Special vol. 4, pts. 1 and 2 of *Ontario Geological Survey*. Toronto: Ontario Geological Survey, 1991–92.

Woodrow, Maureen. *Challenges to Sustainability in Northern Ontario*. Toronto: Environmental Commission of Ontario, 2002.

Canada

Advisory Committee on Reconstruction. *Final Report of the Housing and Community Planning Subcommittee*. Ottawa, ON: King's Printer, 1946.

Agricultural Economics Research Council. *Potentials for Agricultural Development in the Sudbury Regional Municipality*. Ottawa, ON: Agricultural Economics Research Council, 1979.

Barlow, A.E. "Report on the Origin, Geological Relations and Composition of the Nickel and Copper Deposits in the Sudbury Mining District, Ontario, Canada." *Geological Survey of Canada*, Annual Report no. 873 (1904), 1–236.

Bell, Robert. "Report on the Sudbury Mining District." *Geological Survey of Canada*, Annual Report. n.s. 5, part 1, 5F–95F, 1891.

Canada. *Sessional Papers*. No. 48, Appendix 9 (1882), 35–36.

Canada. *Sessional Papers*. No. 27 (1883), 15.

The Canada Gazette 17 (April 17, 1858), 676–77.

Commissioner of Crown Lands of Canada. *Annual Report for the Year 1856*. Toronto, ON: Stewart Derbishire and George Desbarats, 1857.

———. *Annual Report for the Year 1866*. Toronto, ON: Hunter, Rose & Co., 1867.

———. *Annual Report for the Year 1872*. Toronto, ON: Hunter, Rose & Co., 1873.

Department of Indian Affairs and Northern Development. *Registered Indian Population by Sex and Residence 2000*. Ottawa, ON: Department of Indian Affairs and Northern Development, Information Management Branch, March 2001.

Energy, Mines and Resources, Canada. *Canada's Nonferrous Metals Industry: Nickel and Copper—A Special Report*. Ottawa, ON: Energy, Mines and Resources, Canada, 1984.

Folger, Capt. W.W., and Lieut. B.W. Buckingham. *Report to the Secretary of the United States Navy upon Nickel and Copper Deposits of Sudbury*. Ottawa, ON: n.p., 1898. Reprint.

Gee, G.W., and B.F. Findlay. *The Sudbury Tornado—August 20, 1970*, Technical Memoranda 764. Ottawa, ON: Atmospheric Environment Service, Department of the Environment, 1972, 1–35.

Geological Survey of Canada. *Report of Progress for the Year 1848–49*. Montreal, QC: Lovell and Gibson, 1849.

Julig, Patrick, J., ed. *The Sheguiandah Site: Archeological, Geological and Paleobotanical Studies at a Paleoindian Site on Manitoulin Island, Ontario*. Mercury Series. Archeological Survey of Canada, Paper 161. Ottawa, ON: Canadian Museum of Civilization, 2002.

———. "Archeological Conclusions from the Sheguiandah Site Research." In Julig, *The Sheguiandah Site*, 297–314.

Lindeman, E. *Moose Mountain Iron-Bearing District Ont*. Ottawa, ON: Department of Mines, 1914.

Murray, A. "Report for the Year 1856." In *Geological Survey of Canada, Report of Progress for the Years 1853–54–55–56*. Appendix No. 52. Toronto: John Lovell, 1857.

Natural Resources Canada. *Canada Minerals Yearbook*. Ottawa, ON: Natural Resources Canada, 1996–2011.

Nicholson, Norman L. *The Boundaries of Canada, Its Provinces and Territories*. Geographical Branch Memoir 2. Ottawa, ON: Department of Mines and Technical Surveys, 1964.

Parliament of Canada. *History of Nipissing, Sudbury, and Nickel Belt Federal Ridings since 1867*. Ottawa, ON: Parliament of Canada, 2011.

Phillips, David. *The Climates of Canada*. Ottawa, ON: Minister of Supply and Services Canada, 1990.

Royal Commission on Industrial Relations. *Report of Commission*. Ottawa, ON: Committee of the Privy Council, 1919.

Statistics Canada. *Annual Demographic Estimates Subprovincial Areas: 2005 to 2010*. Catalogue no. 91-214-X. Ottawa, ON: n.p., 2011.

———. *Censuses of Canada (1891–2011)*.

———. *Single and Multiple Ethnic Origins for the City of Greater Sudbury*. Catalogue no. 97-562-XCB2006006. Ottawa, ON: Statistics Canada, 2011.

———. *Annual Demographic Estimates: Subprovincial Areas: 2005–2010*.Catalogue no. 91-214-X. Ottawa, ON: Statistics Canada, 2011.

Statutes of Canada. 1852–53, 16 Vic., c. 176, 720–24.

Statutes of Canada. 1857, 20 Vic., c. 60, 283–303.

Thomson, Don W. *Men and Meridians,* Vol. 1. Ottawa, ON: Queen's Printer, 1966.

Wright, J.V. *Ontario Prehistory: An Eleven-Thousand-Year Archeological Outline*. Ottawa, ON: National Museum of Canada, National Museum of Man, 1972.

Other

Paley, William S., Honorable Chairman. *President's Materials Policy Commission*. Washington, DC: Government Printing Office, 1952.

INTERVIEWS

Julig, P. Professor and Anthropologist. Interview by author. July 26, 2000.

Michel, Arnold. Owner of A & J Home Hardware. Interview by author. September 10, 2003.

Erola, Judy. Former Member of Parliament and Director, International Nickel Company of Canada. Interview by author. August 10, 2008.

Bray, Matt. Professor and historian. Interview by author. October 16, 2010.

DeStefano, Dick. Executive Director of Sudbury Area Mining Supply and Services Association. Interviews by author. August 4, 2010, September 3, 2011, and January 23, 2012.

Lautenbach, Bill. General Manager of Growth and Development, City of Greater Sudbury. Interview by author. August 26, 2011.

NEWSPAPER ARTICLES

The Sudbury Star

"10 Dwellings May Be Erected in Kingsmount." February 26, 1930.

"A Helping Hand." November 28, 2012, A1.

"Adam Was a Happy-Go-Lucky Kid." May 7, 2009.

"Against All Odds: Northern Cancer Centre Flourishes." March 31, 1992.

"Agreement Awaits Approval." January 23, 2012.

"Annexation Adds 171 McKim Acres to Size of Sudbury." January 17, 1951.

"Anniversary of Church Recalls Early Struggle." June 20, 1923.

"Art and the City." August 30, 2008.

"As the Community Sees It." December 5, 1958.

"Baseball Found Fertile Ground in Sudbury's Early Years." May 31, 1983.

"Bitter Standoff Grinds Down Company Town." June 6, 2010.

"Black Smoke Hid the Sun." May 5, 2007.

"Boom Goes Sudbury." May 31, 2012.

"Boréal Is on to Promising Projects." March 28, 2008.

"Brains vs. Bins—Blast, Batter, Bury Bulky Building?" January 10, 1956.

"A Brand New Home." January 28, 2012.

"The Buck Stops Here." November 25, 2011.

"A Building to Call Their Own." accessed July 20, 2011, http://www.thesudburystar.com/ArticleDisplay,aspx?e=2865797.

"Business Establishments Give Superior Service to Residents." *Sudbury Daily Star*, August 5, 1950.

"Built in 1908, Has Featured the Famous." November 25, 1948.

"Caisse Populaire in Flour Mill No Ordinary Financial Institution." August 5, 1989.

"Call Centres Major Force in City." March 28, 2008.

"Calls Erupt for Clement's resignation." July 21, 2009.

"Caring beyond Tomorrow." December 12, 2007.

"Celebrates 50 Years." May 29, 2010.

"Changes Would Cause More Urban Sprawl." February 6, 2012.

"City Celebrates Aboriginal Culture." September 30, 2010.

"City's Retail Sector Booming." March 28, 2008.

"A City Suffers Great Loss." *Sudbury Daily Star*, January 13, 1945.

"Cliffs Picks Sudbury." May 8, 2012.

"Cohesive Leadership Is Key in Difficult Municipal Year." January 8, 1976.

"College du Sacré-Coeur." March 8, 1978.

"A Company with Global Ambitions." September 19, 2009.

"Controversial Union Figure, Weir Reid, 53, Dies Monday." August 3, 1971.

"Co-operation Sets Development Pace." August 5, 1950.

"CPR Releases Last Group of Lots for Sale." June 4, 1927.

"Cradle of Life." November 21, 2009.

"Creighton Reunion Being Planned." April 26, 1988.

"Davies' Estate Is Enormous." By Rick Bartolucci. December 13, 1997.

"Deeds 110 Acres Bordering Lake to Parks Board." February 17, 1926.

"District Due for Vigorous Expansion." June 1, 1927.

"Dollar Day Is Tag Day for St. Joseph's Hospital." May 16, 1923.

"Donovan Residents Take Great Pride in District." *Sudbury Daily Star*, June 22, 1950.

"Downtown Needs Rethought." January 23, 2012.

"Eddie Santi a Dedicated Man." December 13, 2008.

"End of an Era." June 14, 2003.

"Faith in Nickel District Founded Minnow Lake." June 13, 1938.

"Falconbridge Head, H.J. Fraser Dies." February 3, 1969.

"Farming Fame of Nickel District Spreads." December 30, 1955.

"First Nation Aims for Excellence." May 19, 2012.

"Fishing Hole Shows Promise." April 21, 2012.

"The Flour Mill: A Depressed Area That Flourished." August 5, 1989.

"Flour Mill Has Deep Francophone Roots." August 23, 2008.

"Forgotten Cemeteries." October 7, 2006.

"Former Convent Gets Reprieve." August 20, 2008.

"Francophone Businesses Flourished in Early Days." November 3, 2008.

"Full Production Planned for Lockerby Mine." January 30, 2012.

"Games Leave Legacy." July 25, 2011.

"Gatchells Did Live in Sudbury Subdivision." August 14, 1993.

"Gatchell Subdivision Shows Faith in Nickel District." June 13, 1928.

"Gatchell Then and Now." January 21, 1989.

"Goats Are Gone, Understanding of 2001 'Better.'" April 28, 1981.

"Good Services Provided." *Sudbury Daily Star*, June 23, 1950.

"Gordon Blasts Mother Inco." March 14, 1979.

"Gordon 'Left His Mark' on Sudbury." June 14, 2003.

"Government to Blame for Vale's Intrusion." June 3, 2010.

"Grey Nuns Built Hospital in 1895." May 31, 1983.

"Groups Want Cut of Mining Tax." July 6, 2011.

"Growth Plan Short on Details." Accessed July 21, 2011. http://www.thesudburystar.com/ArticleDisplay.aspx?archive=true&e=225.

"Happy Valley Residents Not Happy." June 3, 1970.

"Has Grown with Sudbury." June 23, 1923.

"Held Record as Mayor, Bill Beaton Dies at 59." April 2, 1956.

"Historic Flag Celebrates 35th Anniversary." September 28, 2010.

"Historical Highlights." February 7, 1993.

"Horsemen Blast Cash Grab." March 20, 2012.

"Hundreds Pay Final Tribute to Con. Beaton." April 6, 1956.

"Inside-Out City." July 30, 2011.

"James Gordon." By Sarah Lashbrook. February 26, 2011.

"Jesuits Brought First University to Area." May 31, 1983.

"Joint Ventures." Accessed February 5, 2011, http://www.thesudburystar.com/ArticleDisplay.aspx?archive=true&e=261.

"Junction Creek Key to City." June 25, 2011.

"Kingsmount Residents Win First Round over Building." May 30, 1938.

"Labour Council's Legacy Runs Deep in Sudbury." May 19, 2007.

"Legend Surrounds Motives of 'Count' Building Grotto." July 5, 1958.

"Let's Be Friends." November 12, 2011.

"Let's Hear a Viable Plan for Sudbury." October 11, 2010.

"Living with Lakes Recognized." May 8, 2012.

"Local YMCA Celebrating 75 Years." December 24, 2011.

"Lockerby Up and Running." May 1, 2012.

"Logging at Wahnapitae before the Century Dawned." By Gary Peck. January 7, 1978.

"Main Downtown Mall Gets Multi-Million-Dollar Facelift." April 26, 1988.

"Mall Led Retail Revolution." October 20, 2007.

"Many Fine Homes in Sudbury District." June 13, 1938.

"Many Interests Plan Erection of Residences." April 23, 1930.

"May Prohibit Apartments in Kingsmount." May 13, 1938.

"Mayor Gordon Strikes Back at Games Critics." October 19, 1977.

"Marguerite Grace Lougheed: 1929–2006." May 8, 2010.

"Merger Puts Focus Back on Mining." By Stan Sudol. November 11, 2005.

"Mine Mill and the Mounties." November 11, 1993.

"Mine-Mill Declared a Communist Organization." February 13, 1962.

"Mine Mill to Build New Union Hall." November 4, 2009.

"Mines Assessment Irks Mayor." March 12, 1970.

"Mines Less of a Tax Motherlode for Sudbury." October 15, 1988.

"Mining Industry Begins Slow Recovery." March 26, 2010.

"More Action Needed to Improve Lake." July 16, 2007.

"Mrs. W.H. Howey—Sudbury's First Historian." August 4, 1953.

"Nearly Nine Years since Secret Meeting of City Council Started the Whole Thing."
    November 16, 1959.

"New Church of the Epiphany Rising from the Ashes." April 26, 1988.

"New City Mayor to Seek Mandate." January 7, 1976.

"New Civic Thinking behind Plans for Donovan Heights." November 7, 1956.

"New Ukrainian Church Based on Christianity." December 19, 1928.

"Nickel Belt Was to Be Iron Belt but Sellwood Bubble Burst in 1919." *Sudbury Daily Star*,
    June 14, 1951.

"Nickel Park." April 28, 1923.

"Nobody Voted for That Contract." July 10, 2010.

"Northern Plans Fail to Hit the Mark." July 21, 2011.

"Not Happy Campers." June 8, 2012.

"The 'Old Count' Gave Sudbury Some Novel Names." By Joe Pundit. October 22, 1956.

"Once Upon a Time, Back in 1951 …" February 8, 1958.

"Optimism High as Plant Close to Completion." September 26, 1960.

"Ottawa Post for Faintuck: Was First Sudbury Planner." June 9, 1957.

"Our 100th Year." October 29, 2008.

"Pay Tribute to Late Publisher for His Good Works." November 18, 1952.

"Pick Polish-Born Planner for Sudbury Area Post." August 27, 1957.

"Pine Street Water Tower Doomed." July 16, 2011.

"Pioneer Woman, Mrs. W.H. Howey, Called by Death." November 11, 1936.

"Please Make It Rain." May 29, 2010.

"Polish Hall Celebrates 60th Birthday." May 28, 2012.

"Power Broker." July 19, 2003.

"Private Sector Urged to Play Leading Role." April 26, 1988.

"Public Art Spaces Help Define Cities." August 8, 2008.

"Quadra, FNX Both Thriving." November 11, 2010. http://www.thesudburytar.com/ArticleDisplay.aspx?e=2841299.

"Questions Arise from Sale." June 26, 2010.

"Railway Shops Must Stay in Capreol by Agreement." August 4, 1951.

"Rap 'Company' Towns." November 9, 1958.

"Rather Good Name Than Great Wealth: Theme of Eulogy Spoken for Mayor." October 7, 1966.

"RCMP Files Tell Tales of Bombings and Espionage." November 11, 1993.

"Ritchie: A Tower in Early Mining." December 1, 1979.

"Robert C. Stanley." *Sudbury Daily Star*, February 16, 1951.

"Royal Visit Edition." Friday, June 2, 1939.

"Saga of Lily Creek Plaza Finally Ended by Cabinet." May 4, 1976.

"Sagamok, Quadra Ink Deal." January 18, 2012.

"Searching for the Universe's Secrets." May 19, 2012.

"Settlement First Started around Old Brick Yard." June 23, 1950.

"Sign of the Times Makes for Boring Core." February 6, 2012.

"Sisters Played Crucial Role at Hospital." July 18, 2008.

"Sisters of St. Joseph Open Convent on Saturday." December 5, 1923.

"Slag Makes Sudbury Unique." July 14, 2002.

"Some Camps in Preliminary Preparation." October 23, 1936.

"Spirit of Creighton Mine Lives On." By Tom Davies. November 26, 1991.

"Steel to City Businesses: You're Either on Our Side or Vale's." September 21, 2009.

"Sudbury Becomes City of 75 000 as McKim, Frood, Half of Neelon Absorbed in Amalgamation Move." November 16, 1959.

"Sudbury Business Moving into the Big Leagues."Accessed January 23, 2011, http://www.thesudburystar.com/ArticleDisplay.aspx?e=2942129.

"Sudbury's Business, Retail Community Is Thriving." November 25, 2011.

"Sudbury Council Approves Law Restricting Building Outside Commercial Area." August 3, 1955.

"Sudbury in Good Position." November 28, 2011.

"Sudbury Loses Its Greatest Philanthropist; Ricard Dies at 96." June 4, 2003.

"Sudbury Luckiest City." January 23, 2012, A1 and A4.

"Sudbury Needs to Be Aggressive." June 9, 2012.

"Sudbury to Play Leading Role in Treatment of Cancer." April 27, 1988.

"Sudbury Remembers." June 18, 2009.

"Sudbury Worst City in North America Records Disclose." July 17, 1935.

"Sun Sets on Old Inco." October 12, 2005.

"Tax Revenue Available for Nickel Belt Needs without Amalgamation." June 16, 1958.

"Think Tank Bullish on Sudbury in 2012." January 14, 2012.

"Thomas Frood and His Faith in the New Ontario." April 8, 1978.

"To Make Early Start on Memorial Park." February 17, 1926.

"Town of Creighton Mine Rejuvenated by Reviving Industrial Activity." September 29, 1934.

"Township Can't Tax Inco Iron Plant." February 13, 1957.

"Tragedy at Stobie Mine a Reminder of Our Spirit." June 10, 2011.

"Ukrainian Seniors' Centre Will Be Ready by Fall." April 26, 1988.

"Uncertainty Looming." April 28, 2012.

"Union Leaders Upbeat about Changes at Inco." October 26, 2006.

"Ursa Major Minerals Set to Resume Production." Accessed February 5, 2011, http://www
.thesudburystar.com/ArticleDisplay.aspx?archive=true&e=218.

"Vale and Inco Don't Go Together." June 7, 2010.

"Vale Strike Marks Threshold of Brave New World for Company, Union: Horn." December 21,
2009.

"Value of a Strike." July 14, 2010.

"Victoria Mines Soon to Be Lost." November 10, 1952.

"Wahnapitae in the Late 19th Century Termed 'a Hustling and Bustling Village.'" By Gary Peck.
December 3, 1977.

"Waives Claims on Land, Gets 198 Lots." *Sudbury Daily Star*. March 17, 1951.

"The Way Ahead Lies North." November 14, 2009.

"We Want to Be a City of Learning." October 20, 2011.

"West End Section of City Opened When Homes Built Near Brewery." *Sudbury Daily Star*, July 7,
1950.

"When Sudbury Village of Shacks, Tents." March 4, 1922.

"Will Sellwood See Its Second Boom?" November 26, 2011.

"William Joseph Bell, a Lumber Baron with an Obsession for Creating Parks." April 1, 1976.

"W.J. Bell Dies at Home after Long Illness." January 12, 1945.

"Xstrata Working on Innovative Mining Projects." Mining Supplement, May 10, 2012, 4.

Other

"'Alarming' Findings in Road Audit." **Northern Life**, August 16, 2012, 1 and 5.

"A New Mayor: A New Era: Old Rules." *Northern Life*, October 28, 2010.

"Bartolucci Calls Cliffs Agreement 'Historic.'" *Northern Life*, May 10, 2012.

"Black Slag Pile Will Soon Be Green." *Northern Life*, August 7, 2007.

"Caruso Club Welcomes First Female President." *Northern Life*, December 30, 2010.

"City's Slot Share Up in the Air." *Northern Life*, March 20, 2012.

"Cornetist Was a Pioneer." *Globe and Mail*, May 4, 1978.

"CVRD Boss Makes Nice with Union Officials." *Northern Life*, October 27, 2006.

"Favourite Watering Hole Closes in the Donovan." *Northern Life*, February 9, 2000.

"First Nations Fuming over Smelter Decision." *Northern Life*, May 10, 2012.

"Goodbye Inco, 'bem-vindos' to Sudbury CVRD." *Northern Life*, October 25, 2006.

"Happily Ever After." *Northern Life*, October 12, 2005.

"It Doesn't Pay to Take Little Things for Granted." *Northern Life*, June 6, 1993.

"It's the Right Thing to Do." *Sudbury Mining Week 2012* (Sudbury, ON: Northern Life, 2012), 9.

"Labour Minister Urges Deadlocked Parties to Continue Talks." *Northern Life*, July 2, 2010.

"Metal Market Boom Coming: City Poised to Reap Rewards." *Northern Life*, September 1, 2011.

"No Easy Way to Sum Up Jim Gordon." *Northern Life*, June 26, 2003.

"OGS Expects to Be Swamped with Applications for Positions." *Northern Life*, April 1, 1992.

"Our Management of Resources Second Rate." *Northern Life*, May 2, 2007.

"Regional Councillors Say Good-Bye to Their Boss." *Northern Life*, November 28, 1997.

"Remembering a Community Builder." By Heidi Ulrichsen. *Northern Life*, January 10, 2011. http://www.northernlife.ca/news/localNews/2011/01grassby11011.aspx.

"Scaled-Down 2001 Carries on with Job Mandate." *Northern Life*, March 20, 1985.

"SRDC Fears Project Will Lose Momentum." *Northern Life*, April 21, 1993.

"Sudbury 2001: Legend of Job Creation Program Lives on, but Spirit That Fuelled It Seems Dead." *Northern Life*, May 29, 1996.

"Sudbury Economy on the Move." *Northern Life*, May 15, 2012.

"Sudbury in No-Win Situation," *Northern Life*, June 12, 2012.

"Tales of Lives Lived." *Northern Life*, August 2, 2005.

"Unions Happy with Takeover." *Northern Life*, October 12, 2005.

"Vale Pumps $2B into Environmental Project." *Northern Life,* November 25, 2011. http://www.northernlife.ca/news/localNews/2011/11/25-Vale-AER-project.

"With a Future, Appropriate to Shed Light on Worthington's Past." *Northern Life*, May 9, 2007.

"A Tribute to Tom Davies." **Northern Ontario Business**, December, 1997.

"Last Chance for the Grand Theatre." *Northern Ontario Business*, February, 1989.

"Laurentian Is Leaner and Fitter: Daniel." *Northern Ontario Business*, February 1989.

"Too Busy to Retire." By Ricard Baxter. *Northern Ontario Business*, December 1984.

"Xstrata PLC Sparking Merger Mania." By Ian Ross. *Northern Ontario Business*, July 18, 2008.

"Appreciating the Perspiration & Inspiration of Jim Gordon." **South Side Story**, July 2003.

"The Count of Sudbury." *South Side Story*, March 1995.

"Frank Dennie 1880–1991." *South Side Story*, January 2005.

"Great Lives Lived in Greater Sudbury." *South Side Story*, January 2005.

"William Edge Mason: 1882–1948." *South Side Story*, January 2005.

"The N.W. Corner of Elm and Elgin Streets." By Ray Thoms. **Snap Sudbury** 1, no. 1, 2008.

"The Best Investment in the Land Is Land Itself: Donovan Sub-Division." **Sudbury Journal**, March 19, 1908.

"The Fate of Sudbury's Streetcars." **Sudbury Sun**, April 12, 2000.

"Minnow Lake during the 1930s." *Sudbury Sun*, July 5, 2000.

"Bots Go Down the Mine Shaft." **Toronto Star**, August 6, 2001.

"The Death of a Prison Brings Cries of Protest." *Toronto Star*, November 20, 1974.

"Iconic Inco Rides Off into the Sunset." *Toronto Star*, October 23, 2006.

"Recent Wave of Takeovers Raises Foreign Eyebrows." *Toronto Star*, October 24, 2006.

"Robeson's 'Beloved Friends in Canada.'" *Toronto Star*, April 9, 1998, G10.

"This Town's Only Big Enough for One of Us." *Toronto Star*, October 15, 2005.

"Vale Inco Mulls Deep Cuts at Sudbury Mine." *Toronto Star,* accessed December 8, 2010. http://www.thestar.com/business/article/7522989.

"Never a Hockey Fan Like Maxie." **Toronto Weekly Star**, January 3, 1953.

"Raiding the Icebox: Behind Its Warm Front, the United States Made Cold Calculations to Subdue Canada." By Peter Carlson. **Washington Post**, December 30, 2005.

Inco Triangle

"The 1175-Foot Dam at Big Eddy." *Inco Triangle*, December 1946.

"$1 200 000 for Levack Housing." *Inco Triangle*, July 1952.

"Copper Cliff Fifty Years Ago." By F.P. Bernhard. *Inco Triangle*, August 1949.

"Copper Cliff Club for 30 Years a Hub of Good Entertainment." *Inco Triangle*, July 1946.

"The Copper Cliff Mine, Now Market Street." *Inco Triangle*, March 1951.

"Inco Water Storage around Bisco Has System of 11 Dams." *Inco Triangle*, December 1946.

"Fine Range of Facilities at Inco Employees' Club." *Inco Triangle*, February 1938.

"First House in Cliff Built 60 Years Ago." *Inco Triangle*, January 1945.

"Five Plants in Inco's Huronian Hydro-Electric Power System." *Inco Triangle*, August 1952.

"Full Speed Ahead on New Housing at Levack." *Inco Triangle*, October 1952.

"Garson Proud of Handsome Employee Club." *Inco Triangle*, March 1950.

"Green Thumb." *Inco Triangle*, August/September 1972.

"High Falls School and Scholars." *Inco Triangle*, January 1951.

"How Iron Became the Fourteenth Element on the Inco Product Team." *Inco Triangle*, January 1965.

"Inco Iron Ore Goes to Mills by Rail, Boat." *Inco Triangle*, October 1970.

"Inside Inco's Power Plants." *Inco Triangle*, February 1972.

"Levack District Takes Pride in Its Beautiful New High School." *Inco Triangle*, February 1959.

"Levack Is Proud of Smart New Curling Centre." *Inco Triangle*, February 1953.

"Moldering Ruins Recall Early Days of the Nickel Industry." *Inco Triangle*, June 1959.

"Mother Nature's Dark Secrets Probed by the 'Flying X-Ray.'" *Inco Triangle*, January 1953.

"New Curling Rink Proves Very Popular." *Inco Triangle*, February 1950.

"Plant's Capacity Will Be Tripled." *Inco Triangle*, November 1960.

"Presided over Transformation of Swamp into Inco Park at Copper Cliff." *Inco Triangle*, January 1957.

"Process of New Iron Ore Plant Makes History." *Inco Triangle*, October 1953.

"Rare Historical Pictures Taken in 1892 Show First Smelting Operations at Copper Cliff." *Inco Triangle*, July 1960.

"Recall Highlights at Popular Centre Now in 40th Year." *Inco Triangle*, June 1955.

"Reminiscences of the Early Days." *Inco Triangle*, January 1953.

"Reminiscences of Sudbury, 1886." *Inco Triangle*, February 1951.

"Robert C. Stanley." *Inco Triangle*, March 1951.

"Sequestered High Falls Is Nerve-Centre of Industry." *Inco Triangle*, August 1937.

"Six Different Housing Styles In '49 Program." *Inco Triangle*, July 1949.

"Stanley Stadium." *Inco Triangle*, February 1967.

"Sudbury Almost Changed Name One Day in 1891." *Inco Triangle*, January 1949.

"Three Stacks of Copper Cliff World Famous." *Inco Triangle*, September 1951.

"T-i-m-b-e-r!!!" *Inco Triangle*, August 1944.

"U.S. Govt. to Buy 120 Million Pounds of Inco Metallic Nickel." *Inco Triangle*, July 1953.

ONLINE ARTICLES

"CFB Falconbridge." Ontario abandoned places.com. Accessed February 2, 2011. http://www
.ontarioabandonedplaces.com/articles/cfbfalconbridge.html.

"CFS Falconbridge." Ghosttownpix.com. Accessed January 31, 2011. http://www.ghosttownpix
.com/ontario/towns/falconbridge.html.

"Cooperative Freshwater Ecology Unit," Laurentian University. Accessed August 30, 2012. http://
www3.laurentian.ca/livingwithlakes/about/cooperative-freshwater-ecology-unit.

"Earth Impact Database." Impact Structures as Sorted by Age and Diameter. PASSC.net. Accessed
August 28, 2012. http://www.passc.net/Earth Impact Database.

"Here's What's Wrong with Municipal Mining Revenue in Ontario." Republic of Mining. Last
modified June 4, 2008. http://www.republicofmining.com/2008/06/04.

"Inco Building the World's Leading Nickel Company." Vale. Accessed October 31, 2005. http://
www.inco.com/about/history/default.aspx.

"The Lougheed Family." Community Builder Awards. Accessed February 8, 2010. http://www
.cbawards.ca/Winners/2008

"The Lougheed Family." Lougheeds.ca. Accessed February 8, 2010. http://www.lougheeds.com/
history.asp.

"Personalities & Biographies: Judy Erola." CKSO.com. Accessed December 14, 2009. http://www
.ckso.com/personalities_judyerola.html.

"Re-Greening and Beautification." Greater Sudbury.ca. Accessed May 26, 2012. http://www
.greatersudbury.ca/cms/index.cfm?app=div_earthcare&lang=en&currID=6855.

"Riding Profile for Sudbury." Election prediction project. Accessed August 13, 2011. http://www
.electionprediction.org/1999_ontario/north/sudbury.html.

"Sellwood." Ghosttownpix.com. Accessed February 5, 2011. http://www.ghosttownpix.com/
ontario/towns/sellwood.html.

"Sellwood." Ontarioabandonedplaces.com. Accessed February 5, 2011. http://www
.ontarioabandonedplaces.com/sellwood/sellwood.asp.

"Sudbury Water Tower Redevelopment Project." Accessed August 12, 2011. http://
sudburywatertower.com/.

"Wallbridge Mining Company Limited." Wallbridge Mining Co. Ltd. Accessed December 1,
2010. http://www.wallbridgemining.com/s/North Range.asp.

"Wallbridge Mining Company Limited." Wallbridge Mining Co. Ltd. Accessed December 1,
2010. http://www.wallbridgemining.com/s/Sudbury.asp.

Wikipedia. "CFS Falconbridge." Accessed January 31, 2011. http://en.wikipedia.org/wiki/
CFS Falconbridge.

Wikipedia. "Mid-Canada Communications." Accessed December 9, 2009. http://en.wikipedia
.org/wiki/Mid-Canada Communications.

Wikipedia. "Nickel Belt (provincial electoral district)." Accessed June 5, 2008. http://en.wikipedia.
org/wiki/Nickel_Belt_%28provincial_electoral_district%29.

Wikipedia. "Sudbury (provincial electoral district)." Accessed June 5, 2008. http://en.wikipedia.
org/wiki/Sudbury_(provincial_electoral_district).

Wikipedia. "Sudbury East." Accessed June 5, 2008. http://en.wikipediaorg/wikiSudbury_East.

THESES

Battaglini, Sandra P. "'Don't Go Down the Mine, Mamma!' Women in Production Jobs at Inco During World War II, 1942–1945." M.A. thesis, Laurentian University, 1996.

Colussi, James. "The Rise and Fall of the British America Nickel Corporation 1913–24." M.A. thesis, Laurentian University, 1989.

Dennie, Donald. "Sudbury 1883–1946: A Social Historical Study of Property and Class." Ph.D. thesis, Carleton University, 1989.

De Santi, N. "The Spatial Organization of the Sudbury Transit System in the Region of Sudbury." Honours B.A. thesis, Laurentian University, 1997.

Farquhar, John A. "The Historical Evolution of the Territorial District Boundaries in Northern Ontario with Reference to their Suitability at Present." Honours B.A. thesis, Laurentian University, 1972.

Goltz, Eileen Alice. "Genesis and Growth of a Company Town: Copper Cliff: 1886–1920." M.A. thesis, Laurentian University, 1983.

———. "The Exercise of Power in a Company Town: Copper Cliff, 1886–1980." Ph.D. thesis, University of Guelph, 1989.

Hallsworth, Gwenda. "A Good Paying Business": Lumbering on the North Shore of Lake Huron, 1850–1910 with Particular Reference to the Sudbury District." M.A. thesis, Laurentian University, 1983.

Harris, Judith E. "Well-Being in Sudbury 1931–1971: A Social Indicator Analysis." M.Sc. thesis, University of Guelph, 1977.

Krats, Peter V.K. "The Sudbury Area to the Great Depression: Regional Development on the Northern Resource Frontier." Ph.D. thesis, University of Western Ontario, 1988.

Lang, John B. "A Lion in a Den of Daniels: A History of the International Union of Mine Mill and Smelter Workers in Sudbury Ontario, 1942–1962." M.A. thesis, University of Guelph, 1970.

Liu, Kam-biu. "Postglacial Vegetational History of Northern Ontario: A Palynological Study." Ph.D. thesis, University of Toronto, 1982.

Miller, Jason A. "Divided We Stand: A Study of the Development of the Conflict between the International Union of Mine, Mill and Smelter Workers and the United Steelworkers of America in Sudbury, Ontario, 1942–1969." M.A. thesis, McMaster University, 2003.

Nicholson, Thomas Henry. "A Sordid Boon: The Business of State and the State of Labour at the Canadian Copper Company, 1890 to 1918." M.A. thesis, Queen's University, 1991.

Reilly, Karey. "Les Italiens de Copper Cliff, 1886–1914." Honours B.A. thesis, Laurentian University, 1994.

Sajatovic, Stephen Michael. "The Borgia Area Redevelopment Project: A Case Study in Urban Renewal." Honours B.A. thesis, Laurentian University, 1973.

Stephenson, Robert A. "To Strike—or Not to Strike?" Honours B.A. thesis, Laurentian University, 1978.

Zembrzycki, Stacy, R. "Memory, Identity, and the Challenge of Community among Ukrainians in the Sudbury Region, 1901–1919." Ph.D. thesis, Carleton University, 2007.

BOOKS AND REPORTS

Ambrose, Linda M. *Glad Tidings Tabernacle: 70 Years of Pentecostal Ministry in Sudbury, 1937–2007.* Sudbury, ON: Glad Tidings Tabernacle, 2007.

Arnopoulos, Sheila McLeod. *Voices from French Ontario.* Montreal, QC: McGill-Queen's University Press, 1982.

Asher, G.M. *Henry Hudson the Navigator.* New York: Burt Franklin, 1963.

Association of Ontario Land Surveyors. *Annual Report.* Toronto: Association of Ontario Land Surveyors, 1915.

Bajc, A.F., and P.J. Barnett. *Quaternary Geology and Geomorphology of the Sudbury Region,* Field Trip A5 Guidebook. Sudbury, ON: Geological Association of Canada–Mineral Association of Canada, 1999.

Beck, Kaili, Chris Bowes, Gary Kinsman, Mercedes Steedman, and Peter Suschnigg, eds. *Mine Mill Fights Back.* Sudbury, ON: Mine Mill /CAW Local 598, 2005.

Berton, Pierre. *The Impossible Railway: The Building of the Canadian Pacific.* New York: Alfred A. Knopf, 1972.

Bertulli, Margaret, and Rae Swan, eds. *A Bit of the Cliff: A Brief History of the Town of Copper Cliff, Ontario 1901–1972.* Copper Cliff, ON: Copper Cliff Museum, 1982.

Bishop, Morris. *White Men Came to the St. Lawrence.* Montreal, QC: McGill University, 1961.

———. *Champlain: The Life of Fortitude.* Toronto, ON: McClelland and Stewart, 1963.

Black, Trent, Christine Blondin, William J. Campbell, Mike Hrytsak, Frederick Malin, and Wendy Phipps, eds. *Nickel Centre Yesterdays.* Sudbury, ON: Northern Heritage Nickel Centre, 1974.

BMO Financial Group. *Sudbury: Mining for Economic Growth.* Toronto ON: BMO Capital Markets Economic Research, May 31, 2012.

Bonin, Lionel, and Gwenda Hallsworth. *Street Names of Downtown Sudbury.* Sudbury, ON: Your Scrivener Press, 1997.

Bouchard, Jeannette. *Seven Decades of Caring.* Sudbury, ON: Laurentian University Press, 1984.

Brasch, Hans. *A Miner's Chronicle: Inco Ltd. and the Unions 1944–1997.* Sudbury, ON: Hans Brasch, 1997.

Bray, Matt. *Science North: Evolution of a Dream.* Sudbury, ON: Science North, 2011.

———, ed. *Laurentian University: A History.* Montreal, QC: McGill-Queen's University Press, 2010.

Brown, Ron. *Ghost Towns of Ontario,* vol. 2. Toronto, ON: Cannon Books, 1983.

Bryce, George. *The Remarkable History of the Hudson's Bay Company.* Toronto, ON: William Briggs, 1910.

Buchanan, K.T. *An Archeological Survey of the Sudbury Area and a Site Near Lake of the Mountains.* Report No. 6. Sudbury, ON: Laurentian University, 1979.

Butterfield, Consul Willshire. *History of Brûlé's Discoveries and Explorations, 1610–1626.* Cleveland, OH: Helman-Taylor, 1898.

Cadieux, Lorenzo. *Frédéric Romanet du Caillaud «Comte» de Sudbury (1847–1919).* Montreal, QC: Les Éditions Bellarmin, 1971.

Canadian Pacific Railway. *CPR Plan of Sudbury.* Montreal, QC: Canadian Pacific Railway, December 29, 1886.

Clement, Wallace. *Hardrock Mining: Industrial Relations and Technological Changes at Inco*. Toronto, ON: McClelland and Stewart, 1981.

Closs, Gudrun Jahns. *Canadian Mosaic: Life Stories of Post-War German-Speaking Immigrants to Sudbury*. Sudbury, ON: German-Canadian Association, 2003.

Coniston Historical Group. *The Coniston Story*. Coniston, ON: Coniston Historical Group, 1983.

Copper Cliff Italian Heritage Group. *Up the Hill: The Italians of Copper Cliff*. Sudbury, ON: Journal Printing, 1997.

Cranston, J. Herbert. *Etienne Brûlé: Immortal Scoundrel*. Toronto, ON: Ryerson Press, 1949.

Cumming, W.P., R.A. Skelton, and D.B. Quinn, eds. *The Discovery of North America*. London: Elek, 1971.

Dawson, K.C.A. *Prehistory of Northern Ontario*. Thunder Bay, ON: Thunder Bay Historical Museum Society, 1983.

Dean, W.G., ed. *Economic Atlas of Ontario*. Toronto, ON: University of Toronto Press, 1969.

Dennie, Donald. *Á L'Ombre de L'Inco: Étude de la transition d'une communauté canadienne-française de la région de Sudbury 1890–1972*. Ottawa, ON: Les Presses de l'Université d'Ottawa, 2001.

Department of History, Laurentian University. *Index to The Sudbury Star*, Vols. 1 and 2. Sudbury, ON: Laurentian University, 1980.

Deverell, John, and the Latin American Working Group. *Falconbridge: Portrait of a Canadian Mining Multinational*. Toronto, ON: James Lorimer, 1975.

Dickson, Robert. *Sudbury Iron Bridge*. Sudbury, ON: Canned Collective Works, 1978.

Dorian, Charles. *The First 75 Years: A Headline History of Sudbury Canada*. Devon, ON: Arthur H. Stockwell, 1958.

Evans, Robert. *An Eye on Everything*. Sudbury, ON: Laurentian University Press, 1966.

Eyles, Nick. *Ontario Rocks: Three Billion Years of Environmental Change*. Markham, ON: Fitzhenry & Whiteside, 2002.

Falconbridge Nickel Company. *Celebrating 75 Years: 1928–2003*. Falconbridge, ON: Falconbridge Nickel Company, 2003.

Fortin, Jim, ed. *There Were No Strangers: A History of the Village of Creighton Mine*. Walden, ON: Anderson Farm Museum, 1989.

Fox, Captaine Luke. *North-West Fox*. London: B. Alsop and Tho. Fawcet, 1635.

Galbraith, John S. *The Hudson's Bay Company as an Imperial Factor, 1821–1869*. Berkeley: University of California Press, 1957.

Gatchell History Committee. *Memories of Gatchell*. Sudbury, ON: Gatchell History Committee, 1997.

Gentilcore, R. Louis, ed. *Historical Atlas of Canada*. Volume II: *The Land Transformed 1800–1891*. Toronto, ON: University of Toronto Press, 1987.

Gilchrist, Gilbert. *As Strong as Steel*. Sudbury, ON: USWA, 1999.

Gunn, John M., ed. *Restoration and Recovery of an Industrial Region*. New York: Springer-Verlag, 1995.

Hallsworth, Gwenda. *A Brief History of Laurentian University*. Sudbury, ON: Laurentian University, 1985.

Hambleton, Jack. *Fire in the Valley*. Toronto, ON: Longmans, Green & Company, 1960.

Harpelle, Ronald, Varpu Lindstrom, and Alexis Pogorelskin, eds. *Karelian Exodus: Finnish*

Communities in North America and Soviet Karelia during the Depression Era. Beaverton, ON: Aspasia Books, 2004.

Harris, R. Cole, ed. *Historical Atlas of Canada.* Vol. 1, *From the Beginning to 1800.* Toronto, ON: University of Toronto Press, 1987.

Hayes, Kenneth. "Be Not Afraid of Greatness or Sudbury: A Cosmic Accident." In *Sudbury: Life in a Northern Town,* edited by Shawn Van Sluys et al., 16–25. Sudbury, ON: Musagetes and Laurentian Architecture, 2011.

Heidenreich, C.E. *Explorations and Mapping of Samuel de Champlain, 1603–1632.* Supplement No. 2 to *Canadian Cartographer* 13, 1976.

Héroux, L.P. *Aperçu Sur Les Origines de Sudbury.* La Société Historique du Nouvel-Ontario, no. 2 (1943).

Higgins, Edwin G. *Whitefish Lake Ojibway Memories.* Cobalt, ON: Highway Book Shop, 1982.

Howey, Florence. R. *Pioneering on the C.P.R.* Ottawa, ON: Mutual Press, 1938.

International Nickel Company of Canada Limited. *Map of Copper Cliff.* Copper Cliff, ON: International Nickel Company of Canada Limited, August, 1932.

———. Research Library. *History of Nickel Extraction from Sudbury Ores: Part I (1846–1920) and Part II (1921–1956).* Copper Cliff, ON: International Nickel Company of Canada Limited, 1956.

———. *A Submission by the International Nickel Company of Canada, Limited, on the Report of the Ontario Committee on Taxation.* Copper Cliff, ON: International Nickel Company of Canada, Limited, 1968.

International Union of Mine, Mill and Smelter Workers (Canada). *Submission Before the Select Committee of the Ontario Legislature on Mining.* Sudbury, ON: International Union of Mine, Mill and Smelter Workers (Canada), 1964.

Iuele-Colilli, Diana. *Italian Faces: Images of the Italian Community of Sudbury.* Welland, ON: Soleil, 2000.

James, Captaine Thomas. *The Strange and Dangerous Voyage.* New York: Da Capo Press, 1968.

Julig, P.J. *Laurentian University Field School at Lavase Site, North Bay Ontario, 1996.* Report No. 27 of the Archeological Survey of Laurentian University. Sudbury, ON: Laurentian University, 1998.

Kaattari, Ray. *Voices from the Past: Garson Remembers.* Garson, ON: Garson Historical Group, 1992.

Karni, Michael, G., Olavi Koivukangas, and Edward W. Laine, eds. *Finns in North America.* Turku, Finland: Institute of Migration, 1984.

Kechnie, Margaret, and Marge Reitsma-Street, eds. *Changing Lives: Women in Northern Ontario.* Toronto, ON: Dundurn Press, 1996.

Kerr, D.G.G., ed. *A Historical Atlas of Canada.* Toronto: Thomas Nelson & Sons, 1960.

Khatter, Kapil. *Sudbury Human Health Risk Assessment Briefing.* Toronto, ON: Environmental Defence, September 22, 2008.

Kinsman, Gary, Dieter K. Buse, and Mercedes Steedman, eds. *Whose National Security? Canadian State Surveillance and the Creation of Enemies.* Toronto, ON: Between the Lines, 2000.

Knowles, John D. *The Sudbury Streetcars.* Sudbury, ON: Nickel Belt Rails, 1983.

Koivukangas, Olavi, ed. *Entering Multiculturalism: Finnish Experience Abroad.* Turku, Finland: Institute of Migration, 2002.

Ladell, John L. *They Left Their Mark: Surveyors and Their Role in the Settlement of Ontario*. Toronto, ON: Dundurn Press, 1993.

Landry, Denis. *Azilda, Comme Je L'ai Connu: Document Historique 1890–1972*. Azilda, ON: Denis Landry, 2001.

Larmour, Adelle. *Until the End: Memoirs of Sinter-plant Activist Jean L. Gagnon*. North Bay, ON: Wynterblue Publishing Canada, 2010.

Laurentian University Museum and Art Centre. *Bell Rock—Laurentian University Museum and Art Centre*. Sudbury, ON: Laurentian University Museum and Art Centre, 1990.

Le Club 50 de Chelmsford. *Chelmsford 1883–1983*. Chelmsford, ON: Le Club 50 de Chelmsford, 1983.

Leadbeater, David, ed. *Mining Town Crisis: Globalization, Labour and Resistance in Sudbury*. Halifax, NS: Fernwood Publishing, 2008.

LeBelle, Wayne F. *Valley East 1850–2002*. Field, ON: WFL Communications, 2002.

LeBourdais, D.M. *Sudbury Basin: The Story of Nickel*. Toronto, ON: Ryerson Press, 1953.

Lower, A.R.M. *The North American Assault on the Canadian Forest: A History of the Lumber Trade between Canada and the United States*. Toronto, ON: Ryerson Press, 1938.

MacDowell, Laurel Sefton. *"Remember Kirkland Lake": The History and Effects of the Kirkland Lake Gold Miners' Strike, 1941–1942*. Toronto, ON: University of Toronto Press, 1983.

MacLelland, Sister Bonnie. "A Message to the Citizens of Sudbury." Press Release, June 17, 2010. Sisters of St. Joseph of Sault Ste. Marie.

Main, O.W. *The Canadian Nickel Industry: A Study in Market Control and Public Policy*, Canadian Studies in Economics No. 4. Toronto, ON: University of Toronto Press, 1955.

McCharles, Aeneus. *Bemocked of Destiny*. Toronto: William Briggs, 1908.

McKechnie, D. and G.R. Stock. *Report to the Tourism and Convention Committee on the Historic Sites in the Sudbury District*. Sudbury, ON: Chamber of Commerce, n.d.

Mika, Nick and Helma. *Places in Ontario: Their Name, Origins and History*, Parts 1–3. Belleville, ON: Mika Publishing, 1977, 1981, and 1983.

———. *The Shaping of Ontario from Exploration to Confederation*. Belleville, ON: Mika Publishing, 1985.

Nelles, H.V. *The Politics of Development: Forests, Mines & Hydro-electric Power in Ontario, 1849–1941*. Toronto: Macmillan of Canada, 1975.

Newell, Dianne. *Technology on the Frontier*. Vancouver: University of British Columbia Press, 1986.

Newman, Peter, C. *Company of Adventurers*, Vol. 1. Markham, ON: Penguin Books of Canada, 1985.

Nissilä, Eino. *Pioneers of Long Lake*. Sudbury, ON: Eino Nissilä, 1987.

Northern Ontario Local Training and Adjustment Boards. *An Aging Population in Northern Ontario*. Thunder Bay, ON: Northern Ontario Local Training and Adjustment Boards, October 31, 2002.

Oleson, Tryggvi J. *Early Voyages and Northern Approaches 1000–1632*. Toronto: McClelland and Stewart, 1963.

Ontario Motor League—Nickel Belt Club. *Sixty Golden Years 1915–1975*. Sudbury, ON: Nickel Belt Club, 1975.

Ontario North Economic Development Corporation. *Pan Northern Mining Supply and Services Strategy*. Sudbury, ON: Ontario North Economic Development Corporation, 2009.

Ontario Prospectors Association. *Ontario Mining & Exploration Directory 2011*. Thunder Bay, ON: Ontario Prospectors Association.

Otis, Charles Pomeroy. *Voyages of Samuel de Champlain*. New York: Burt Franklin, 1882.

Peake, F.A. *The Church of the Epiphany, Sudbury, Ontario: A Century of Anglican Witness*. Sudbury, ON: Church of the Epiphany, 1982.

Peake, F.A., and R.P. Horne. *The Religious Tradition in Sudbury 1883–1983*. Sudbury, ON: Downtown Churches Association, 1983.

Peters, T.H. *The Use of Vegetation to Stabilize Mine Tailings at Copper Cliff*. Copper Cliff, ON: International Nickel, 1969.

Pilon, Claire. *Le Moulin à Fleur*. Sudbury, ON: Claire Pilon, 1983.

Potvin Air Management Consulting. *Air Quality Trends: City of Greater Sudbury, Ontario 1953–2002*. Sudbury, ON: Potvin Air Management Consulting, 2004.

Preston, Robert. *The Defence of the Undefended Border: Planning for War in North America 1867–1939*. Montreal: McGill-Queen's University Press, 1977.

Quimby, George Irvin. *Indian Life in the Upper Great Lakes*. Chicago: University of Chicago Press, 1960.

Radecki, Henry. *The 1978–79 Strike at Inco: The Effects on Families*. Sudbury, ON: Laurentian University, 1979.

Rayburn, Alan. *Place Names of Ontario*. Toronto: University of Toronto Press, 1997.

Raymonde, Alphonse, S.J. *Paroisse Sainte-Anne de Sudbury 1883–1953*. Sudbury, ON: La Société Historique du Nouvel-Ontario, no. 26 (1953).

Ricard, Alma. *Fondation Baxter & Alma Ricard*. Sudbury, ON: Fondation Baxter & Alma Ricard, December 15, 1998.

Rich, E.E. *The Fur Trade and the Northwest to 1857*. Toronto, ON: McClelland and Stewart, 1976.

Ross, Nicola. *Healing the Landscape: Celebrating Sudbury's Reclamation Story*. Sudbury, ON: Vegetation Enhancement Technical Advisory Committee, 2001.

Rudmin, Floyd. *Bordering on Aggression: Evidence of U.S. Military Preparations against Canada*. Hull, QC: Voyageur Publications, 1993.

Saarinen, Oiva, W. "Finnish Adaptation and Cultural Maintenance: The Sudbury Experience." In *Entering Multiculturalism: Finnish Experience Abroad*, edited by Olavi Koivukangas, 199–215. Turku, Finland: Institute of Migration, 2002.

———. *Between a Rock and a hard Place: A Historical Geography of the Finns in the Sudbury Area*. Waterloo, ON: Wilfrid Laurier University Press, 1999.

Saarinen, Oiva W., and Gerry Tapper. "Sudbury in the Great Depression: The Tumultuous Years." In Harpelle, Lindstrom, and Pogorelsk, In *Karelian Exodus: Finnish Communities in North America and Soviet Karelia during the Depression Era*, 48–66.

St. Joseph's Hospital. *Golden Jubilee 1896–1946*. Sudbury, ON: St. Joseph's Hospital, 1946.

Savageau, David with Ralph D'Agostino. *Places Rated Almanac*. New York: IDG Books Worldwide, 2000.

Scott, Geoffrey A.J. *Canada's Vegetation: A World Perspective*. Montreal, QC: McGill-Queen's University Press, 1995.

Smyth, David William. *A Map of the Province of Upper Canada*, 2nd edition. London: W. Faden, 1813.

Solski, Mike, and John Smaller. *Mine Mill: The History of the International Union of Mine, Mill and Smelter Workers in Canada Since 1895*. Ottawa, ON: Steel Rail Publishing, 1984.

Southern, Frank. *The Sudbury Incident*. Toronto, ON: York Publishing, 1978.

Steedman, Mercedes, Peter Suschnigg, and Dieter K. Buse, eds. *Hard Lessons: The Mine Mill Union in the Canadian Labour Movement*. Toronto, ON: Dundurn Press, 1995.

Stephenson, Robert, Michael Gauvreau, Tom Kiley, Marie Lalonde, Nancy Pellis, and Mira Zirojevic, eds. *A Guide to the Golden Age: Mining in Sudbury, 1886–1977*. Sudbury, ON: Laurentian University, Department of History, 1979.

Sudbury & District Historical Society. *Industrial Communities of the Sudbury Basin: Copper Cliff, Victoria Mines, Mond and Coniston*. Sudbury, ON: Sudbury & District Historical Society, 1986.

Sudbury Area Risk Assessment Group (SARA). *Sudbury Area Risk Assessment, Final Report*. Guelph, ON: SARA Group, 2008.

Sudbury Area Risk Assessment. Vol. 1, *Background, Study Organization and 2001 Soils Survey*. Sudbury, ON: SARA, January 2008.

———. Vol. 2, *Human Health Risk Assessment*. Sudbury, ON: SARA, May 2008.

———. Vol. 3, *Ecological Risk Assessment*. Sudbury, ON: SARA, March 2009.

———. *Summary of Vol 2: Human Health Risk Assessment*. Sudbury, ON: SARA, May 2008.

———. *Summary of Vol. 3: Ecological Risk Assessment*. Sudbury, ON: SARA, March 2009.

Sudbury Centennial Foundation. *To Our City/À Notre Ville*. Sudbury, ON: Sudbury Centennial Foundation, 1983.

Sudbury Secondary School Reunion Committee. *Sudbury Secondary School: 100 Years Alumni Reunion Celebration July 31–August 3, 2008*. Sudbury, ON: Sudbury Secondary School Alumni Association, 2008.

Sudbury 2001. *A Conference on Economic Development: April 6, 7 and 8, 1978*. Sudbury, ON: Sudbury 2001, 1978.

———. *Foundations*. Sudbury, ON: Sudbury 2001, 1983.

———. *Proceedings*. Sudbury, ON: Sudbury 2001, 1983.

———. *Retrospect: Prospect*, Sudbury, ON: Sudbury 2001, 1981.

Swift, Jamie, and The Development Education Centre. *The Big Nickel: Inco at Home and Abroad*. Kitchener, ON: Between the Lines, 1977.

Tapper, G.O., ed. *Wa-Shai-Ma-Gog: Memories of Fairbank Lake and Surrounding Area*. Sudbury, ON: Fairbank Lake Camp Owners Association, 2000.

Tapper, G.O., and Oiva W. Saarinen, eds. *Better Known as Beaver Lake: An History of Lorne Township and Surrounding Area*. Walden, ON: Walden Public Library, 1998.

———, eds. *Beaver Lake II: Sisu, Stumps, and Sugar Lumps—A Way of Life*. Sudbury, ON: Laurentian University, 2003.

Tester, Jim. *The Shaping of Sudbury: A Labour View.* Sudbury, ON: Sudbury & District Historical Society, 1979.

Thompson, John F., and Norman Beasley. *For the Years to Come: A Story of International Nickel of Canada.* Toronto: Longmans, Green & Company, 1960.

Thoms, Ray, and Kathy Pearsal. *Sudbury.* Boston, MA: Mills Press Book, 1994.

Thorpe, T. *A Review of the Logging and Pulp and Paper Operations in the Sudbury District during the Years 1901–1950.* Sudbury, ON: Ontario Department of Lands and Forests, 1950.

Thwaites, Reuben Gold, ed. *The Jesuit Relations and Allied Documents.* Cleveland, OH: Burrows Bros., 1896–1901.

Trudel, Marcel. *Atlas de la Nouvelle France: An Atlas of New France.* Quebec City, QC: Les Presses de l'Université Laval, 1968.

Urban Aboriginal Task Force. *Sudbury Final Report.* Sudbury, ON: Ontario Federation of Indian Friendship Centres, 2007.

Van Sluys, Shawn, Kenneth Hayes, and Jocelyn Laurence, eds. *Sudbury: Life in a Northern Town.* Sudbury, ON: Laurentian Architecture, 2011.

Wallace, C.M., and Ashley Thomson, eds. *Sudbury: Rail Town to Regional Capital.* Toronto, ON: Dundurn Press, 1993.

Wallbridge Mining Company. *Annual Report 2009.* Sudbury, ON: Wallbridge Mining Company, 2010.

Wilson, Dale. *Algoma Eastern Railway.* Sudbury, ON: Nickel Belt Rails, 1979.

Winter, J.R. *Sudbury: An Economic Survey.* Sudbury, ON: Sudbury & District Industrial Commission, 1967.

Yeates, Maurice. *Main Street Canada: Windsor to Quebec City.* Toronto, ON: Macmillan Company of Canada, 1975.

Zaslow, Morris. *The Opening of the Canadian North 1870–1914.* Toronto, ON: McClelland and Stewart, 1971.

———. *Reading the Rocks: The Story of the Geological Survey of Canada 1842–1972.* Toronto, ON: Macmillan Company of Canada, 1975.

ARTICLES

Addison, William D, Gregory R. Brumpton, Daniela A. Vallini, Neal J. McNaughton, Don W. Davis, Stephen A. Kissin, Philip W. Fralick, and Anne L. Hammond. "Discovery of Distal Ejecta From the 1850 Ma Sudbury Impact Event." *Geology* 33, no. 3 (2005): 193–96.

"An Alternatives Interview with Elie W. Martel." *Alternatives* 2, no. 3 (1979): 10.

Barlow, A.E. "Report of the Nickel and Copper Deposits of the Sudbury Mining District." *Geological Survey of Canada* 14 (1904): Part H, 191.

———. "On the Origin and Relations of the Nickel and Copper Deposits of Sudbury, Ontario, Canada." *Economic Geology* 1 (1906): 454–66.

"Baxter Ricard." In *Greater Sudbury 1883–2008*, edited by Vicki Gilhula, 65. Sudbury, ON: Laurentian Media Magazine Group, 2008.

Becker, Luann, Jeffrey L. Bada, Randall E. Winans, Jerry E. Hunt, Ted E. Bunch, and Bevan M. French. "Fullerenes in the 1.85-Billion-Year-Old Sudbury Impact Structure." *Science*, n.s. 265, no. 5172 (1994): 642–45.

Berton, Pierre. "Dionne Quintuplets." In *The Canadian Encyclopedia*, vol. 1. 2nd ed. Edmonton, AB: Hurtig Publishers, 1988, 599.

Bodsworth, Fred. "The Unknown Giant of Canadian Mining—Thayer Lindsley." *Maclean's*, August 15, 1951, 7–9, 44–47.

Brandt, Gail Cuthbert. "The Development of French-Canadian Social Institutions in Sudbury, Ontario, 1883–1920)." *Laurentian University Review* 11, no. 2 (1979): 12–13.

Bray, Matthew. "The Province of Ontario and the Problem of Sulphur Fumes Emissions in the Sudbury District: An Historical Perspective." *Laurentian University Review* 16, no. 2 (1984): 81–90.

———. "1910–1920." In Wallace and Thomson, *Sudbury: Rail Town to Regional Capital*, 86–112.

———. "Samuel J. Ritchie." In *Dictionary of Canadian Biography*, vol. 13 (1901–1910), edited by Ramsay Cook and Jean Hamelin, 873–76. Toronto, ON: University of Toronto Press, 1994.

———. "Thomas Frood." In Cook and Hamelin, *Dictionary of Canadian Biography*, vol. 14 (1911 to 1920), 378. Toronto, ON: University of Toronto Press, 1998.

———. "The Founding of Laurentian University." In *Laurentian University: A History*, edited by Matt Bray. Montreal, QC: McGill-Queen's University Press, 2010, 17–32.

Bray, Matthew, and Angus Gilbert. "The Mond–International Nickel Merger of 1929: A Case Study in Entrepreneurial Failure." *Canadian Historical Review* 76, no. 1 (1995): 19–42.

Brown, L. Carson. "Elliot Lake: The World's Uranium Capital." *Canadian Geographical Journal* 75, no. 4 (1967): 120–33.

Buse, Dieter K. "The 1970s." In Wallace and Thomson, *Sudbury: Rail Town to Regional Capital*, 242–74.

———. "Weir Reid and Mine Mill: An Alternative Union's Cultural Endeavours." In Steedman, Suschnigg, and Buse, *Hard Lessons: The Mine Mill Union in the Canadian Labour Movement*, 269–322.

Chadwick, B., P. Claeys, and B. Simonson. "New Evidence for a Large Paleoproterozoic Impact: Spherules in a Dolomite Layer in the Ketilidian Orogen, South Greenland." *Journal of the Geological Society* 158, no. 2 (2001): 331–40.

Cornwall, Claudia. "These Mayors Mean Business." *Canadian Reader's Digest*, August 2000, 88–93.

Delaplante, Don. "Sudbury: Melting Pot for Men and Ore." *Maclean's*, April 15, 1951, 12–13 and 52–54.

Dence, M.R. "Meteorite Impact Craters and the Structure of the Sudbury Basin." *Geological Association of Canada*, Special Paper Number 10 (1972): 7–18.

Dence, M.R., and J. Popelar. "Evidence for an Impact Origin for Lake Wanapitei Ontario." In *New Developments in Sudbury Geology*, Special Paper no. 10, edited by J. V. Guy-Bray, 117–24. Toronto, ON: Geological Association of Canada, 1972.

Dennie, Donald. "Le local 902 du Mine Mill: Les dix premières années (1949–1959) du syndicate des travailleurs de la ville et du district de Sudbury." In Steedman, Suschnigg, and Buse, *Hard Lessons: The Mine Mill Union in the Canadian Labour Movement*, 50–67.

———. "French Ontario: Two Realities." In Leadbeater, *Mining Town Crisis; Globalization, Labour and Resistance in Sudbury*, 200–201.

Dietz, Robert S. "Sudbury Structure as an Astrobleme." *Journal of Geology* 72 (1964): 412–34.

Executive and Operating Staffs. "The Operations and Plants of International Nickel Company of Canada Limited." *Canadian Mining Journal* 58 (November 1937): 581–89 and 67 (May 1946): 309–18.

"The Falconbridge Story." *Canadian Mining Journal* 80, no. 6 (1959): 103–230.

"Featured Funeral Homes of Sudbury." *Canadian Funeral News*. November 1978, n.p.

Filion, Pierre. "Potentials and Weaknesses of Strategic Community Development Planning: A Sudbury Case Study." *Canadian Journal of Regional Science* 11, no. 3 (1988): 393–411.

French, Bevan, M. "Sudbury Structure, Ontario: Some Petrographic Evidence for an Origin by Meteorite Impact." In *Shock Metamorphism of Natural Materials*, edited by Bevan M. French and N.M. Short. Baltimore, MD: Mono Book Corporation, 1968: 383–412.

———. "Shock-Metamorphic Features in the Sudbury Structure Ontario: A Review." *Geological Association of Canada*, Special Paper no. 10 (1972): 19–28.

"From Creosote to Critters." *Green Living*, Summer 2011, 8.

Gagnebin, A.P. *Shaping the Natural Environment*, address given to the Canadian Manufacturers' Association, June 2, 1969, 13.

Gaudreau, Guy. "Les activités forestières dan deux communautés agricoles du Nouvel-Ontario, 1900–1920." *Revue d'Histoire de L'Amérique Française* 54, no. 4 (2001): 528–29.

———. "The Origins of Laurentian University. " In Bray, *Laurentian University: A History*, 3–16.

Gervais, Gaétan. "Sudbury, 1883–1914." In Sudbury Centennial Foundation, *To Our City/À Notre Ville*, 17–31.

Gilbert, A.D. "The 1920s." In Wallace and Thomson, *Sudbury: Rail Town to Regional Capital*, 116.

Gray, James H. "Big Nickel." *Maclean's*, October 1, 1947, 23, 49 and 53–54.

Grieve, R.A.F. "The Sudbury Structure: Additional Constraints on Its Origin and Evolution." *Abstracts of the 25th Lunar Planetary Science Convention*," Houston, TX, March 14–18 (1994), 477–78.

Gunn, J.M., and W. Keller. "Urban Lakes: Integrators of Environmental Damage and Recovery." In Gunn, *Restoration and Recovery of an Industrial Region*, 257–69.

Guy-Bray, J.V. Introduction. In *New Developments in Sudbury Geology*, Special Paper no. 10, edited by J.V. Guy-Bray. Toronto, ON: Geological Association of Canada, 1972, 1–5.

Hallsworth, Gwenda, and Peter Hallsworth. "The 1960s." In Wallace and Thomson, *Sudbury: Rail Town to Regional Capital*, 215–41.

Hayes, Kenneth. "Be Not Afraid of Greatness or Sudbury: A Cosmic Accident." In *Sudbury: Life in a Northern Town*, edited by Shawn Van Stuys, Kenneth Hayes, and Jocelyn Laurence, 16–25. Sudbury, ON: Musagetes and Laurentian Architecture.

Humphreys, David. *Nickel: An Industry in Transition*. Paper presented at the World Stainless Steel Conference, Dusseldorf, September 17–19 (2006), 1–10.

"Judy Erola: The One-Woman Energy Resource Who's Minister of State for Mines." *Chatelaine*, April 1981, 52, 146–53.

Katary, Narasim, to O.W. Saarinen, October 12, 2012. Private Collection.

Keck, Jennifer, and Mary Powell. "Working at Inco: Women in a Downsizing Male Industry." In *Changing Lives,* edited by Margaret Kechnie and Marge Reitsma-Street, 147–61. Toronto, ON: Dundurn Press, 1996.

"Laberge, J.A." In *Greater Sudbury 1883–2008: The Story of Our Times,* edited by Vicki Gilhula, 134. Sudbury, ON: Laurentian Media, 2008.

Lang, John B. "One Hundred Years of Mine Mill." In Steedman, Suschnigg, and Buse, *Hard Lessons: The Mine Mill Union in the Canadian Labour Movement,* 13–20.

"The Last Angry Socialist." *Time,* May 3, 1971, 9.

Leadbeater, David. Preface. In Leadbeater, *Mining Town Crisis: Globalization, Labour and Resistance in Sudbury,* 7–10.

———. Introduction. In Leadbeater, *Mining Town Crisis: Globalization, Labour and Resistance in Sudbury,* 11–48.

Lewis, Gertrud Jaron. "German-Speaking Immigrants in the Sudbury Region," *Polyphony* 5, no. 1 (1983): 82–85.

Lindström, Varpu. "Finnish Women's Experience in Northern Ontario Lumber Camps, 1920–1939." In Kechnie and Reitsma-Street, *Changing Lives,* 107–22.

———. "Central Organization of the Loyal Finns in Canada." *Polyphony* 3, no. 2 (1981): 97–103.

"Lougheed Funeral Homes, Sudbury, Ontario." *Canadian Funeral Homes,* January 2000.

Makinen, W.H. "The Mond Nickel Company and the Communities of Victoria Mines and Mond." In *Industrial Communities of the Sudbury Basin: Copper Cliff, Victoria Mines, Mond and Coniston,* edited by F.A. Peake, 23–44. Sudbury, ON: Sudbury and District Historical Society, 1986.

Martel, Elie. "The Name of the Game Is Power: Labour's Struggle for Health and Safety Legislation." In Steedman, Suschnigg, and Buse, *Hard Lessons: The Mine Mill Union in the Canadian Labour Movement,* 195–209.

"Memorial of Horace John Fraser." *American Mineralogist* 55 (1970): 554–61.

Michalak, Wieslaw. "Economic Changes of the Canadian Urban System 1971–1981." *Alberta Geographer* 22 (1986): 65.

Michelutti, B., and M. Weiseman. "Engineered Wetlands as a Tailings Rehabilitation Strategy." In Gunn, *Restoration and Recovery of an Industrial Region,* 135–41.

Mika, Nick and Helma. "Historic Sites of Ontario," In *Encyclopedia of Ontario,* vol. 1, Belleville, ON: Mika Publishing Company, 1974, 180.

"Milestones." *Time,* August 16, 1971, 49.

Mount, Graeme S. "The 1940s." In Wallace and Thomson, *Sudbury: Rail Town to Regional Capital,* 168–89.

Naldrett, Anthony J. "From Impact to Riches: Evolution of Geological Understanding as Seen at Sudbury, Canada." *GSA Today,* February 2003, 4–7.

Newman, Peter C. "Sudbury's Sunny Renaissance." *Maclean's,* April 1, 1991, 40.

Parent, Huguette. "Le Township de Hanmer 1904–1969." *La Société Historique du Nouvel-Ontario,* Documents Historiques No. 70, 12–13. Sudbury, ON: Université de Sudbury, 1979.

Pearson, David A.B, and J. Roger Pitblado. "Geological and Geographic Setting." In Gunn, *Restoration and Recovery of an Industrial Region*, 10–11.

Peters, T.H. "Rehabilitation of Mine Tailings: A Case of Complete Ecosystem Reconstruction and Revegetation of Industrially Stressed Lands in the Sudbury Area, Ontario, Canada." In *Effects of Pollutants at the Ecosystem Level*, edited by P.J. Sheehan et al., 403–21. New York: John Wiley & Sons, 1984.

———. "Revegetation of the Copper Cliff Tailings Area." In Gunn, *Restoration and Recovery of an Industrial Region*, 123–33.

Pope, K.O., S.W. Kieffer, and D.E. Ames. "Empirical and Theoretical Comparisons of the Chicxulub and Sudbury Impact Structures." *Meteoritics & Planetary Science* 39, no. 1 (2004): 97–116.

Potvin, Raymond R., and John J. Negusanti. "Declining Industrial Emissions, Improving Air Quality, and Reduced Damage to Vegetation." In Gunn, *Restoration and Recovery of an Industrial Region*, 54–56.

Puro, M.J., W.B. Kipkie, R.A. Knapp, T.J. McDonald, and R.A. Stuparyk. "Inco's Copper Cliff Tailings Area." In *Proceedings of Sudbury'95–Mining and the Environment*, vol. 1, edited by Thomas P. Hynes and Marcia C. Blanchette. Ottawa, ON: CANMET, 1995, 181–91.

Radecki, Henry. "Polish Immigrants in Sudbury, 1883–1980." *Polyphony* 5, no. 1 (1983): 49–58.

Ray, D. Michael. "The Spatial Structure of Economic and Cultural Differences: A Factoral Ecology of Canada." *Papers in Regional Science* 23, no. 1 (1969): 7–24.

Reid, Ruth. "We're Still Here." In Steedman, Suschnigg, and Buse, *Hard Lessons: The Mine Mill Union in the Canadian Labour Movement*, 151–57.

Richards, Howard J. "Lands and Policies: Attitudes and Controls in the Alienation of Lands in Ontario during the First Century of Settlement." *Ontario History* 50 (1958): 193–209.

Roberts, Leslie. "Sudbury Looks to the Future." *Maclean's*, March 15, 1931, 13, 48–50.

Rogers, J.W. "A History of Continents in the Past Three Billion Years," *Journal of Geology* 104 (1996), 91–107.

Ross, Val. "The Arrogance of Inco." *Canadian Business*, May 1979, 44–55, 116–42.

Rousell, Don, H., Harold T. Gibson, and Ian R. Jonasson. "The Tectonic, Magmatic and Mineralization History of the Sudbury Structure." *Exploration and Mining Geology* 6, no. 1 (1997): 1–22.

Saarinen, Oiva W. "Planning and Other Developmental Influences on the Spatial Organization of Urban Settlement in the Sudbury Area." *Laurentian University Review* 3, no. 3 (1971): 38–70.

———. "Ethnicity and the Cultural Mosaic in the Sudbury Area." *Polyphony* 5, no. 1 (1983): 86–92.

———. "Sudbury: A Historical Case Study of Multiple Urban-Economic Transformation." *Ontario History* 82, no. 1 (1990): 55–81.

———. "Finnish Adaptation and Cultural Maintenance: The Sudbury Experience." In *Entering Multiculturalism: Finnish Experience Abroad*, edited by Olavi Koivukangas, 199–215. Turku, Finland: Institute of Migration, 2002.

Saarinen, O.W., and G.O. Tapper, "The Beaver Lake Finnish-Canadian Community: A Case Study of Ethnic Transition as Influenced by the Variables of Time and Spatial Networks, ca. 1907–1983." In *Finns in North America*, edited by Michael G. Karni, Olavi Koivukangas, and Edward W. Laine, 166–200. Turku, Finland: Institute of Migration, 1988.

Sebert, L.M. "The Land Surveys of Ontario 1750–1980." *Cartographica* 17, no. 3 (1980): 65–106.

Simmons, James, and Brian Speck. *Spatial Patterns of Social Change: The Return of the Great Factor Analysis.* Research Paper No. 160. Toronto, ON: Centre for Urban and Community Studies, University of Toronto, 1986.

Simpich, Frederick. "Ontario, Next Door." *National Geographic*, August 1932, 131–84.

Slack, John, F., and William F. Cannon. "Extraterrestrial Demise of Banded Iron Formations 1.85 Billion Years Ago." *Geology*, November 2009, 1011–14.

Slade, Gord. "One Man, 27 Different Jobs." In *Falconbridge 75 Years*, 23–24. Falconbridge, ON: Falconbridge Ltd., 2003.

Speers, E.C. "The Age Relation and Origin of the Common Sudbury Breccia." *Journal of Geology* 65 (1957): 497–514.

Spragge, George W. "The Districts of Upper Canada." *Ontario History* 39 (1947): 91–100.

———. "Colonization Roads in Canada West, 1859–1867." *Ontario History* 44 (1957): 1–17.

Steedman, Mercedes. "The Red Petticoat Brigade: Mine Mill Women's Auxiliaries and the Threat from Within, 1940s–70s." In *Whose National Security? Canadian State Surveillance and the Creation of Enemies*, edited by Gary Kinsman, Dieter K. Buse, and Mercedes Steedman, 55–72. Toronto, ON: Between the Lines, 2000.

Stefura, Mary. "The Ukrainian Co-operative Movement in Sudbury." *Polyphony* 2, no. 1 (1979): 45–46.

———. "Ukrainians in the Sudbury Region." *Polyphony* 5, no. 1 (1983): 71–81.

Stelter, Gilbert A. "Origins of a Company Town: Sudbury in the Nineteenth Century." *Laurentian University Review* 3, no. 3 (1971): 3–37.

"The Sudbury Community Adjustment Project: It Works." *Matrix* 1, no. 1 (1988): 1.

Suschnigg, Peter. "Talking Peace But Preparing for War." In *Mine Mill Fights Back: Mine Mill/CAW Local 598 Strike 2000–2001 Sudbury*, edited by Kaili Beck et al., 74–93. Sudbury, ON: Mine Mill/CAW Local 598, 2005.

Tester, Jim. "We're Still Here." In Steedman, Suschnigg, and Buse, *Hard Lessons: The Mine Mill Union in the Canadian Labour Movement*, 141–92.

Teller, J.T. "Proglacial Lakes along the Southern Margin of the Laurentide Ice Sheet." In *North America and Adjacent Oceans during the Last Glaciation*, edited by W.F. Ruddiman and H.E. Wright, DNAG vol. K-3, 39–69. Geological Society of America, 1987.

Thomson, Ashley. "The 1890s." In Wallace and Thomson, *Sudbury: Rail Town to Regional Capital*, 33–57.

Thompson, J.E., and H. Williams. "The Myth of the Sudbury Lopolith." *Canadian Mining Journal* 80, no. 3 (1959): 57–62.

United Steelworkers of America. "Trade Unionism in the Nickel Industry: The Sudbury Story." *Information* 12, no. 3 (1964): 1–11.

Visentin, Maurizio, A. "The Italians of Sudbury." *Polyphony* 5, no. 1 (1983): 30–36.

Vlahovich, Gordana, and S. Moutsatsos. "Serbians in Sudbury." *Polyphony* 5, no. 1 (1983): 118–22.

Wallace, C.M. "Sudbury: The Northern Experiment with Regional Government." *Laurentian University Review* 27, no. 2 (1985): 87–191.

———. "The 1880s." In Wallace and Thomson, *Sudbury: Rail Town to Regional Capital*, 11–32.

———. "The 1930s." In Wallace and Thomson, *Sudbury: Rail Town to Regional Capital*, 138–67.

"Why the Leafs Stink." *Maclean's*, April 14, 2008, 49.

Winterhalder, Keith. "Environmental Degradation and Rehabilitation in the Sudbury Area." *Laurentian University Review* 16, no. 2 (1984): 15–47.

———. "Early History of Human Activities in the Sudbury Area and Ecological Damage to the Landscape." In *Restoration and Recovery of an Industrial Region*, edited by John M. Gunn, 17–31. New York: Springer-Verlag, 1995.

Zhao, Guochun., M. Sun, M. Wilde, and I. Sanzhong "A Paleo-Mesoproterozoic Supercontinent: Assembly, Growth and Breakup." *Earth-Science Reviews* 67 (2004): 91–123.

Zucchi, John. "Società Italiana di Copper Cliff." *Polyphony* 2, no. 1 (1979): 29–30.

Index